Ali:

This is a gift of appreciation
and inspiration as a result
of ^your guidance.

Respectively yours,

Bill

03/03/98

SELENIUM
in the
ENVIRONMENT

BOOKS IN SOILS, PLANTS, AND THE ENVIRONMENT

Soil Biochemistry, Volume 1, edited by A. D. McLaren and G. H. Peterson
Soil Biochemistry, Volume 2, edited by A. D. McLaren and J. Skujiņš
Soil Biochemistry, Volume 3, edited by E. A. Paul and A. D. McLaren
Soil Biochemistry, Volume 4, edited by E. A. Paul and A. D. McLaren
Soil Biochemistry, Volume 5, edited by E. A. Paul and J. N. Ladd
Soil Biochemistry, Volume 6, edited by Jean-Marc Bollag and G. Stotzky
Soil Biochemistry, Volume 7, edited by G. Stotzky and Jean-Marc Bollag
Soil Biochemistry, Volume 8, edited by Jean-Marc Bollag and G. Stotzky

Organic Chemicals in the Soil Environment, Volumes 1 and 2, edited by C. A. I. Goring and J. W. Hamaker
Humic Substances in the Environment, M. Schnitzer and S. U. Khan
Microbial Life in the Soil: An Introduction, T. Hattori
Principles of Soil Chemistry, Kim H. Tan
Soil Analysis: Instrumental Techniques and Related Procedures, edited by Keith A. Smith
Soil Reclamation Processes: Microbiological Analyses and Applications, edited by Robert L. Tate III and Donald A. Klein
Symbiotic Nitrogen Fixation Technology, edited by Gerald H. Elkan
Soil-Water Interactions: Mechanisms and Applications, edited by Shingo Iwata, Toshio Tabuchi, and Benno P. Warkentin
Soil Analysis: Modern Instrumental Techniques, Second Edition, edited by Keith A. Smith
Soil Analysis: Physical Methods, edited by Keith A. Smith and Chris E. Mullins
Growth and Mineral Nutrition of Field Crops, N. K. Fageria, V. C. Baligar, and Charles Allan Jones
Semiarid Lands and Deserts: Soil Resource and Reclamation, edited by J. Skujiņš
Plant Roots: The Hidden Half, edited by Yoav Waisel, Amram Eshel, and Uzi Kafkafi
Plant Biochemical Regulators, edited by Harold W. Gausman
Maximizing Crop Yields, N. K. Fageria
Transgenic Plants: Fundamentals and Applications, edited by Andrew Hiatt
Soil Microbial Ecology: Applications in Agricultural and Environmental Management, edited by F. Blaine Metting, Jr.
Principles of Soil Chemistry: Second Edition, Kim H. Tan

Water Flow in Soils, edited by Tsuyoshi Miyazaki
Handbook of Plant and Crop Stress, edited by Mohammad Pessarakli
Genetic Improvement of Field Crops, edited by Gustavo A. Slafer
Agricultural Field Experiments: Design and Analysis, Roger G. Petersen
Environmental Soil Science, Kim H. Tan
Mechanisms of Plant Growth and Improved Productivity: Modern Approaches, edited by Amarjit S. Basra
Selenium in the Environment, edited by W. T. Frankenberger, Jr., and Sally Benson
Plant–Environment Interactions, edited by Robert E. Wilkinson

Additional Volumes in Preparation

Handbook of Plant and Crop Physiology, edited by Mohammad Pessarakli

Handbook of Phytoalexin Metabolism and Action, edited by M. Daniel and R. P. Purkayastha

Seed Development and Germination, edited by Jaime Kigel and Gad Galili

Stored Grain Ecosystems, edited by D. S. Jayas, N. D. G. White, and W. E. Muir

Nitrogen Fertilization in the Environment, edited by Peter Bacon

Soil–Water Interactions: Mechanisms and Applications, Second Editon, Revised and Expanded, Shingo Iwata, Toshio Tabuchi, and Benno P. Warkentin

Selenium in the Environment

edited by

W. T. FRANKENBERGER JR.
University of California
Riverside, California

SALLY BENSON
Lawrence Berkeley Laboratory
Berkeley, California

Marcel Dekker, Inc.   New York • Basel • Hong Kong

Library of Congress Cataloging-in-Publication Data

Selenium in the environment / edited by W. T. Frankenberger, Jr., and Sally Benson.
   p.   cm. — (Books in soils, plants, and the environment)
   Includes bibliographical references and index.
   ISBN 0-8247-8993-8
   1. Selenium—Environmental aspects.   I. Frankenberger, W. T. (William T.).   II. Benson, Sally.   III. Series.
QH545.S45S46   1994
574.5'222—dc20   94-11106
                                                                                         CIP

The publisher offers discounts on this book when ordered in bulk quantities. For more information, write to Special Sales/Professional Marketing at the address below.

This book is printed on acid-free paper.

Copyright © 1994 by Marcel Dekker, Inc.   All Rights Reserved.

Neither this book nor any part may be reproduced or transmitted in any form or by any means, electronic or mechanical, including photocopying, microfilming, and recording, or by any information storage and retrieval system, without permission in writing from the publisher.

Marcel Dekker, Inc.
270 Madison Avenue, New York, New York 10016

Current printing (last digit):
10 9 8 7 6 5 4 3 2 1

PRINTED IN THE UNITED STATES OF AMERICA

Preface

Selenium intrigues modern scientists because of the environmental consequences of its double-edged behavior. Depending on its concentration and chemical form, selenium functions as an essential element or potent toxicant to humans, livestock, plants, waterfowl, and certain bacteria. As a result, scientists have searched for effective techniques to augment this important element in selenium-deficient lands while at the same time they have sought to identify plants and soils with dangerously high levels of selenium to protect multiple receptors from contamination, reproductive deformities, and even death.

This book provides a worldwide compilation of scientific studies on selenium, spanning the breadth of disciplines necessary for a global assessment of selenium's complex environmental behavior. Chapter 1 discusses the global importance and cycling of selenium, its worldwide distribution, and the factors controlling its fate and transport within and between major environmental media. Chapter 2 focuses on selenium's role in plant and animal health. Chapter 3 discusses the distribution of two endemic diseases in China, Keshan disease and Kaschin-Beck disease, and their relationships to selenium adsorption, volatilization, and speciation in different types of soil.

Dramatic evidence of selenium poisoning at Kesterson Reservoir in California in 1983 spurred a new wave of multidisciplinary research on selenium by wildlife biologists, toxicologists, microbiologists, geologists, and agricultural and chemical engineers. The results of their research, documented in several chapters in this book, provide an excellent case study of how eminent scientists from many disciplines can work quickly to evaluate an urgent environmental problem, propose practical remediation strategies that address immediate needs but also have worldwide applicability, and advance the theoretical knowledge base to prevent future occurrences of the problem. Chapter 4 begins with an ecological risk assessment of Kesterson Reservoir. Subsequent chapters address biochemical and geological cycling of selenium, microbial transformations, vegetative uptake, toxicity, and remedial treatments, including geochemical mobilization and chemical fixation.

Chapter 5 through 9 discuss selenium speciation, transformation, and transport, kinetics and seasonal cycling in aquatic biota, accumulation and colonization by plants, geochemistry in groundwater systems, and geologic origin from the California Coast Range to the West-Central San Joaquin Valley of California. The reader can trace the movement of selenium in soils of the San Joaquin Valley, one of the world's most productive agricultural areas. In the valley, irrigation water applied to agricultural soils dissolves and carries selenium and associated solutes into buried drains, where the water is collected and carried north to Kesterson Reservoir, a storage reservoir that served as a feeding and/or breeding area for migratory and resident waterfowl in the Pacific flyway. At Kesterson, dissolved selenium undergoes complex transformations from the selenate form delivered with the drainage water to a variety of forms with differing mobility and toxic characteristics. Some of the selenium is incorporated into plants and algae at the base of the food web and has led to death and deformities of waterfowl. Some is converted to gaseous forms that are dissipated into the atmosphere. Some is converted to insoluble and benign forms when the drainage water seeps into the underlying groundwater aquifer. Once the plight of the waterfowl was discovered, research efforts shifted to redefining "safe" levels of selenium in various habitats, finding cost-effective measures for removing selenium from agricultural drainage water, and detoxifying selenium-contaminated soils and sediments.

Several chapters devoted to the discussion on Kesterson Reservoir focus on selenium management and remediation strategies, including agroforestry and vegetation management techniques (Chapters 10, 12, and 13); the development of an algal-bacterial selenium removal system (Chapter 11); enhancement of plant and microbial volatilization processes (Chapters 14 and 15); microbial reduction (Chapter 16); and characterization of a selenate-respiring bacterium capable of removing selenium oxides from San Joaquin Valley drainage water (Chapter 17).

Scientific pursuit of knowledge about selenium has paralleled the trends taking place throughout the environmental sciences. We learned quickly that total elemental concentrations are not reliable indicators of environmental toxicity. Extensive efforts were devoted to developing selenium-specific speciation and soil-fractionation techniques. Scientists also learned more about the important role that microorganisms play in the transformation of selenium—anaerobic bacteria reduce selenium from soluble, toxic forms to sparingly soluble nontoxic forms while fungi convert soluble and, to a lesser extent, sparingly soluble forms to a gas that can be dissipated into the atmosphere. Competitive and synergistic effects between chemical constituents were also recognized. Under some circumstances, the presence of nitrate may interfere with microbial reduction of selenate; sulfate may compete with selenate uptake by plants; boron and selenium may interact to decrease toxicity to waterfowl. In unearthing the complex behavior of selenium in the environment, scientists found that it is not an element to fear but one that requires control. The spectrum of selenium's biological and geochemical behavior about which scientists have much more to learn has provided the impetus to expand our knowledge of methods for controlling environmental exposure to selenium.

The period since 1983 has seen dynamic growth in understanding selenium, due largely to the teamwork of a worldwide community of researchers who have worked in parallel, shared knowledge, and competed for the next exciting revelation. The notable progress they have made, documented in this volume, would not have been possible without the generous support from state and federal agencies and private industry to universities and government research laboratories.

Appreciation is expressed to Deborah Silva for her editorial assistance.

W. T. Frankenberger, Jr.
Sally Benson

Contents

Preface		*iii*
Contributors		*xi*
1	Global Importance and Global Cycling of Selenium Philip M. Haygarth	1
2	Selenium in Plant and Animal Nutrition H. F. Mayland	29
3	Adsorption, Volatilization, and Speciation of Selenium in Different Types of Soils in China J. A. Tan, W. Y. Wang, D. C. Wang, and S. F. Hou	47
4	Kesterson Reservoir—Past, Present, and Future: An Ecological Risk Assessment Harry M. Ohlendorf and Gary M. Santolo	69
5	Field Investigations of Selenium Speciation, Transformation, and Transport in Soils from Kesterson Reservoir and Lahontan Valley Tetsu K. Tokunaga, Peter T. Zawislanski, Paul W. Johannis, Douglas S. Lipton, and Sally Benson	119

6	Geologic Origin and Pathways of Selenium from the California Coast Ranges to the West-Central San Joaquin Valley Theresa S. Presser	139
7	Distribution and Mobility of Selenium in Groundwater in the Western San Joaquin Valley of California Steven J. Deverel, John L. Fio, and Neil M. Dubrovsky	157
8	Chemical Oxidation-Reduction Controls on Selenium Mobility in Groundwater Systems Arthur F. White and Neil M. Dubrovsky	185
9	Kinetics of Selenium Uptake and Loss and Seasonal Cycling of Selenium by the Aquatic Microbial Community in the Kesterson Wetlands Alexander J. Horne	223
10	Agroforestry Farming System for the Management of Selenium and Salt on Irrigated Farmland Vashek Cervinka	237
11	The Algal-Bacterial Selenium Removal System: Mechanisms and Field Study Tryg J. Lundquist, Matthew B. Gerhardt, F. Bailey Green, R. Blake Tresan, Robert D. Newman, and William J. Oswald	251
12	Selenium Accumulation and Colonization of Plants in Soils with Elevated Selenium and Salinity Lin Wu	279
13	Vegetation Management Strategies for Remediation of Selenium-Contaminated Soils David R. Parker and Albert L. Page	327
14	Selenium Volatilization by Plants Norman Terry and Adel M. Zayed	343
15	Microbial Volatilization of Selenium from Soils and Sediments W. T. Frankenberger, Jr., and Ulrich Karlson	369

CONTENTS

16 *Biogeochemical Transformations of Selenium in Anoxic Environments* 389
    Ronald S. Oremland

17 *Biochemistry of Selenium Metabolism by* Thauera selenatis *gen. nov. sp. nov. and Use of the Organism for Bioremediation of Selenium Oxyanions in San Joaquin Valley Drainage Water* 421
    Joan M. Macy

Index 445

Contributors

Sally Benson, Ph.D. Earth Sciences Division, Lawrence Berkeley Laboratory, Berkeley, California

Vashek Cervinka, Ph.D. Agricultural Resources Branch, California Department of Food and Agriculture, Sacramento, California

Steven J. Deverel, Ph.D. Supervisory Hydrologist, Water Resources Division, U.S. Geological Survey, Sacramento, California

Neil M. Dubrovsky, Ph.D. Supervisory Hydrologist, California District Water Resources Division, U.S. Geological Survey, Sacramento, California

John L. Fio Research Hydrologist, California District Water Resources Division, U.S. Geological Survey, Sacramento, California

W. T. Frankenberger, Jr. Professor of Soil Microbiology and Biochemistry, Department of Soil and Environmental Sciences, University of California, Riverside, California

Matthew B. Gerhardt, Ph.D., P.E. Senior Engineer, Brown and Caldwell, Pleasant Hill, California

CONTRIBUTORS

F. Bailey Green, Ph.D. Senior Research Associate, Environmental Engineering and Health Sciences Laboratory, University of California, Berkeley, Richmond, California

Philip M. Haygarth, Ph.D. Soil and Environmental Scientist, Soil Science Group, North Wyke Research Station, Institute of Grassland and Environmental Research, Okehampton, Devon, England

Alexander J. Horne, Ph.D. Professor of Applied Ecology, Department of Civil Engineering, University of California, Berkeley, California

S. F. Hou Associate Professor, Department of Chemical Geography, Institute of Geography, Chinese Academy of Sciences, Beijing, People's Republic of China

Paul W. Johannis Research Associate, Earth Sciences Division, Lawrence Berkeley Laboratory, Berkeley, California

Ulrich Karlson Senior Scientist, Department of Marine Ecology and Microbiology, National Environmental Research Institute, Roskilde, Denmark

Douglas S. Lipton, Ph.D. Senior Associate Scientist, Levine-Fricke, Inc., Emeryville, California

Tryg J. Lundquist, M.S. Environmental Engineering & Health Sciences Laboratory, University of California, Berkeley, Richmond, California

Joan M. Macy, Ph.D. Department of Animal Science, University of California, Davis, California

H. F. Mayland, Ph.D. Research Soil Scientist, Soil and Water Management Research Unit, Agricultural Research Service, U.S. Department of Agriculture, Kimberly, Idaho

Robert D. Newman Junior Associate Engineer, Algal Research Group, University of California, Berkeley, Richmond, California

Harry M. Ohlendorf, Ph.D. Environmental Scientist, CH2M HILL, Inc., Sacramento, California

Ronald S. Oremland, Ph.D. Project Chief, Microbial Biogeochemistry, Water Resources Division, U.S. Geological Survey, Menlo Park, California

CONTRIBUTORS

William J. Oswald, Ph.D., P.E. Professor Emeritus of Environmental Engineering, College of Engineering and School of Public Health, University of California, Berkeley, Richmond, California

Albert L. Page, Ph.D. Professor of Soil Chemistry, Department of Soil and Environmental Sciences, University of California, Riverside, California

David R. Parker, Ph.D. Assistant Professor of Soil Chemistry, Department of Soil and Environmental Sciences, University of California, Riverside, California

Theresa S. Presser Research Chemist, National Research Program, U.S. Geological Survey, Menlo Park, California

Gary M. Santolo Environmental Scientist, CH2M HILL, Inc., Sacramento, California

J. A. Tan Professor, Department of Chemical Geography, Institute of Geography, Chinese Academy of Sciences, Beijing, People's Republic of China

Norman Terry, Ph.D. Professor of Environmental Plant Biology, Department of Plant Biology, University of California, Berkeley, California

Tetsu K. Tokunaga, Ph.D. Staff Scientist, Earth Sciences Division, Lawrence Berkeley Laboratory, Berkeley, California

R. Blake Tresan Environmental Engineering and Health Sciences Laboratory, University of California, Berkeley, Richmond, California

D. C. Wang Assistant Professor, Department of Chemical Geography, Institute of Geography, Chinese Academy of Sciences, Beijing, People's Republic of China

W. Y. Wang Associate Professor, Department of Chemical Geography, Institute of Geography, Chinese Academy of Sciences, Beijing, People's Republic of China

Arthur F. White, Ph.D. Geochemist, Water Resources Division, U.S. Geological Survey, Menlo Park, California

Lin Wu, Ph.D. Professor, Department of Environmental Horticulture, University of California, Davis, California

Peter T. Zawislanski Senior Research Associate, Earth Sciences Division, Lawrence Berkeley Laboratory, Berkeley, California

Adel M. Zayed, Ph.D. Postdoctoral Researcher, Department of Plant Biology, University of California, Berkeley, California

SELENIUM
in the
ENVIRONMENT

1
Global Importance and Global Cycling of Selenium

Philip M. Haygarth

*Institute of Grassland and Environmental Research
Okehampton, Devon, England*

I. INTRODUCTION

In a global context, selenium (Se) is a complex but interesting element. Its properties make the boundaries between animal toxicity and deficiency relatively narrow, and, indeed, both phenomena are common around the globe. There is also a contrasting distribution and rate of transfer between different environmental compartments. Most recently, human interference in the global cycling of Se has added to this heterogeneity. This chapter introduces the global significance and distribution of Se and considers the factors that control its fate and transport within and between major environmental media.

II. THE GLOBAL IMPORTANCE OF SELENIUM

A. Properties and Applications of Selenium

Selenium (Se) has chemical and physical properties that are intermediate between those of metals and nonmetals and belongs in Group VIA of the periodic table. Pure Se is allotropic and exists as gray hexagonal, red monoclinic, and vitreous amorphous forms. The physicochemical properties of Se are summarized in Table 1. It has a valence of 2– in combination with hydrogen or metals, and in oxygen compounds it exists in the +4 and

+6 oxidation states. Such properties give rise to an array of Se compounds that are common in the environment, some of which are summarized in Table 2. Six stable isotopes of Se occur with varying degrees of abundance: ^{74}Se (0.87%), ^{76}Se (9.02%), ^{77}Se (7.58%), ^{78}Se (23.52%), ^{80}Se (49.82%), and ^{82}Se (9.19%), and there are a number of short-lived isotopes, of which ^{75}Se is most commonly used in neutron activation, radiology, and tracer applications [1].

More than 1600 tonnes of Se is produced annually from mine production [1,2], primarily as a by-product of copper refining. Over 80% of this Se is derived from anodic mud deposited during the electrolytic refining of copper [13]. Selenium is used most commonly in the glass industry to counteract coloration due to iron oxides and can also be used to color glass red [1]. Glassmaking constitutes about 20% of its overall industrial use. Selenium is also used for pigments in plastics and paints, because it is resistant to heat, light, weathering, and various chemicals. It also has properties that can increase wear resistance, especially important in the rubber industry. Selenium is an antioxidant, which makes it useful for inclusion in inks, mineral and vegetable oils, and lubricants. There are a wide range of electronics applications because of its photoelectric and semiconductor properties. In the pharmaceutical industry, selenium mono- and disulfide are used in the treatment of dandruff and its sulfides have also been used to treat the fungal infection *Tinea versicolor*. Small amounts of [^{75}Se]selenomethionine have been used as a diagnostic

Table 1 Physicochemical Properties of Selenium

Property	Selenium
Atomic number	34
Atomic mass	78.96
Density, g/cm^3	4.79[a]
Melting point, °C	217
Boiling point, °C	685.4
Atomic radius, μm	0.117
Hardness, relative units	2[a]
Electronegativity, relative units (Li = 1)	2.4
Latent heat of fusion, J/g (cal/g)	6.91 (16.5)
Heat of vaporization, J/g	272.98 (65.2)
Thermal conductivity, W/(m·°C)	0.293–0.766

[a]Hexagonal modification
Source: Modified from Ref. 1.

Table 2 Compounds of Selenium Common in the Environment

Name	Formula	Where found
Selenides (−II)	Se^{2-}	Reducing environments, e.g., soils. Forms metal complexes; highly immobile.
Dimethylselenide (DMSe)	$(CH_3)_2Se$	Gas formed by volatilization from soil bacteria and fungi.
Dimethyldiselenide (DMdSe)	$(CH_3)_2Se_2$	Gas formed by volatilization from plants.
Dimethylselenone/methyl methylselenite	$(CH_3)_2SeO_2$	Volatile metabolite, possibly formed as a final intermediate prior to reduction to DMSe.
Hydrogen selenide	H_2Se	Gas, unstable in moist air; decomposes to Se^0 in water.
Elemental selenium (0)	Se^0	Stable in reducing environments; (a) red crystalline alpha and beta monoclinic; (b) red glossy or black amorphous forms, all insoluble in water and oxidation/reduction very slow.
Selenite (+IV)	SeO_3^{2-}	Soluble form, common in mildly oxidizing conditions, e.g., soils or air particles.
Trimethylselenonium ($TMSe^+$)	$(CH_3)_3Se^+$	$TMSe^+$ is an important urinary metabolite of dietary Se and is made rapidly unavailable to plants by fixation and volatilization.
Selenous acid	H_2SeO_3	Selenous acid is protonated in acid/neutral conditions. Se(IV) is easily reduced to Se^0 by ascorbic acid (vitamin C) or sulfur dioxide in acidic environments by microorganisms. Readily available by Fe oxides, amorphous Fe hydroxides, and aluminum sesquioxides in soils.
Selenium dioxide	SeO_2	Gas SeO_2 formed as a product of fossil fuel combustion (sublimation temperature, 300°C); dissolves in water to form selenous acid.
	$HSeO_3^-$	Common soils.
Selenate (+VI)	SeO_4^{2-}	SE(VI) is stable in well-oxidized environments, and very mobile in soils, hence easily available to plants. Slowly converted to more reduced forms: not as strongly absorbed as SE(IV).
Selenic acid	$H_2SeO_4^-$	
	$HSeO_4^-$	Common in soils.

scanning and labeling material because it concentrates in the liver, pancreas, and other important organs that are difficult to study by conventional X-rays. The uses of Se are reviewed by Newland [1] and Fishbein [3] and are summarized in Table 3.

B. Passive Anthropogenic Interference in the Global Se Cycle

In addition to mining production of Se, the element is also mobilized as a by-product of industrial activity. This "passive" release has by far the major influence on the global cycle of Se, releasing 76,000–88,000 tonnes/yr globally [2]. Nriagu [2] combined industrial mobilization with release from mining (1600 tonnes) to estimate total industrial mobilization and obtained a *biospheric enrichment factor* (BEF) by dividing the combined

Table 3 Selected Selenium Compounds and Their Uses

Compound	Use
Elemental selenium	Rectifiers, photoelectric cells, blasting caps; in xerography; stainless steel; dehydrogenation catalyst
Sodium selenate	As insecticide; in glass manufacture; in medicinals to control animal diseases
Sodium selenite	In glass manufacture; as soil additive for selenium-deficient areas
Selenium diethyldithiocarbamate	Fungicide; vulcanizing agent
Selenium disulfide	In veterinary medicine
Selenium dioxide	Catalyst for oxidation, hydrogenation, or dehydrogenation of organic compounds
Selenium monosulfide	In veterinary medicine
Selenium hexafluoride	As gaseous electric insulator
Selenium oxychloride	Solvent for sulfur, selenium, tellurium, rubber, bakelite, gums, resins, glue, asphalt, and other materials
Aluminum selenide	Preparation of hydrogen selenide for semiconductors
Ammonium selenite	Manufacture of red glass
Cadmium selenide	Photoconductors, photoelectric cells, rectifiers
Cupric selenate	In coloring copper and copper alloys
Tungsten diselenide	In lubricants

Source: Modified from Ref. 3.

total by the estimated total "natural" release (4500 tonnes/yr). This gives a BEF value of 17, and since this BEF value is substantially greater than 1, it indicates a high magnitude of human interference in the cycle. For comparison purposes, BEF values for other elements are As, 1.7; Ni, 4.4; Cd, 9.6; Hg and Pb, 20; and Cu, 24. Other global inventories on the anthropogenic emissions of Se have been published by Nriagu and Pacyna [4] and Nriagu [5]. The key industries involved are primarily coal and oil combustion, pyrometallurgical nonferrous metal production (principally Cu and Ni, but also Pb, Zn, and Cd), secondary nonferrous metal production, steel and iron manufacturing, municipal and sewage sludge refuse incineration, and production of phosphate fertilizers.

Recent investigations concerning the Se content of herbage, systematically sampled since the mid-1800s, have revealed changes in the long-term global cycling of Se resulting from such emissions [6]. Samples were taken from an undisturbed grassland at a semirural site at Rothamsted in southeast England, bulked at 5-year intervals between 1861 and 1990. Total Se concentrations changed considerably, primarily due to changes in industrial emissions of the element, particularly coal combustion (see Figure 1). Highest Se levels were between 1940 and 1970, but more recently levels have declined. This is thought to reflect a change in the dominance of domestic coal burning to industrial coal burning, a more efficient combustion process that probably emits less Se. Another cause of the reduction is a shift away from coal, which is generally a high Se emission source, to sources that emit less Se, such as oil, gas, and nuclear power [6]. Moreover, this demonstrates that human interference is having a major impact on the Se cycle in the natural ecosystem.

C. Selenium and Diet

1. Benefits of Selenium

Recently, there has been an increase in popular interest in the use of Se as a dietary supplement [e.g., 7]. Fishbein [8] gives detailed consideration to the benefits of Se and lists over 20 medical complaints associated with a low Se intake. Among these are protein energy malnutrition, hemolytic anemia, cardiomyopathy (Keshan disease), hypertension, ischemic heart disease, alcoholic cirrhosis, cystic fibrosis, infertility, cancer, arthritis, muscular dystrophy, and multiple sclerosis. Human intake levels in a low-intake area of China where Keshan disease is common are on the order of 11 µg per person per day; levels of "adequate" Se intake were considered to be in the range of 110 µg per person per day, whereas "high" Se intake may be up to 750 µg per person per day. Chronic selenosis (Se poisoning)

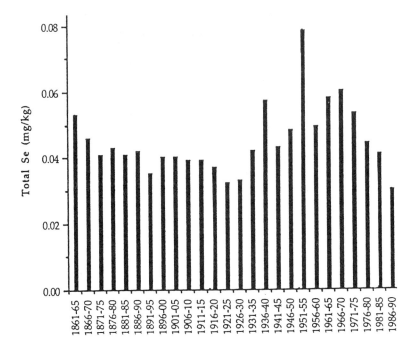

Figure 1 Long-term changes in anthropogenic Se emissions as reflected from retrospective analysis of archived herbage from Park Grass, Rothamsted, U.K.

is associated with an intake of nearly 5 mg/day. There is also experimental evidence for Se anticarcinogenicity, determined by tumor size in rats and mice. In the breast, the percent reduction associated with Se supplementation was 28–62%; in the colon, the reduction ranged from 7 to 54%; in the lungs, the reduction was 100%; and in liver, 35–83% [9]. Selenium and anticarcinogenicity are also considered by Underwood [9].

The metabolic and functional effects of Se deficiency are considered by Diplock [10]. The work describes how Se is an integral part of glutathione peroxidase, which catalyzes the reduction of lipid hydroperoxides and hydrogen peroxide. This involves a complex interaction between vitamin E, Se, and polyunsaturated fatty acids. Diplock suggests three aspects of the interaction that may be important:

1. Selenium in glutathione peroxidase controls intracellular levels of hydrogen peroxide that influence active oxygen metabolite formation and serve as initiators of lipid peroxidation. Se is closely related to

superoxide dismutases, which control intracellular levels of the superoxide anion.
2. Through these antioxidative properties, vitamin E controls the formation of lipid hydroperoxides.
3. Glutathione peroxidase catalyzes the reduction of lipid hydroperoxides formed from membrane lipids, without detriment to the cellular economy.

Geographical surveys of Se in human health are commonly determined by levels in urine, blood, serum, and scalp hair [11,12]. Most recently, Maksimovic et al. [12] reported serious Se deficiency in the Yugoslav population. Considerable interest in the benefits of Se in the metabolism of grazing livestock has developed and will be discussed below.

2. Selenium Toxicity

Only excretion is influenced by Se nutritional status; absorption is not regulated by the body, which means that toxicity can arise from a high level of Se in dietary intake [13]. The major sources of Se toxicity in humans are occupational (through inhalation or dermal contact), uncontrolled self-medication, and common diet [13]. Robin [14] surveyed workers in a by-products department of a copper refinery who were exposed to high levels of Se and observed cutaneous (eye, skin, nose irritation) and digestive (metallic taste, epigastric pain, weight loss, and decreased appetite) disorders. "Garlic" breath is also common, denoting the exhalation of dimethylselenide gas and related compounds.

Self-medication of one tablet containing 2 mg of sodium selenite per day for more than 2 years elevated blood concentrations to about 0.179 mg/L and hair Se to 830 ng/g. This caused thickened, fragile nails and garlic odor. Daily dietary supplements of 30 mg of Se for 77 days were associated with hair loss, fingernail and fingertip disorders (including fingertip inflammation and nail bed purulence), vomiting, fatigue, and a blood serum level of 0.528 mg/L [13]. Clearly, such means of self-medication need to be regulated.

High levels of dietary intake of Se have been associated mostly with people farming over the seleniferous soils of South Dakota [16]. The symptoms are gastrointestinal disorders and yellowish skin discoloration, thought to be due to a hepatic dysfunction. Seafood is another known source of high dietary Se, but it is unlikely that Se toxicity can be attributed to such sources [13]. With the trend away from subsistence agriculture and the prevalence of worldwide food distribution channels in the modern economy, it can be speculated that diet-derived Se toxicity may become less common due to the dilution effect of food distribution.

D. Selenium in Agriculture

There is widespread concern about Se contamination of wildlife populations, especially around the San Joaquin Valley of central California, where much recent Se research has focused on Kesterson Reservoir, in the Kesterson National Wildlife Refuge (Merced County), which consists of 12 shallow evaporation ponds (ca. 500 ha each when filled) for agricultural drainage waters from the western San Joaquin Valley. Selenium concentrations entering Kesterson along the San Luis Drain have been determined to be as high as 0.3 mg/L [15]. Selenium contamination in this environment arises from irrigation of otherwise nonarable land in arid areas. Irrigation in arid regions leads to salt build-up [16], and in order to maintain an acceptable salt balance, excess water needs to be applied to flush off the drainage water, in which trace elements and Se are unwanted passengers. Samples of mosquitofish (*Gambusia affinis*) contained levels of Se nearly 100 times greater than those of the nearby Volta Wildlife Area, which did not receive drainwater [15]. Avian populations have also been affected by Se toxicity, with noted embryonic mortality and external anomalies, particularly in mallards (*Anas platyrhynchos*) [15,17]. Deformities include missing or abnormal eyes, beaks, wings, legs, and feet. Brain, heart, liver, and skeletal abnormalities have also been observed. Other wildlife affected by Se toxicity include invertebrates, amphibians and reptiles, and mammals [18]. Problems associated with agricultural drainage waters in the western United States are the subject of some of the chapters that follow.

The other significant effect on agriculture attributable to Se is its role in the regulation of reproductive performance and general health in livestock agriculture. The problem of livestock Se deficiency has become a major focus in contemporary Se research and has brought together scientists from many disciplines, including the veterinary, chemical, soil, plant, and geochemical sciences. Reviews on this issue and on Se in agriculture and the environment have been presented by Oldfield [19] and Girling [20].

The popular term for livestock Se deficiency is "white muscle disease," although nowadays the syndrome is known to encompass more than muscular disorders. Levels of Se intake required for adequate animal nutrition range between 0.04 and 0.1 mg/kg dry matter, depending on the animal species and the level of vitamin E in the diet [20]. Sheep, cattle, and swine generally require 0.1 mg/kg intake, but chickens and turkeys require approximately 0.15–0.2 mg/kg. Toxicity is far less common but is generally associated with levels an order of magnitude higher [20,21]. Cortese [22] presented evidence to link Se with reproductive performance in dairy cattle. Ropstad et al. [23] noted a seasonal variation in the Se status of dairy cows, with highest levels in the winter associated with indoor feeding of Se-rich concentrates. They concluded that further supplementation was not required, provided Se-rich concentrates were used. Their results complement the findings of Slaweta et al. [24], who reported a significant winter increase in total and Se-dependent glutathione peroxidase activity in bovine semen.

Anderson et al. [25] surveyed 329 farms around Britain and found that 47% were probably unable to provide grazing livestock with sufficient Se to maintain blood levels greater than 0.075 mg/L, attributing the increased incidence of Se deficiency (white muscle disease) to the increasing use of home-grown feeds in Britain. They used erythrocyte glutathione peroxidase activity to indicate animal Se status; Wuyi et al. [26] proposed wool as an accurate and less costly measure of sheep Se status. Frøslie et al. [27] surveyed Se in animal feedstuffs in Norway by analyzing a range of domestic and imported feeds, mixed concentrates, and liver samples. Animals fed roughages and feeds of domestic origin suffered from the greatest Se deficiency. Grain products were found to be especially associated with low Se.

Remedial practices commonly employed to counter the potential of animal Se deficiency include injections, dietary supplements, salt licks, and drenches [e.g., 23,26]. These mechanisms are used with a reasonable degree of success, although they are often expensive and time-consuming. Such difficulties probably cause farms to neglect to dose stock regularly and can lead to recurrences of the problem. In an attempt to understand the syndrome, recent studies on Se deficiency in agriculture have adopted a mass balance approach [28–30], to be discussed in detail in the section on soil–plant cycling that follows.

III. THE GLOBAL CYCLING OF SELENIUM

A. Overview of the Selenium Cycle

Past approaches to Se cycling have focused on geological and geochemical phenomena, which are clearly essential in governing the natural or background concentrations of the element in localized environments. This approach was the basis of the classic work published by Rosenfield and Beath in 1964 [31] for the United States and later by Webb et al. [32] for the

United Kingdom. Generally, parent materials known to have the highest Se concentration are black shales (around 600 ppm) and phosphate rocks (1–300 ppm), both potentially giving rise to seleniferous soils and food chain Se toxicity [31]. Other typical crustal concentrations are limestones (0.08 ppm), sandstones (0.05 ppm), and igneous rock (0.004–1.5 ppm).

An overview of the global cycle pathways for Se is presented in Figure 2. Nriagu [33] has provided a comprehensive global inventory of Se in principal environmental compartments, together with a citation of the typical concentrations, which is summarized in Table 4. However, an understanding of compartmental concentrations has only preliminary use in considering the global Se cycle, because it is the fluxes, driven by both natural (biogenic) and human (anthropogenic) processes, that are of environmental interest, as they are responsible for an ongoing redistribution of Se throughout the global system. Such fate and transport processes have become the focus of current research, and the following subsections

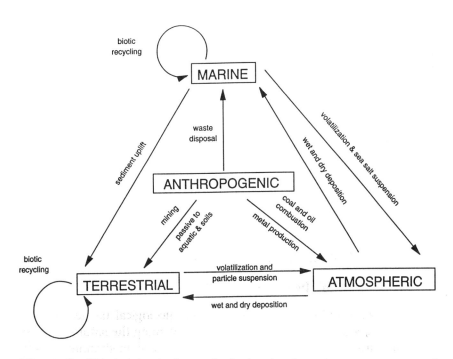

Figure 2 Global cycling pathways for Se showing the major compartments and flux pathways. For quantitative information on the fluxes, see Table 5.

Table 4 Global Inventories of Selenium in Selected Environmental Compartments

Reservoir	Reservoir mass (g)	Average selenium conc. (mg/kg)	Selenium pool (g)
Lithosphere (down to 45 km)	57×10^{24}	0.05	2.8×10^{18}
Soils (down to 1.0 m)	3.3×10^{20}	0.4	1.3×10^{14}
Soil organic matter	1.6×10^{18}	0.2	3.2×10^{11}
Fossil fuel deposits			
Coal	1.0×10^{19}	3.4	3.4×10^{13}
Oil shale	4.6×10^{19}	2.3	1.1×10^{14}
Crude oil	2.3×10^{17}	0.2	4.6×10^{10}
Terrestrial biomass			
Plants	1.2×10^{18}	0.05	6.0×10^{10}
Animals	1.0×10^{15}	0.15	1.5×10^{8}
Forest litter	1.2×10^{17}	0.08	9.6×10^{9}
Oceans			
Dissolved, surface mixed layer	2.8×10^{22}	30 ng/kg	8.4×10^{11}
Dissolved, deep ocean	1.4×10^{24}	95 ng/kg	1.3×10^{14}
Suspended particulates, mixed layer	7×10^{16}	3	2.1×10^{11}
Pore water in sediments	3.3×10^{23}	0.3 µg/kg	9.9×10^{13}
Rivers			
Dissolved	3.4×10^{19}	60 ng/kg	2.0×10^{9}
Suspended load	1.5×10^{19}	0.8	2.4×10^{12}
Shallow groundwater	4×10^{8}	0.2 µg/kg	8.0×10^{8}
Polar ice	2×10^{22}	20 ng/kg	4.0×10^{11}
Selenium reserves	—	—	112×10^{9}
Selenium resources	—	—	291×10^{9}

Source: Nriagu [33], Table 1, p. 330.

provide a summary of the issues, that, in my opinion, represent the key processes and themes in contemporary Se research.

B. Surface–Atmosphere Fluxes

Transformations of Se through the atmosphere represent a dynamic and critical stage in the global Se cycle. Although concentrations are low (typically around 1 ng/m^3 air), the physical and chemical transformations can be very rapid, affecting ecosystems worldwide. For example, the residence time for particulate Se is relatively short [34], on the order of a few weeks, but in relation to other compartments the turnover of Se in the atmosphere is extremely rapid. Nriagu [33] estimated the mean residence time of Se in the atmosphere as being 45 days. Particle-bound Se may have to be transported thousands of kilometers within this short period, before it is redeposited onto the global surface. The physical constituents of atmospheric Se are the particle phase, predominantly less than 1 μm in diameter [35], and gaseous forms [36]. Gaseous atmospheric Se is considered to be dimethylselenide, although dimethyldiselenide and dimethylselenone have also been reported [37]. Atkinson et al. [34] suggested that dimethylselenide is unstable in the atmosphere because it reacts with OH, NO_3, and O_3 within a few hours of emission. Gaseous atmospheric Se can also bond to particulate material for long-range transport.

1. Volatilization

Volatilization of Se occurs as a result of microbial methylation of dimethyl forms of Se from soil, plant, and water surfaces. Most of the studies on the volatilization process are derived from experimental procedures involving spiking of media with stable Se or ^{75}Se [e.g., 38–41]. Bacteria and fungi responsible for the process are probably numerous, and *Penicillium* and *Alternaria* are examples that have been identified in recent research [42,43].

Factors influencing microbial volatilization are essentially those that (1) affect microbial activity, such as the carbon source, temperature, and oxygen conditions; and (2) affect the availability of Se to the microorganism, such as irreversible adsorption, precipitation, and complexation.

Determination of the net flux of volatile Se from grassland surfaces has been undertaken by Haygarth et al. [44] using a vertical gradient method. The flux rate was estimated to be on the order of 100–200 $\mu g/(m^2 \cdot yr)$ (0.07–0.14% of that in soil), and this result agreed with estimations by mass balance from soil lysimeters. On a global scale, biomethylation of volatile Se from the land biota on the terrestrial surface has been estimated to contribute 1200 tonnes/yr to the atmosphere [33]. Biogenic volatile fluxes from the ocean to the atmosphere are considerably greater, being on the order of 5000–8000 tonnes/yr [33,45].

2. Particle Suspension

Physical removal of Se associated with particulate material from global surfaces has attracted an unduly small amount of research, despite its significance as a transport pathway in certain circumstances. Wind erosion of Se bound to particulate material has been estimated to be responsible for a loss of up to 0.5 mg Se per kilogram of soil on the earth surface, and in areas particularly susceptible to wind erosion this loss may be considerably higher. On a global scale, this flux has been estimated to be on the order of 180 tonnes/yr [5]. Suspension of sea salts from the marine system is also a physical means of surface atmospheric cycling, estimated at 550 tonnes Se/yr [5].

3. Deposition

Deposition of Se from the atmosphere to the global surface occurs in both wet and dry forms. On a global scale, deposition to land contributes 6600 tonnes/yr, and deposition to the marine system, 8700 tonnes/yr [33].

On a more localized scale, Haygarth et al. [6,44,51] have conducted a number of studies on the importance of the deposition process as a means of input to soils and plants in the United Kingdom. They found that

1. Deposition has contributed to a 15% increase in soil Se concentration in the last century.
2. Deposition contributes between 33% and 82% of plant leaf Se uptake.
3. Deposition is heavily influenced by geographical proximity to an emission source, with highest levels associated with industrial and coastal zones.
4. The deposition process is considerably higher during the winter months, probably reflecting the prevalence of wet removal processes combined with climatological characteristics, temperature inversions, and occult deposition (see below), with higher winter combustion and emission rates.

a. Wet Deposition "Wet deposition" refers to rainout and washout of all forms of atmospheric Se. Wet deposition delivers particulate and gaseous forms to the surface, some of which may have been converted to soluble species during transport. Cawse [46 and references cited therein] indicated that in the U.K. climate wet deposition accounted for 76–93% of Se deposition to a bulk deposition sampler over terrestrial surfaces, and, indeed, >70% of this was in a soluble form. A washout factor (W) can be calculated to express the concentration of Se in rain relative to that in air. Cawse [46] grouped elements according to the W value, and Se was classed as having a low W (<500), similar to that of Al, Cr, Fe, and Pb, indicating a local or near-ground source of emission. A process related to wet deposition that is important in montane environments is occult deposition, where exchange between the atmosphere and the terrestrial surface occurs in low cloud situations [47,48]. In the United Kingdom, occult deposition is speculated to be higher during the winter months [52]. The chemical content of clouds is known to be high, and there is reason to believe that occult deposition may be a significant transfer pathway for Se. However, it has never been studied directly for this element and is obviously an area that requires further research.

b. Dry Deposition Dry deposition [49] is the exchange of particulate and gaseous material between the atmosphere and the global surface. It is a function of two factors:

1. The physicochemical characteristics of the surface. Surface roughness (and its subsequent effects on local wind speed and dynamics) play an important role [50,52]. For example, a forest canopy is a more efficient trapper of atmospheric chemicals than grassland.
2. Particle size (gases, of course, excepted).

Deposition velocity (V_g, in centimeters per second) is derived empirically and commonly employed as a constant to aid in prediction of the rate of dry deposition (micrograms per square centimeter per second) divided by the concentration in air (micrograms per cubic centimeter). V_g of Se gases onto surfaces has, to date, not been determined, and indeed it is becoming clear that over terrestrial surfaces there is a net emission of gaseous Se [44]. There have been several determinations of the deposition flux of particulate material. Cawse [46] estimated the V_g of Se to a sampling device to be about 0.1 cm/s. More recently, deposition to soil under various crop treatments was determined, and V_g was calculated as being 0.018–0.028 cm/s [6]. V_g of Se particulates to grassland has been estimated to be 0.0068–0.034 cm/s [44,51].

C. Cycling Through Soil–Plant Systems

Mobilization of selenium within soil–plant systems is a well-researched but highly complex subject [53]. Most research has been driven by attempts to understand the soil–plant fluxes of the element in grassland agricultural systems that are Se-deficient. This discussion concentrates on mass balance (inputs and outputs), soil mobilization, and plant uptake.

1. Mass Balance in Agricultural Soil–Plant Systems

Given the Se-deficiency problem, clearly the processes that control levels of Se in agricultural soil are important. Whether these are natural processes or occur as a result of agricultural management, they can be summarized by the mass balance equation [after 28–30]

$$\text{Se total} = (Se_p + Se_a + Se_f + Se_s) - (Se_{cr} + Se_l + Se_v)$$

where the subscripts are p, parent material; a, atmospheric deposition; f, fertilizers; s, sewage sludge; cr, crop Se offtake; l, Se leaching; and v, Se volatilization.

Consideration of the Se mass balance in agricultural soils, with respect to these transport pathways, should become the focus for research into Se agricultural deficiencies. Processes such as atmospheric deposition and volatilization and the importance of soil parent material have been discussed earlier in this chapter. Figure 3 provides an illustrative appraisal of some of the key components of Se mass balance, some of which are discussed below.

a. Influence of Grazing Management Pastures may become Se-deficient as a direct result of grazing management. Essentially, deficiency will occur when animal ingestion and absorption exceeds the rate of endogenous excretion (ca. 100 days half-life [55]). For example, if young lambs are removed from the pasture at ages less than 100 days, then equilibrium between gut Se absorption and endogenous excretion will not have been attained (animal in a state of net Se accumulation), and removal of the animal to market will result in a net loss of Se from the grassland. Pasture deficiency will be exacerbated by high stocking rates or grazing by larger animals with high rates of ingestion. Losses will also occur by removal of animal products such as sheep wool or cow's milk. Therefore, animal deficiencies can occur when the rate of supply from the soil to the pasture does not match the rate of removal by agricultural grazing practices.

b. Fertilizer Inputs Addition of Se to the grassland via fertilizer may elevate plant and soil Se levels. Additions may be either to the soil or by spraying onto the foliage for direct uptake. Concerning additions of Se to soil, Bisjerb and Gissel-Nielsen [56] reported that addition of 70 g of Se (sodium selenite) per hectare (or 0.02 µg/g to the plough layer) was required to remove Se deficiencies in New Zealand pastures. Kabata-Pendias and Pendias [57] reported that concentrations of Se in phosphate fertilizers ranged from 0.5 to 25 mg/kg and limestones had only 1 mg/kg. Moreover, Senesi et al. [58] surveyed the Se content of fertilizers and reported that ammonium nitrate and triple superphosphate were high in Se, with concentrations up to 10 and 13.25 mg/kg, respectively. Gissel-Nielsen [59] surveyed the Se content of various fertilizers. None of the other fertilizers surveyed (ammonium sulfate, calcium nitrate, urea, calcium cyanamide, superphosphate, potassium sulfate, NP compounds, and NPK compounds) contained Se.

Gissel-Nielsen [60] conducted a series of field experiments examining the soil application of selenized fertilizers to barley on 21 Danish farms. Selenite was added at amounts corresponding to 0, 60, and 120 g of Se per

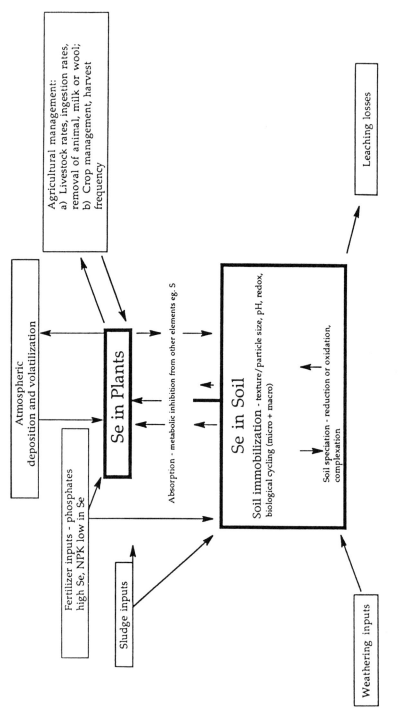

Figure 3 Major components in the mass balance of Se in soil–plant systems.

hectare. After 3 years it was obvious that the additions of the fertilizer caused no biologically significant difference in the Se concentration of the barley. This can be explained largely in terms of rapid soil immobilization and competition for uptake with other anions (particularly sulfate and phosphate) (see below), implying that addition of Se to soils is an inefficient means of controlling livestock deficiency problems.

Direct application of Se to foliar surfaces has been shown to be more effective than soil applications, clearly avoiding the problem of soil immobilization. Gissel-Nielsen [60–62] added 0.5–50 g Se/ha by spraying barley using a high-pressure atomizer. A linear correlation between the amount of Se added and the Se concentration of the mature barley was obtained. Uptake to the barley grain was about twice as great as uptake to the straw, and it was found that application of 5 g Se/ha was sufficient to raise the grain concentration above 0.05 ppm. A more extensive field experiment, covering 21 Danish farms, confirmed the initial findings, leading Gissel-Nielsen to conclude that "foliar application appears to be a safe, inexpensive measure to prevent Se deficiency in livestock." Limitations to foliar application practices are poor recoveries (4.2–8.1% of Se added) and variable results, both of which have been attributed to factors such as local humidity, time elapsed between spraying and rainfall events, and the simultaneous addition of other, inhibiting compounds.

c. Sewage Sludge The spreading of domestic and industrially derived sewage sludge onto agricultural land is known to pose a significant mechanism of input to agricultural soil. In the United Kingdom, sludge dumping at sea will soon cease, so the transfer onto agricultural soil will rise. Chaney [63] reported that typical Se concentrations of sludge in the United States are 1.7–17.2 µg/g with a median value of 5. Kabata-Pendias and Pendias [57] cited a typical global range of 2–9 µg/g. Sauerbeck [64] reviewed sludges in the United Kingdom, quoting a range of 1–10 µg/g, with a common value of 3 µg/g. The maximum permissive Se concentration considered acceptable for application to agricultural lands is 25 µg/g. Sauerbeck also quoted the normal background soil Se level as 0.5 µg/g, with a maximum permissive soil Se level of 3 µg/g (notably higher in France, at 10 µg/g).

Although there is little doubt that land application of sludge increases total Se levels, evidence for transfer into plants and livestock is less clear. Garnett et al. [65] showed that sludges contain high quantities of nitrilotriacetic acid, which increases solubilities of metals in soil, although Se was not one of the metals studied in the work. In a study of accumulations of Se in crops (barley, swiss chard, and radish) grown on sludge-treated soil,

Logan et al. [66] found mobility and transfer from the sludge-treated soil to be minimal. Most significantly, there was little or no measurable uptake of Se by the barley, chard, or radish crops.

d. *Leaching* Bar-Yosef [67] showed that leaching of Se from calcium kaolinite increased as pH increased from 5.6 to 8.7. Yläranta [68] found that <0.2% Se was leached through an artificial column of clay in fine sandy soils, whereas up to 84% was leached from peat. Gissel-Nielsen and Hamdy [69] added selenite to soil columns and found that movement was greatest in sandy soils and in soils to which lime had been added but decreased with the addition of organic matter. They concluded that losses from the rooting zone by leaching would be insignificant in the short-term Se cycle in the environment.

2. Mobilization of Se Through Soil

Selenium is readily immobilized in various soil compartments, making it unavailable for root uptake and obviously contributing to Se deficiencies in grassland and livestock. Mayland et al. [70] defined mobile soil Se as those forms that are moved by water under toxic soil conditions, being available for plant uptake. Selenium immobilization is controlled by adsorption onto mineral material or organic matter, complexation and precipitation with other chemicals, and incorporation into organic matter.

a. *Adsorption-desorption* Kaolinite, montmorillonite, and calcite (for calcareous soils) have been shown to be the important minerals for Se adsorption [70 and references cited therein]. In addition, Hamdy and Gissel-Nielsen [71] found that the vermiculite clay mineral was a sink for selenium. Weerasooriya et al. [72] found that goethite and gibbsite were the main sinks for selenite, most readily adsorbed in soils of pH range 2.0–3.0. Bar-Yosef and Meek [73] studied adsorption of both selenate and selenite into kaolinite and montmorillonite and found that adsorption decreased with increasing pH, becoming negligible at pH 8. Adsorption onto kaolinite was greater by selenite than selenate.

Hydroxide ions are important in modifying Se adsorption capacity [70]. For example, Hamdy and Gissel-Nielsen [71] found iron hydrous oxides to be important in the fixation of Se. The major part of the selenite added to the Fe_2O_3 system was in a poorly available form, suggesting that Se fixation takes place both by exchange reactions and as ferric selenite precipitates. Vuori et al. [74] studied the sorption of selenate on Finnish agricultural soils and found that sorption correlated positively with clay content, iron content, and surface area of the particles and negatively with sulfuronic acid–extractable phosphorus. Yläranta [75] studied selenate and selenite adsorption in relation to soil texture, finding that immobilization was greater in clay (77% of Se added) than peat (39% of Se added) and lowest in fine sandy soil (34% of Se added).

Rajan and Watkinson [76] found that low concentrations of selenite exchanged with adsorbed sulfate, silicate, and hydroxyl ions into clay. Selenite adsorption was found to reach a saturation level and could be desorbed by phosphate, which could be adsorbed to greater strength. Christensen et al. [77] observed that selenite became fixed to fine clay particles (ca. 64–65%) in preference to silt (ca. 45–61%) or sand (<5%), 1 h after addition. Time was also shown to be an important factor, as the amount increased 24 h after addition. The work also demonstrated the importance of adsorption onto organic matter, as the amount adsorbed was halved after the soil was treated with hydrogen peroxide.

b. Turnover in the Soil Organic Material The role of living organic matter in the soil Se cycle has attracted a disproportionately small amount of research. The biological half-life $t_{1/2}$(biol) for Se in most animals (mammals, birds, fish, insects, and annelids) is in the range of 10–28 days; the $t_{1/2}$(biol) for earthworms, *Lumbricus terrestris*, is longer, 64 days [78–81]. Soil bacteria and fungi are also known to have an affinity for accumulating soil Se, although this has been studied primarily with respect to volatilization processes rather than immobilization per se [e.g., 41,42].

Mobility of organic Se compounds, such as seleno amino acids, proteins, and selenomethionine, in soil is a neglected research area. Recent research by Gustafsson and Johnsson [82] has pioneered studies on speciation of Se in association with microbial decomposition and the degree of soil humification. Their novel work found that in Swedish forest soils most Se was associated with hydrophobic fulvates. This is an area of Se cycling in soils that has been neglected and is likely to attract more research in the future.

c. Precipitation–dissolution Reactions Precipitates of Se are extremely important in controlling the chemical mobility of Se in soils. Se has a particular affinity for precipitation with various metals, and this complexation is governed by interaction between redox potential and soil pH. Manganese and calcium selenites are common precipitates in oxic acid soils, with dissolution at higher pH. Lead and barium selenites occur in oxic alkaline soils, with dissolution in more reducing conditions. Coprecipitation with iron oxides and gypsum is another common means of immobilizing selenite in soil solution. Other complexes such as copper and mercury selenide immobilizes Se in reducing conditions across the pH range. It has been postulated that the chemical mobility of Se is favored by sodium hydrogen selenite, potassium ammonium hydride with Se attached, manganese selenate, and nickel selenate [53].

3. Plant Uptake of Selenium

Mechanisms of Se uptake by plants have attracted a small amount of research. Understanding of the mechanisms is relatively poor. Clearly, the processes governing availability in the soil are important (discussed above), but assuming availability is achieved, factors governing adsorption from the rhizosphere require consideration.

Anions are generally taken up directly from the soil solution. Mycorrhizal associations, providing a bridge between plant roots and the soil, are known to be important mechanisms in anion uptake. However, other uptake mechanisms such as carbonic acid exchange and contact exchange reactions may be important where the Se is in a complex that has a net positive charge.

Plant uptake of Se has been shown to be dependent on biochemical interactions with other elements. Interaction with sulfur is particularly significant, as Se has been shown to be taken up and incorporated into Se-methylselenocysteine and selenohomocysteine as a substitute for S-amino acids [83]. Gissel-Nielsen [79] showed that the interaction can work in the opposite direction, whereby S uptake inhibits adsorption of Se into barley and red clover; the effect is more marked with selenate than selenite. Milchunas et al. [84] demonstrated that atmospheric SO_2 uptake can reduce soil Se uptake by *agropyron smithii* from a plant concentration of 2.10 µg/g (control plot) to 1.79 µg/g (SO_2 fumigated plot). Interactions between S and Se have also been reported by Hurd-Karrer [85], Rosenfield and Beath [31], and Singh and Singh [86]. Other chemicals that have been reported to reduce plant Se uptake are ammonia, phosphorus (phosphatic salts increase Se uptake, but superphosphate reduces Se uptake due to the presence of sulfate), and barium chloride [87].

4. Significance of Soil–Plant Transformations

Recent studies have determined the mass balance and flux pathways for Se in a grassland soil–plant system [54]. Over 99% of Se was defined as being fixed, that is, immobile and unavailable for plant uptake, with a flux into the mobile or labile pool on the order of 1 µg/kg soil per year (assuming a soil Se concentration of 0.68 ppm). With a soil depth of 20 cm and density of 1 g/cm^3, this is a flux into the labile pool of about 200 µg/yr. Approximately half of this is leached away to groundwater, and the other half is volatilized to the atmosphere by soil microorganisms. Only 8 µg/(m^2·yr) is taken from the labile pool into grassland plants.

Using this information, Haygarth [54] was able to calculate the recycling time for Se in grassland soil to be on the order of 500–700 yr. On an agricultural scale, this slow cycle rate represents a very large "loss" or "waste" of immobilized Se in the soil, illustrating the potential for livestock Se deficiency brought about by poor soil–plant transfer in grassland systems.

D. Aquatic and Marine Pathways

The most significant pathways, globally the largest fluxes in the Se cycle, are the flux from land along aquatic pathways into and within the marine system. A comprehensive study on these pathways has been undertaken by Nriagu [33], who estimates that transfer from land along aquatic pathways into oceans is a flux of about 14,000 tonnes/yr. Nriagu also suggests that a major transport pathway may be sediment uplift, a geological phenomenon resulting from continental thrusting causing long-term transfer of Se from ocean sediments to land. According to Nriagu's estimates, this pathway is the largest in the global Se cycle, at 15,000 tonnes/yr [33], but it is felt that this estimated rate of transfer is an overestimation. A more realistic scenario may be considerably lower than Nriagu's estimate, with a much slower turnaround in deep ocean sediments. I was unable to find quantitative information on the influence of anthropogenic releases on the marine system, but it is generally acknowledged that atmospheric and waste dumping will constitute significant pathways.

Transfer of Se from the terrestrial ecosystem into groundwater and the aquatic systems has been implicated as a major pathway of loss from the terrestrial ecosystem. Recent studies have indicated that significant proportions of the labile Se in soil and of the Se deposited atmospherically onto soils are rapidly leached into groundwater, representing a major loss mechanism [54]. Processes governing the leaching of Se from soil systems are poorly understood and warrant more research.

Of the worldwide Se load of rivers, 85% is in particulate form [88], indicating that loss of particulate-bound colloidal Se may be an important mechanism in the global Se cycle, presumably especially prevalent in tropical regions. Takayanagi and Wong [89] studied the waters of the Chesapeake Bay and found that the contribution of colloidal inorganic Se to dissolved inorganic Se decreased with increasing salinity.

Studies on the speciation of Se in the marine environment have been undertaken by Cutter and Bruland [90], who postulated the existence of three species of dissolved selenium: selenite, selenate, and organic selenide. The latter is thought to consist of seleno-amino acids in peptides and to coincide with the maxima of primary productivity. Nriagu [33] observed that biological assimilation of Se was on the order of 16,000 tonnes/yr, which exceeds the total input, suggesting that the turnover of Se in the productive zones is very rapid.

The mean residence time for Se in the mixed layer of ocean is estimated to be around 70 yr, whereas the mean life in the deep ocean is about 1100 [33]. This system therefore represents the major sink in the global cycle of Se. Selenium in the marine system has also been reviewed by Siu and Berman [91].

IV. CONCLUSIONS

A. General Summary

A summary of the global mobilization pathways for Se is presented in Table 5 together with collated data on fluxes. Clearly, the largest pathway is the total release of Se as a result of "passive" anthropogenic activity, particularly combustion and metal production activities. The remaining three sources (Table 5) each have particular features that contribute to their importance.

The marine system poses the largest natural cycle pathway, with volatilization and recycling through biota representing the main pathways. On a global scale, the fluxes through the atmosphere are smaller, but these are of elevated importance because of the rapidity of turnover and transport through this system. The terrestrial system is important because terrestrial pathways are involved most directly with human activities such as agricultural crop and livestock production.

B. Selenium in the 21st Century

Paradigms in Se research have altered over the last 20 years, with a move away from a geochemical approach toward that of environmental chemistry and a focus on problem solving. There are still many problems to be resolved, particularly in the health and agricultural sciences.

Further investigations are required regarding the metabolism and effects of Se in animals and humans. In humans, the potential for Se to have a role in cancer prevention continues to receive attention. At the time of writing, a contemporary popular article in the Toronto *Globe and Mail* acknowledges the links between selenium and cancer prevention, illustrating that this is a timely and topical issue [93]. A fuller understanding of the proposed mechanisms is essential. The uses of Se in the diet of grazing

Table 5 Global Selenium Mobilization (tonnes per year) and Pathways

Source	Pathway	Se Flux (tonnes per year)	Reference
Anthropogenic	Mining production	1,600	1[d],2
	Release to atmosphere[a]	7,300	2,4,92
	Release to aquatic pathways	41,000	2,4
	Release to terrestrial system	41,000	2,4
	Release to marine system	?	
	Total	76,000–88,000	2[d]
Atmospheric	Wet deposition to marine system[b]	7,395	33[d]
	Dry deposition to marine system[b]	1,305	33[d]
	Wet deposition to terrestrial system[b]	5,610	33[d]
	Dry deposition to terrestrial system[b]	990	33[d]
Marine	Volatilization to atmosphere	6,700	33[d], 45
	Sea salt suspension	550	5
	Into marine biota	16,000	33
	Sediment transfer to land[c]	<15,000	33
Terrestrial	Volatilization to atmosphere	1,200	33
	Particle suspension	180	5
	Via rivers, particulate, to marine pathways	12,000	33
	Via rivers, dissolved to marine pathways	2,000	33

[a]Mean for cited values.
[b]Assumes 85% is wet deposition.
[c]This is probably an overestimation: see discussion in text.
[d]With modifications.

animals will continue to be a major issue, and progress will be enhanced provided scientists from many disciplines work together.

Multidisciplinary research will also help advance the understanding of the fate and transport of Se in the environment. This chapter has presented examples in which system mass balances have aided an understanding of the mechanisms that, for example, contribute to soil immobilization of Se and therefore to plant and livestock Se deficiency. Soil, as a major medium, will persist as a focus for research. However, previous approaches to Se soil science have centered on chemistry, such as the interaction between redox potential and pH and adsorption–desorption behavior. Clearly, these phenomena are critical in increasing understanding, but it appears that alternative, perhaps equally important, processes such as the mechanisms of plant uptake, the role of soil biota, and organic material have received an unduly small amount of attention. These issues should be addressed into the next century.

Transfers between the environmental compartments are, of course, fundamental in assessing the cycling of elements through the environment. Exchanges between surfaces and the atmosphere are starting to attract attention, and the magnitude of particle suspension, volatilization, and occult deposition are all aspects that have been neglected. Transfer from soil to the aquatic and ultimately marine systems is clearly a major vector (Table 5), yet very little is known about the rates and mechanisms of Se leaching from agricultural soils into groundwater. Budgets and understanding of transformations of Se within the marine system are also required to further knowledge about selenium.

The types of systems that will receive attention are those where the problems occur. With Se deficiency, the system of focus is grassland agriculture. In warmer climes, where irrigation waters are prevalent, concentration of Se in drainage waters has and will continue to receive a considerable proportion of the Se research attention in the United States. Research conducted around Kesterson Reservoir, California, has dominated the scientific literature in the last five years and is an excellent example in which problem-solving research is being employed successfully. Driven by concern over avian populations and toxicity to wildlife in general, knowledge has increased regarding cycling processes such as microbial volatilization.

The future of the Se cycle is clearly in the hands of anthropogenic activity. Humans are persistently releasing Se into the environment that would otherwise have been immobilized for millions of years, and the rate of Se release is much in excess of any natural pathway. This trend is set to continue and will pose fresh challenges to scientists. Problem solving will, out of necessity, be the focus of Se research in the next century.

ACKNOWLEDGMENTS

I wish to thank Dr. Kevin C. Jones for support and encouragement with selenium work and Anne Frain, Linda Heard, and Heather Dennis for helping to prepare the manuscript.

REFERENCES

1. L. W. Newland, *Handbook of Environmental Chemistry*, Springer-Verlag, New York, 1982, pp. 45–57.

2. J. O. Nriagu, *Heavy Metals in the Environment*, Vol. 1, CEP Consultants, Edinburgh, 1991, pp. 1–5.
3. L. Fishbein, *Fundam. Appl. Toxicol.* 3: 411 (1983).
4. J. O. Nriagu and J. M. Pacyna, *Nature* 333: 134 (1988).
5. J. O. Nriagu, *Nature* 338: 47 (1989).
6. P. M. Haygarth, A. I. Cooke, K. C. Jones, A. F. Harrison, and A. E. Johnston, *J. Geophys. Res. (Atmos.)* 98(D9): 16769–16776 (1993).
7. A. Li Wan Po and T. Maguire, *Pharm. J.* April 28, 513 (1990).
8. L. Fishbein, *Toxicol. Environ. Chem.* 12: 1 (1986).
9. E. J. Underwood, *Phil. Trans. Roy. Soc. Lond.* B288: 5 (1979).
10. A. T. Diplock, *Phil. Trans. Roy. Soc. Lond.* B 294: 105 (1981).
11. H. J. Robberecht and H. A. Deelstra, *Clin. Chim. Acta* 1984: 107.
12. Z. Maksimovic, V. Jovic, I. Djujic, and M. Rsumovic, *Environ. Geochem. Health* 14(4): 107 (1992).
13. World Health Organisation, Review of potentially harmful substances— arsenic, mercury and selenium, Reports and Studies No. 28, USEPA, Geneva, 1986.
14. J. P. Robin, *Heavy Metals in the Environment*, Vol. 2, CEP Consultants, Edinburgh, 1991, pp. 283–285.
15. H. M. Ohlendorf, D. J. Hoffman, M. K. Saiki, and T. W. Aldrich, *Sci. Total Environ.* 52: 49 (1986).
16. H. M. Ohlendorf, R. L. Hothem, C. M. Bunck, and K. C. Marois, *Arch. Environ. Contam. Toxicol.* 19: 495 (1990).
17. H. M. Ohlendorf and J. P. Skorupa, *Proceedings of the Fourth International Symposium on Uses of Selenium and Tellurium*, Selenium-Tellerium Development Association Inc., Darien, 1989, pp. 314–338.
18. H. M. Ohlendorf, *Selenium in Agriculture and the Environment*, (L. W. Jacobs, Ed.), SSSA Spec. Publ. No. 23, 1989, pp. 133–177.
19. J. E. Oldfield, *Environ. Geochem. Health* 14(3): 81 (1992).
20. C. A. Girling, *Agric. Ecosyst. Environ.* 11: 27 (1984).
21. L. F. James, K. E. Panter, H. F. Mayland, M. R. Miller, and D. C. Baker, *Selenium in Agriculture and the Environment*, (L. W. Jacobs, Ed.), SSSA Spec. Publ. No. 23, 1989, pp. 123–131.
22. V. Cortese, *Agri-Pract.* 9(4): 5 (1988).
23. E. Ropstad, O. Østerås, G. Øvernes, and A. Frøslie, *Acta Vet. Scand.* 29(2): 159 (1988).
24. R. Slaweta, T. Laskowska, and E. Szymanska, *Anim. Reproduct. Sci.* 17: 303 (1988).
25. P. H. Anderson, S. Berrett, and D. S. P. Patterson, *Vet. Record* 104: 235 (1979).
26. W. Wuyi, S. van Dorst, and I. Thornton, *Environ. Geochem. Health* 9: 48 (1987).
27. A. Frøslie, J. T. Karlsen, and J. Rygge, *Acta Agric. Scand.* 30:19 (1980).
28. B. J. Alloway, *Heavy Metals in Soils*, Blackie, London, 1990.
29. P. M. Haygarth, K. C. Jones, and A. F. Harrison, *Sci. Total Environ.* 103: 89 (1991).
30. X. Wu and J. Låg, *Acta. Agric. Scand.* 38: 271 (1988).
31. I. Rosenfield and O. Beath, *Selenium, Geobotany, Biochemistry, Toxicity and Nutrition*, Academic, New York, 1964.

32. J. S. Webb, I. Thornton, and K. Fletcher, *Nature* 327(5046) (1966).
33. J. O. Nriagu, *Occurrence and Distribution of Selenium*, (M. Inhat, Ed.), CRC Press, Boca Raton, Fl., 1989, pp. 327–340.
34. R. Atkinson, S. M. Aschmann, D. Hasegawa, E. T. Thompson-Eagle, and W. T. Frankenberger, *Environ. Sci. Technol.* 24: 1326 (1990).
35. R. A. Duce, B. J. Ray, G. L. Hoffman, and P. R. Walsh, *Geophys. Res. Lett.* 3: 339 (1976).
36. B. W. Mosher and R. A. Duce, *J. Geophys. Res.* 88: 6761 (1983).
37. S. Jaing, H. Robberecht, and F. Adams, *Atmos. Environ.* 17: 117 (1983).
38. T. J. Ganje and E. I. Whitehead, *Proc. S.D. Acad. Sci.* 37: 81 (1958).
39. J. W. Doran and M. Alexander, *Soil Sci. Soc. Am. J.* 41: 70 (1977).
40. R. Zieve and P. J. Peterson, *Sci. Total Environ.* 19: 277 (1981).
41. U. Karlson and W. T. Frankenberger, *Sci. Total Environ.* 92: 41 (1990).
42. R. W. Fleming and M. Alexander, *Appl. Microbiol.* 24,D: 424 (1972).
43. E. T. Thompson-Eagle, W. T. Frankenberger, and U. Karlson, *Appl. Environ. Microbiol.* 55: 1406 (1989).
44. P. M. Haygarth, D. Fowler, S. Stürup, B. M. Davison, and K. C. Jones, Determination of gaseous and particulate selenium over a rural grassland in the UK, *Atmos. Environ.*, submitted.
45. B. W. Mosher and R. A. Duce, *J. Geophys. Res.* 92(D11): 13277–13287 (1987).
46. P. A. Cawse, *Pollutant Transport and Fate in Ecosystems*, Blackwell, Oxford, 1987, pp. 89–112.
47. D. Fowler, J. N. Cape, I. D. Leith, T. W. Choularton, M. J. Gay, and A. Jones, *Atmos. Environ.* 22(7): 1355 (1988).
48. D. Fowler, J. N. Cape, and M. H. Unsworth, *Phil. Trans. Roy. Soc. Lond. Ser. B* 324: 247 (1989).
49. D. Fowler, *Phil. Trans. Roy. Soc. Lond. Ser. B* 305: 281 (1984).
50. J. L. Monteith, *Principles of Environmental Physics*, Edward Arnold, London, 1973.
51. P. M. Haygarth, A. F. Harrison, and K. C. Jones, *Environ. Sci. Technol.* 27(13): 2878–2884 (1993).
52. J. A. Garland, *Proc. Roy. Soc. (Lond.) Ser. A* 354: 245 (1977).
53. R. H. Neal, *Heavy Metals in Soils*, Blackie, London, 1990, pp. 235–260.
54. P. M. Haygarth, The role of the atmosphere in the cycling of selenium through soil-plant systems, Ph.D. Thesis, Lancaster Univ., UK, 1992.
55. J. Leibetseder, A. Kment, and M. Skalicky, *Proc. 2nd Int. Symp. Trace Element Metabolism in Animals*, 1974, pp. 581–583.
56. B. Bisjerg and G. Gissel-Nielsen, *Plant Soil* 31: 287 (1969).
57. A. Kabata-Pendias and H. Pendias, *Trace Elements in Soils and Plants*, CRC Press, Boca Raton, Fl., 1984.
58. N. Senesi, M. Polemio, and L. Lorusso, *Commun. Soil Sci. Plant Anal.* 10: 1109 (1979).
59. G. Gissel-Nielsen, *J. Agric. Food Chem.* 19: 564 (1971).
60. G. Gissel-Nielsen, *Risø Report No. 370*, Roskilde (1977).
61. G. Gissel-Nielsen, 7th Int. Colloquium, Hanover, Germany (1974).
62. G. Gissel-Nielsen, *Pflanzenernaehr. Bodenkd.* 1: 97 (1975).

63. R. L. Chaney, Final Report of the Workshop on the International Transportation, Utilization or Disposal of Sewage Sludge Including Recommendations, PNSP/85-01, Pan American Health Organization, Washington, D.C., 1985, pp. 1–56.
64. D. Sauerbeck, *Scientific Basis for Soil Protection in the European Community*, Elsevier, London, 1987, pp. 181–210.
65. K. Garnett, P. W. W. Kirk, R. Perry, and J. N. Lester, *Environ. Pollut. Ser. B* 12: 145 (1986).
66. T. J. Logan, A. C. Chang, A. C. Page, and T. L. Ganje, *J. Environ. Qual.* 164: 349 (1987).
67. B. Bar-Yosef, *Commum. Soil Sci. Plant Anal.* 18(7): 11 (1987).
68. T. Ylåranta, *Ann. Agric. Fenn.* 22: 29 (1983).
69. G. Gissel-Nielsen and A. A. Hamdy, *Z. Pflanzenernaehr. Bodenkd.* 140: 193 (1977).
70. H. F. Mayland, L. P. Gough, and K. C. Stewart, *Proceedings of the 1990 Billings Land Reclamation Symposium on Selenium in Arid and Semi-Arid Environments, Western United States*, U.S. Geol. Survey Circ. 1064, 1990, pp. 57–64.
71. A. A. Hamdy and G. Gissel-Nielsen, *Z. Pflanzenernaehr. Bodenkd.* 140: 63 (1977).
72. S. V. R. Weerasooriya, S. B. Bulumulla, S. A. Tilakaratne Bandara, and M. U. Jayasekara, *Int. J. Environ. Stud.* 33: 111 (1989).
73. B. Bar-Yosef and D. Meek, *Soil Sci.* 144(1): 11 (1987).
74. E. Vuori, J. Vaariskoski, H. Hartikainen, P. Vakkilainen, J. Kumpulainen, and K. Niinvaara, *Agric. Ecosyst. Environ.* 25: 111 (1989).
75. T. Ylåranta, *Ann. Agric. Fenn.* 22: 29 (1983).
76. S. S. S. Rajan and J. H. Watkinson, *Soil Sci. Soc. Am. J.* 40: 51 (1976).
77. B. T. Christensen, F. Bertelsen, and G. Gissel-Nielsen, *Soil Sci.* 40: 641 (1989).
78. G. Gissel-Nielsen, *Ambio* 2: 114 (1973).
79. G. Gissel-Nielsen, *Sci. Food Agric.* 24: 649 (1973).
80. A. C. Evans, *Appl. Biol.* 35(1): 1 (1948).
81. E. Satchel, *Soil Biology*, Academic, London, 1967, pp. 259-322.
82. J. P. Gustafsson and L. Johnsson, *J. Soil Sci.* 43(3): 461 (1993).
83. D. G. Milchunas and W. K. Lauenroth, *Sulphur Deposition, Cycling and Accumulation: The Effects of SO_2 on Grassland*, Springer-Verlag, New York, 1984, pp. 61–95.
84. D. G. Milchunas, W. K. Lauenroth, and J. L. Dodd, *Plant Soil* 72: 117 (1983).
85. A. Hurd-Karrer, *J. Agric. Res.* 49: 343 (1934).
86. M. Singh and N. Singh, *Soil Sci.* 127: 264 (1979).
87. G. A. Fleming, *Applied Soil Trace Elements*, Wiley, London, 1980, pp. 213–234.
88. G. A. Cutter, *Occurrence and Distribution of Selenium*, (M. Inhat, Ed.), CRC Press, Boca Raton, Fl. (1987), pp. 243–263.
89. K. Takayanagi and G. T. F. Wong, *Mar. Biol.* 14: 141 (1984).
90. G. A. Cutter and K. W. Bruland, *Limnol. Oceanogr.* 29(6): 1179 (1984).
91. K. W. M. Siu and S. S. Berman, *Occurrence and Distribution of Selenium*, (M. Inhat, Ed.), CRC Press, Boca Raton, Fl. (1987), pp. 263–293.
92. H. B. Ross, *Tellus* 37B: 78 (1985).
93. "Vitamins reduce cancer risk, study finds," *The Globe and Mail (Toronto, Canada)*, Wednesday, Sept. 14, 1993, p. A13.

2
Selenium in Plant and Animal Nutrition

H. F. Mayland

Agricultural Research Service, U.S. Department of Agriculture
Kimberly, Idaho

INTRODUCTION

Selenium (Se), while not required by plants, is an essential trace element for adequate nutrition and health for fish, birds, animals, and humans. Generally, diets containing 0.1–0.3 mg/kg Se will provide adequate Se for these various animals. However, many soils are incapable of providing that amount to the plants growing on them. Animals consuming low-Se diets will be Se-deficient, grow poorly, or even die. Conversely, there are soils that provide an abundance of soluble Se. Some plants growing on these Se-rich soils may accumulate Se in excess of the 3–15 mg/kg concentration at which animals begin to show Se toxicity symptoms. This multifaceted characteristic of Se makes it imperative that scientists and policy makers recognize the deficiency, adequacy, and toxicity effects of Se on animal health. This chapter presents information about these aspects of Se in the plant and animal system.

HISTORICAL PERSPECTIVE OF SELENIUM NUTRITION

Deficiency

The nutritional value of Se was first recognized in 1957 when it was found to have a complementary role to vitamin E in preventing dietary hepatic necrosis and exudative diathesis in rats and chicks [1,2]. Ensuing reports were published of similar nutritional interactions between Se and vitamin E in birds and animals. In the late 1960s, a specific nutritional requirement for Se was established for chicks [3].

Eventually, Se was shown to be an essential constituent of the biologically important enzyme glutathione peroxidase (SeGSHpx) [4]. SeGSHpx, superoxide dismutase, and catalase convert free radicals to peroxides and then to water and oxygen; whereas, vitamin E scavenges the free radicals and neutralizes their potential damaging effects. Thus, low selenium intake with vitamin E deficiency increases oxidative stress and contributes to the development of oxidative damage.

Combs and Combs [5] reviewed the Se deficiencies affecting fish, laboratory animals, poultry, livestock, and humans. Clinical signs include reduced appetite, growth, production, and reproductive fertility, a general unthriftiness, and muscular weakness. Specific disorders include exudative diathesis and increased embryonic mortality in birds. Nutritional muscular dystrophy is found in birds, fish, and animals. Retained placenta is reported in Se-deficient cows, while mulberry heart disease is noted in pigs. Severe nutritional Se deficiency is associated with endemic juvenile cardiomyopathy (i.e., Keshan disease) in youngsters from a discrete area in China. Selenium may also be involved in the etiology of chondrodystrophic disease (i.e., Kaschin-Beck disease) in young Chinese children.

Toxicity

Marco Polo described a necrotic hoof disease in his horses during his travels in western China in the thirteenth century [6]. He associated the problem with the ingestion of certain plants that were generally avoided by local animals.

In 1560, in Colombia, South America, Father Pedro Simon described hair and hoof loss, tender bone joints, reproduction disorders, and deaths in domestic animals [7]. These disorders were later attributed to Se toxicosis [7]. Father Simon also noted malformation of children and chickens and Indian women giving birth to monsters that were abandoned by their parents. The natives associated the problem with ingestion of foodstuffs grown on certain soils.

The problem was documented again in the mid-nineteenth century by a U.S. Army surgeon, T. W. Madison, who described similar necrotic and sloughed hooves and deaths of horses grazing near Fort Randall, South Dakota [8]. Ranchers associated the toxicosis with the saline seeps and outcrops common to much of the northern Great Plains and named the problem "alkali disease." Alkali disease still occurs in ruminants and monogastrics inhabiting seleniferous areas.

By 1931, researchers identified alkali disease as chronic Se toxicosis (selenosis) characterized by hair and hoof loss and poor growth and reproduction. A second disorder occurring in the area, "blind staggers," results in varying degrees of vision impairment. This neurological dysfunction occurs only among ruminants [8] and was attributed to excess Se in the forage. However, this disorder may be the result of ingesting excess sulfur (S) rather than Se [9].

Selenium toxicosis has been observed in waterfowl inhabiting areas where sediments and aquatic vegetation contain excess Se levels. Ohlendorf [10] described embryocidal deformities in birds feeding on Se-enriched feedstuffs. Earlier reports of these deformities had stimulated investigation of both natural and anthropogenic factors associated with Se cycling.

SELENIUM IN WATER

Selenium naturally enters the food chain through water. It occurs as a minor constituent in water at concentrations ranging from <0.1 to 100 µg Se/L [11]. Few samples exceed the 10 µg Se/L upper limit established by the 1977 Safe Drinking Water Act of the U.S. Environmental Protection Agency [12]. Water derived from Cretaceous geological zones may contain as much as 1000 µg Se/L. Regional rivers that drain such areas may have concentrations approaching 10 µg Se/L [13]. Most of these areas are found in the central plains and western deserts of North America and the interior deserts of other continents. Another area of economic interest is the accumulation of seleniferous drainage waters in the San Joaquin Valley of central California. Selenium may also accumulate in other catchment areas where evaporation concentrates soluble salts.

Small amounts of Se may be found in aerosols that enter the atmosphere and contribute to global cycling of Se. This Se originates from volcanic eruptions and the fine particulates that are generated from fossil fuel combustion and the incineration of municipal wastes. Water used to wash down smoke stacks or precipitators also contains significant levels of Se [13]. Some organic forms are volatilized directly from biological activities. Eventually, these Se compounds are returned to earth. Globally, wet deposition of Se aerosols returns about 1.5 g Se/ha annually.

SELENIUM IN SOILS

Depending on the redox potential of the soil, Se occurs in many different forms. Concentrations in most soils lie within the range of 0.01–2 mg Se/kg. However, some seleniferous soils may contain as much as 38 mg Se/kg as water-soluble selenate. Other soils such as those in Hawaii, Ireland, and the Amazonian rain forest also contain high levels of total Se, but the Se is relatively unavailable to most plants [13]. Inorganic Se forms like SeO_4, SeO_3, and Se^0 have a wide range of solubility in water and subsequent bioavailability to plants and animals (Table 1). Selenium concentrations in plants are related approximately to broad areas described by geology and soils (Figure 1).

Organic forms, including selenomethionine, have been extracted from soils and represent an important source of plant-available Se [14,15]. Selenomethionine is two to four times as available to plants as selenite [16], and its uptake is under metabolic control [17]. Selenocystine is less bioavailable than selenomethionine [16]. In some soils, nearly 50% of the Se may be in organic forms [15]. Identifying these forms will be challenging but is necessary if scientists are to better understand Se cycling.

Table 1 Selenium Solubility in Water and Relative Uptake of Se by Plants from Various Sources Labeled with ^{75}Se in Pot Experiments Using a Loamy Sand Having 2.8% Organic Matter, 5.7 pH, and 0.12 mg Se/kg

Se source	Solubility of Se in cold water[a] (g/L)	Se added to soil (mg/kg)	Uptake relative to added Se (%)		
			Clover	Barley	Mustard
Se	i	2.5	0.005	0.02	0.07
SeO_2	i	0.5	1.0	0.9	1.2
K_2SeO_3	22.4	0.5	1.0	1.1	1.3
Na_2SeO_3	s	0.5	1.0	1.0	1.1
$BaSeO_3$	0.05	0.37	0.9	0.9	0.9
$FeSeO_3$	i	0.35	1.1	1.0	1.1
$CuSeO_3$	i	0.30	0.8	0.8	0.7
K_2SeO_4	390	0.50	24.	12.	24.
$BaSeO_4$	0.03	0.10	63.	27.	61.
$CuSeO_4$	68	0.13	53.	28.	48.

[a]i = insoluble; s = slightly soluble.
Source: Adapted from Mayland et al. [13].

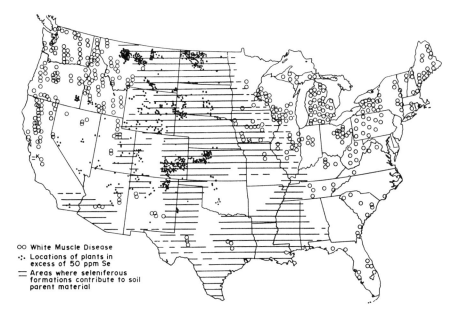

Figure 1 The geographical distribution of Se-rich soils (horizontal lines), locations where plants contain in excess of 50 μg Se/kg (solid dots), and locations of Se deficiency in animals (white muscle disease, clinically termed nutritional muscular dystrophy, open circles). (Adapted from Muth and Allaway [35].)

SELENIUM IN PLANTS

On moderately low Se soil, alfalfa accumulates more Se than many other forage plants [13]. This characteristic may be related to differences in rooting depth and to genetic traits that affect the absorption and translocation of Se to shoots. Sulfur fertilization of legumes will often reduce Se uptake and concentration in the forage [18]. McQuinn et al. [19] estimated that Se concentrations could be increased by 19% in tall fescue (*Festuca arundinacea* Shreb.) through genetic selection. This species is adapted to most of the Se-deficient pastoral areas in the United States. Genetic selection in this forage species promises to increase herbage Se levels and satisfy the Se needs of grazing animals in marginally deficient areas. Similar breeding opportunities may exist in other forages.

Plants exhibit genetic differences in Se uptake when grown on seleniferous soil. Some plants accumulate surprisingly low levels of Se. For example, white clover (*Trifolium repens* L.), buffalograss (*Buchloe dactyloides*

[Nutt.] Engelm.), and grama (*Bouteloua* spp.) are poor accumulators of Se. On the other hand, S-rich plants like the *Brassica* spp. (mustard, cabbage, broccoli, and cauliflower) and other cruciferae are good concentrators of Se [11].

Rosenfeld and Beath [8] identified three plant groups according to their ability to accumulate Se when growing on Se-rich soils. The first two groups of plants were identified by their potential to accumulate moderate or very high concentrations of Se. These are the plants that grow successfully on soil containing high levels of available Se. The presence of these plants and the characteristic dimethylselenide odor are indicative of seleniferous soils. These plants have a different metabolic pathway that shunts Se into nonprotein forms [20]. Nevertheless, a Se requirement has not been shown for any plants.

Plant genera that can accumulate very high concentrations of Se include many species of *Astragalus*, *Machaeranthera*, *Haplopappus*, and *Stanleya*. On a dry weight basis, these species absorb high concentrations of Se, from hundreds to occasionally even thousands of milligrams per kilogram. These plants are found in semiarid environments throughout west central North America (Figures 2 and 3) and other continents. The absence of deep percolation, neutral to alkaline soil pH, and oxidative conditions have allowed much of the soil Se to remain in place. Precipitation in excess of evapotranspiration normally leaches out the soluble Se salts. An exception seems to occur in the Amazonian Plateau, where several members of the Amazonian *Lecythidaceae* family also accumulate high concentrations of Se [7]. Investigations into Se cycling in this area of high precipitation would be interesting.

Plant genera having the potential to accumulate moderately high concentrations of Se include many species of *Aster* and some species of *Astragalus*, *Atriplex*, *Castilleja*, *Grindelia*, *Gutierrezia*, *Machaeranthera*, and *Mentzelia*. They rarely concentrate more than 50–100 mg Se/kg. Nonaccumulator plants make up the third group. It includes grains, grasses, and many forbs that do not usually accumulate more than 50 mg Se/kg when grown on seleniferous soil.

Alfalfa (*Medicago sativa* L.) is commonly grown in seleniferous areas like the Kendrick Reclamation Project area of central Wyoming. A Se survey of alfalfa conducted there during 1988 reported a range of 0.1–40 mg Se/kg with a median of 0.9 mg Se/kg [21]. However, the next year, alfalfa that had contained 17 and 25 mg Se/kg now contained only 0.7 and 0.2 mg Se/kg, respectively. The significant reduction in Se values was attributed to percolation of soluble Se beyond the rooting zone and to dilution in the plant material resulting from increased dry matter production.

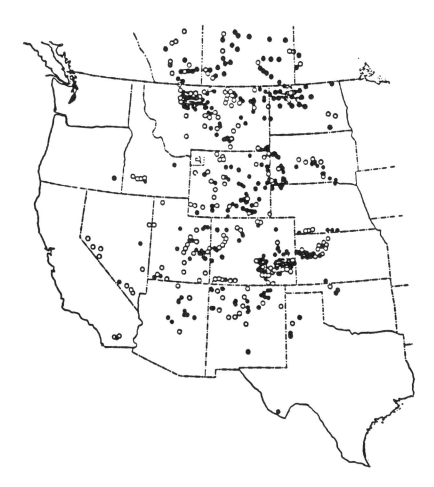

Figure 2 Distribution of seleniferous vegetation in the western United States and Canada. (Adapted from Rosenfeld and Beath [8].) Each open dot represents the place of collection of a plant specimen containing 50–500 mg Se/kg; each solid dot represents specimens containing more than 500 mg Se/kg.

Infrequent incidence of selenosis has been reported on the Kendrick Project in central Wyoming. Tolerance to high Se levels varies considerably among individual animals and birds [22]. In addition, experimental evidence suggests that some animals can accommodate high levels of dietary Se after evidencing some symptoms of chronic toxicosis, such as lameness and hair loss (H. F. Mayland, personal observation).

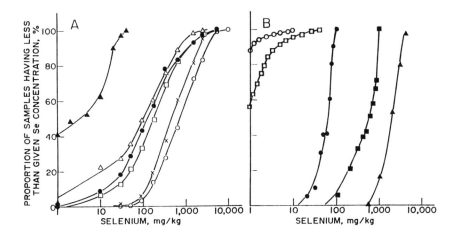

Figure 3 Proportion of samples (percent) having less than given Se concentration. (A) Data for (▲) western wheatgrass (*Pascopyrum smithii* [Rydb.] A. Love) and Sandberg bluegrass (*Poa secunda* Presl.) sampled from seleniferous areas of Montana and Wyoming; (Δ) *Stanleya* spp.; (●) *Xylorhiza* section of *Machaeranthera*; (□) *Astragalus bisulcatus*; (X) *Astragalus pectinatus*; (○) *Oonopsis* section of *Haplopappus*. (B) (●) Vegetative wheat (*Triticum aestivum* L.); (■) *Astragalus bisulcatus*; and (▲) *Astragalus pectinatus* reported in plants from North Dakota. (Data adapted for graphical presentation by Mayland et al. [13]). (○) Big sagebrush (*Artemisia tridentata* Nutt.) and (□) alfalfa (*Medicago sativa* L.) data from Kendrick Project, central Wyoming (adapted from Erdman et al. [36]).

Many different Se compounds have been identified in plants [20]. Much of the Se in nonaccumulating species is found as protein-bound selenomethionine. In contrast, the Se in accumulator plants is mostly water-soluble and found in nonprotein forms like selenium methylselenocysteine. Only trace amounts of the latter compound are found in nonaccumulator species. Selenomethionine, selenocystine, and possibly selenium-methylselenomethionine and selenonium have been detected in nonaccumulators but not in the accumulators tested [23]. The Se metabolites in plants are generally analogues of S compounds. Nevertheless, Se metabolism in nonaccumulator plants cannot be identified from known mechanisms because of scientists' limited understanding of the metabolic pathways for Se in plants [20].

Microorganisms can reduce selenate to elemental Se^0 and Se^{2-}. Many microorganisms, plants, and animals reduce selenite to selenide, giving rise to volatile organic forms. Dimethylselenide is the volatile and odoriferous Se compound that is characteristic of Se accumulator plants. The compound is also detected in the breath of animals and humans respiring excess Se [5,7]. I have detected the aroma of dimethylselenide within an hour of spraying sodium selenite on alfalfa foliage. Obviously, the selenite was rapidly metabolized to the dimethylselenide by the plants or by the microorganisms present on the plants or on ground.

Two methylated Se compounds, dimethylselenide and dimethyldiselenide, are respiratory products of microorganisms, plants, animals, and humans [7, 10, 13, 16]. Hydrogen selenide (H_2Se) is another volatile Se compound. It is highly toxic but under atmospheric conditions quickly decomposes into innocuous Se^0 and water [5].

BIOAVAILABILITY OF SELENIUM IN FEEDSTUFFS

Selenium utilization varies greatly, depending upon the chemical form in which it is fed. Selenium compounds that are insoluble or have low digestibility pass through the digestive tract and are excreted in the feces. Apparent absorption of Se in feedstuffs, inorganic compounds, and Se amino acids is about 70%; however, absorption is highly variable among and within sources [5,24].

Not all Se absorbed is physiologically important. Some is metabolized to methylated forms that are excreted readily. For example, most of the Se in urine is trimethylselenonium. Plants and animals alike metabolize some Se to the volatile dimethylselenide and dimethyldiselenide, which are respired and lost to the atmosphere.

Selenium is normally metabolized to Se analogues of the S-containing amino acids and then to Se-containing polypeptides and proteins. Of these, the Se-dependent enzyme glutathione peroxidase (SeGSHpx) is the only physiologically critical species known [5]. As an analogue of S, Se may be incorporated into some Se proteins that are physiologically inert.

The biological utilization of dietary Se depends on the form and the integrated response of several physiological and metabolic processes of varying complexity. The net response is quantitatively called the bioavailability of the given Se source, an experimentally derived value that must be considered in the context of the measured response. Disease prevention, tissue Se levels, and SeGSHpx activities are criteria of bioavailability. Thus, Se bioavailability values are often expressed on a percentage basis, with the bioavailability of sodium selenite assigned a value of 100%.

Combs and Combs [5] list bioavailability estimates for nearly 300 inorganic and organic Se compounds based on their ability to prevent hepatic necrosis in rats and exudative diathesis in chicks. Assuming sodium selenite to be 100% bioavailable, Se in animal by-product feedstuffs (including fish meal) have low availability (9–25%), while that in various plant products have a much higher bioavailability, of about 80%.

Blood Se concentrations less than 120 ng/L are positively correlated with blood SeGSHpx in humans and other animals. To maintain this concentration, the bioavailability of Se in plant products must be comparable with or greater than that of sodium selenite. However, the Se in foods or feedstuffs of animal origin has a much lower bioavailability [5].

Measures of SeGSHpx activity provide another approach to estimating bioavailability, which generally substantiates the relative value of Se in various feedstuffs determined by the other methods. Bioavailability of Se can be affected by factors other than Se source. A reduction in feed intake often increases the utilization of ingested Se. Dietary fats, and especially unsaturated plant oils, increase the utilization of dietary Se. Suboptimal levels of the S-containing amino acid methionine have been shown to reduce the utilization of Se from selenomethionine [5]. High levels of S, however, have been shown to decrease the metabolism of Se to tissue SeGSHpx in animals and result in increased frequency of skeletal myopathies in lambs and calves. Selenium has been identified as an antagonist against the toxicity of other heavy metals [5,25]. However, it has been shown to exhibit a synergistic effect on lead poisoning in sheep [26].

Studies with common foods and animal feedstuffs have shown that despite variations between bioassay methods and species, some materials, like yeasts, wheat grain, and alfalfa, generally provide highly available sources of Se. Compared to the bioavailability of Se in selenate or selenite, the Se in plant-derived foods and feedstuffs is moderately available, whereas Se in materials of animal origin is poorly available [5].

The toxicity of high levels of Se can be reduced by feeding high levels of protein. Protection is provided by feeding casein, lactalbumin, linseed oil meal, and torula yeast [5]. The active factor in linseed-oil meal is a cyanogenic glycoside. Some heavy metals may increase the biliary elimination of Se but may also potentiate the acute toxicity of trimethylselenonium [5,26]. Supplemental methionine has been shown to protect against Se toxicity by forming readily excreted methylated metabolites like dimethylselenide and trimethylselenonium ion. Large-scale application of these remedies may not be economically practical. Changing diets is often the only realistic solution.

Studies with rats have shown that males are less sensitive than females to intoxication by selenite but more sensitive to methylated forms [5]. Males can adapt to higher levels of Se than females. Genetic strains of chickens and swine have shown heritable traits in response to Se deficiency or toxicity levels, respectively. Animals may be conditioned or acclimated to moderately toxic levels of Se (H. F. Mayland, personal observation). They may adapt to the high Se exposure by increasing their metabolic production of methylated Se compounds, which are readily excreted.

BIOAVAILABILITY OF SELENIUM IN FECES, URINE, AND RESPIRATORY PRODUCTS

Urine is the primary route of Se excretion by monogastric animals. The main route of Se excretion in ruminants, though, depends on the method of administration and the age of the animal [11]. When Se is ingested by ruminants, most of it is excreted in feces. In contrast, Se that is injected either intravenously or subcutaneously into ruminants is excreted mostly in urine. Lambs, and presumably calves, that have not developed rumen function can excrete 65–75% of the orally ingested Se in the urine. As these animals develop functioning rumen systems, microorganisms transform the Se into unavailable forms such as elemental Se, which are then excreted in the feces.

Nearly all of the Se excreted in the feces of ruminants is in an unavailable form, and very little is available for uptake by plants. Research reports summarized by Mayland et al. [13] noted that <0.3% of the Se taken up by plants originated from the Se contained in sheep manure during a 75-day study.

Trimethylselenonium ion (TMSe$^+$) is the primary urinary metabolite. This source is readily absorbed and translocated to leaves and stems of wheat, but not to the grain [27]. However, large differences were observed in Se uptake by barley, wheat, and alfalfa when TMSe$^+$ was applied in a pot study in the greenhouse. Very little of the Se from TMSe$^+$ was absorbed by the plants, and some absorbed TMSe$^+$ was even lost to the atmosphere through volatilization from the plant or perhaps from microbial respiration [27]. Therefore, TMSe$^+$ excreted in animal urine contributes little biologically active Se to plants.

Dimethylselenide is the principal respiratory product of animals ingesting excess Se. Dimethyldiselenide may also be respired, and the proportion of the two compounds is dependent upon the Se source [5]. Dimethylselenide is also respired by plants [13] and accounts for the distinctive odor of Se accumulator plants. These methylated forms are likely absorbed by plants. The Se enrichment of plants growing in Se-free nutrient culture could have occurred by foliar absorption of Se volatilized from adjacent plants growing in selenized nutrient culture [16].

DIAGNOSTICS FOR ADEQUATE SELENIUM NUTRITION IN ANIMALS

Several syndromes in cattle and sheep have been classified as selenium-responsive conditions (Table 2) on the basis of current information [28]. Some of these syndromes are complex because they involve interactions with other nutrients. Scientists have just begun to learn about the involvement of Se with the immune system [29]. Blood levels of over 100 µg Se/L in cattle [30] and 180–230 µg Se/L in swine [31] are needed to maintain optimum immunocompetence. Measures of whole-blood Se and SeGSHpx in the hemoglobin are useful in interpreting the Se nutritional status in

Table 2 Selenium-Responsive Diseases of Cattle and Sheep

Syndrome	Major clinical features
Nutritional myodegeneration (white muscle disease)	Acute onset, stiffness, skeletal and/or cardiac muscles affected. Signs vary from acute death to chronic lameness. Pale, necrotic areas of muscle; Zenker's necrosis present histopathologically.
Retained placenta	Retained placenta.
Abortions, stillbirths	Late third-trimester abortions and stillbirths.
Neonatal weakness	Neonates born weak, with or without gross lesions of nutritional myodegeneration.
Diarrhea	Diarrhea, usually profuse, and weight loss in young and adult cattle.
"Ill thrift" syndrome	Decreased feed efficiency, decreased weight gains, and unthrifty appearance.
Immune system effects	Cell-mediated immune response suppression.
Myodegeneration in adult cattle	Weakness, myodegeneration, myocardial fibrosis, myoglobulinuria.
Infertility	Decreased conception rate, irregular estrous cycles, early embryonic death.

Source: Adapted from Maas and Kohler [28].

Table 3 Selenium Diagnostics for Cattle and Sheep

Category	Whole-blood selenium (mg/kg)	GSHpx μg/(mg heme·min)	Results of Se supplementation
Deficient	0.01–0.04	0–15	Usually beneficial
Marginal	0.05–0.06	15–25	Often beneficial
Normal	0.07–>0.10	25–500	Seldom beneficial

Source: Adapted from Maas and Kohler [28].

cattle and sheep (Table 3). Similar criteria are used in determining Se status of human nutrition [5].

MANAGEMENT OF SELENIUM-RESPONSIVE DISEASES

Producers and veterinarians have several methods for treating Se-deficient animals. The most commonly used therapies in the United States are (1) injectable Se products, (2) salt-mix formulations with supplemental Se, and (3) total-ration formulations with supplemental Se. The U.S. Food and Drug Administration regulations in the United States [32] now permit only 0.1 mg/kg as Se supplementation in the diet. Some practitioners question whether this limit is adequate to meet the nutritional needs of animals [31], especially in areas where feeds may contain high levels of S that reduce Se bioavailability. More information is needed about the bioavailability and cycling of various Se forms in the ecosystem.

Soils, plants, animals, and humans in New Zealand and Finland are deficient to marginally deficient in available Se. These countries have resorted to Se fertilization of crop-producing areas to increase Se concentration in pasture, cereal, and other food crops [13]. To overcome widespread Se deficiencies, these two countries applied selenate fertilizer to crops in the early 1980s. Increasing soil Se levels has effectively increased the general level of Se in feedstuffs for both animals and humans [13,33].

Sulfur deficiencies in the U.S. Pacific Northwest have necessitated S fertilization. The added S has reduced the bioavailability of soil Se and increased the incidence of Se deficiencies in calves and lambs (H. F. Mayland, personal observation).

SELENIUM IN HUMAN NUTRITION AND HEALTH

In the mid-1960s, severe nutritional Se deficiency was identified in specific areas of China [34]. The deficiency was associated with an endemic juvenile cardiomyopathy (i.e., Keshan disease). Selenium deficiency was also implicated in the etiology of a chondrodystrophic disease (i.e., Kaschin-Beck disease) of children in severely Se-deficient parts of China. Such severe deficiencies have not been noted elsewhere.

The efficacy of sodium selenite for the prevention of Keshan disease was evident as early as 1974. Selenium was supplemented to the human population in the affected areas by (1) distribution of selenite tablets, (2) enrichment of table salt with sodium selenite, and (3) foliar sprays of sodium selenite on grain crops. Recent surveys in the affected areas indicate a decline in the incidence of Keshan disease. However, selenite has failed to provide a cure for people already stricken with Keshan disease. Positive results have been obtained for the prevention and cure of Kaschin-Beck disease by supplementation with sodium selenite [34].

Other areas of China are plagued with endemic Se toxicity in humans, characterized by loss of hair and nails [34]. These symptoms are similar to the alkali disease described in the Northern Great Plains and Prairie Provinces of North America. In addition, Kerdel-Vegas [7] earlier reported cases of selenosis in Amazonian peoples. These cases occurred after the consumption of too many nuts of the *Lecythidaceae* family. The ingestion of foods containing excess Se produced symptoms of nausea, vomiting, chills, diarrhea, and breath characteristic of dimethylselenide. In several days there was a loss of body hair and some finger nails and pronounced arthralgia of joints [7]. Hair regrew almost immediately, and few deaths were reported.

SELENIUM TOXICITY IN BIRDS AND ANIMALS

Acute selenosis is characterized by intake of large doses of Se. Clinical signs of toxicosis are dependent on Se source, dose rate, route of administration, and animal species. Acute lethalities associated with Se compounds are greater when the Se is administered parenterally than when it is given orally [5]. Common inorganic Se salts such as sodium selenite, sodium selenate, selenomethionine, and selenodiglutathione are among the more toxic species. Important characteristics associated with the short-term toxicity of these Se forms include their oxidation state and aqueous solubility. The reduced and poorly soluble forms are the least toxic Se compounds.

Subacute or chronic selenosis can occur when birds, fish, and animals are gradually exposed for periods of weeks to moderately high concentrations of Se (5–25 mg/kg) in their diets. The major signs include lesions on the skin, hoof necrosis, loss of long hair, and emaciation. Other signs include anorexia, weight loss, and increases in serum transaminases and alkaline phosphatase [5]. Signs of chronic selenosis may be expected among animals with whole-blood Se concentrations above 2 mg/kg.

Chronic Se toxicosis decreases conception in animals and may cause embryocidal damage in birds and feticidal loss in animals. The method of administration and the Se source both have significant effects on the extent of the toxicosis.

It is not yet possible to rank chronic toxicities of Se compounds by direct comparison [5]. However, sodium selenate and sodium selenite appear to be quite toxic, and selenomethionine appears to have moderate toxicity. The insoluble forms of Se, like elemental Se, exhibit the least long-term toxicity.

SUMMARY

Selenium is an essential element for animal nutrition and health. It serves as the metal cofactor for the biologically important enzyme glutathione peroxidase. Selenium deficiency reduces growth, productivity, and reproduction and even causes death in fish, birds, animals, and humans. Plants, while not requiring Se, absorb it from the soil solution and cycle it to ingesting animals. Plants differ in their Se metabolism, with most food plants converting much of the Se into protein where the Se is readily available to animals.

Animals have a dietary Se requirement of about 0.1 mg/kg in uncomplicated situations. The requirement increases to 0.3 mg/kg when high levels of S or other Se antagonists are present. In many parts of the United States and elsewhere, there is not enough Se in the feedstuffs to provide adequate nutrition for animal health requirements. In these areas Se may be injected into animals, provided as a mineral mix, or supplemented in a complete feed mix. However, the U.S. Food and Drug Administration currently limits such supplementation to 0.1 mg Se/kg in the diet. Several Se-deficient countries are successfully increasing the Se concentration of their feedstuffs by fertilizing pastures and cropland.

Animals develop a chronic selenosis when the Se concentration of the diet increases to levels of 3–15 mg Se/kg. This is a problem in some areas of the United States and elsewhere, where plants grow on seleniferous soils and accumulate excess Se. Animals feeding on these plants may develop moderate to severe health problems. Animal sensitivity to selenosis is dependent upon animal species and preconditioning. Some plants can accumulate Se in excess of 25 mg Se/kg when grown on highly seleniferous soils. Animals consuming these plants often will die of acute stenosis. The actual Se concentration is dose-related.

Current regulations of Se supplementation and management of seleniferous areas are largely driven by the political process. Factually based decisions await more information about the Se cycling in the soil–plant–animal system and the bioavailability of various Se species. Continued progress in understanding Se cycling will require additional methodologies to determine Se speciation.

REFERENCES

1. K. Schwarz, H. G. Bieri, G. M. Briggs, and M. L. Scott, *Proc. Soc. Exp. Biol. Med.* 95: 621 (1957).
2. K. Schwarz and C. M. Foltz, *J. Am. Chem. Soc.* 79: 3292 (1957).
3. J. N. Thompson and M. L. Scott, *J. Nutr.* 100: 797 (1970).
4. J. T. Rotruck, A. L. Pope, H. E. Ganther, A. B. Swanson, D. G. Hafeman, and W. G. Howkstra, *Science* 179: 588 (1973).
5. G. F. Combs, Jr. and S. B. Combs, *The Role of Selenium in Nutrition,* Academic, Orlando, Fla., 1986.
6. R. Latham (Translator), *The Travels of Marco Polo,* The Folio Society, London, 1968, p. 72.
7. F. Kerdel-Vegas, *Econ. Bot.* 20: 187 (1966).
8. I. Rosenfeld and O. A. Beath, *Selenium: Geobotany, Biochemistry, Toxicity and Nutrition,* Academic, New York, 1964.
9. G. J. Beke and R. Hironaka, *Sci. Total Environ.* 101: 281 (1991).
10. H. M. Ohlendorf, in *Selenium in Agriculture and the Environment* (L. W. Jacobs, Ed.), Soil Science Society of America, Madison, Wis., 1989, p. 133.
11. National Academy of Science—National Research Council, *Selenium in Nutrition,* NAS-NRC, Washington, D.C., 1983.
12. U.S. Environmental Protection Agency, *EPA-57019-76-003,* 1977.
13. H. F. Mayland, L. F. James, J. L. Sonderegger, and K. E. Panter, in *Selenium in Agriculture and the Environment* (L. W. Jacobs, Ed.), Soil Science Society of America, Madison, Wis., 1989, p. 15.
14. M. M. Abrams and R. G. Burau, *Commun. Soil Sci. Plant Anal.* 20: 221 (1989).
15. M. M. Abrams, R. G. Burau, and R. J. Zasoski, *Soil Sci. Soc. Am. J.* 54: 979 (1990).
16. M. C. Williams and H. F. Mayland, *J. Range Manage.* 45: 374 (1992).
17. M. M. Abrams, C. Shennan, R. J. Zasoski, and R. G. Burau, *Agron. J.* 82: 1127 (1990).
18. D. T. Westermann and C. W. Robbins, *Agro. J.* 66: 207 (1974).
19. S. D. McQuinn, D. A. Sleper, H. F. Mayland, and G. F. Krause, *Crop Sci.* 31: 617 (1991).

20. A. Shrift, in *Organic Selenium Compounds: Their Chemistry and Biology* (D. L. Klayman and W. H. H. Gunther, Eds.), Wiley-Interscience, New York, 1973, p. 763.
21. J. A. Erdman, R. C. Severson, J. G. Crock, T. F. Harms, and H. F. Mayland, *U.S. Geol. Surv. Circ. 1064*, 1991.
22. F. Lingaas, E. Brun, and A. Froslie, *J. Animal Breeding Genet. 108*: 48 (1991).
23. B. G. Lewis, in *Environmental Biogeochemistry* (J. O. Nriagu, Ed.), Ann Arbor Science, Ann Arbor, Mich., 1976, p. 389.
24. I. Milan, *Occurrence and Distribution of Selenium*, CRC Press, Boca Raton, Fla., 1989.
25. B. Z. Siegel, S. M. Siegel, T. Correa, C. Dagan, G. Galvez, L. Leeloy, A. Padua, and E. Yaeger, *Arch. Environ. Contam. Toxicol. 20*: 241 (1991).
26. H. F. Mayland, J. J. Doyle, and R. P. Sharma, *Biol. Trace Element Res. 10*: 65 (1986).
27. O. E. Olson, E. E. Cary, and W. H. Allaway, *Agron. J. 68*: 805 (1976).
28. J. Maas and L. D. Koller, *Selenium Responsive Diseases in Food Animals*, Veterinary Learning Systems Co., Princeton Junction, N.J., 1985, p. 20.
29. R. J. Turner and J. M. Finch, *Proc. Nutr. Soc. 50*: 275 (1991).
30. J. W. G. Nicholson, R. S. Bush, and J. G. Allen, *Can. J. Animal Sci. 73*: 355 (1993).
31. H. Wuryastuti, H. D. Stowe, R. W. Bull, and E. R. Miller, *J. Animal Sci. 71*: 2464 (1993).
32. U.S. Food and Drug Administration, *Fed. Reg. 58*(175): 47962 (Sept. 13, 1993).
33. P. Ekholm, M. Ylinen, and P. Varo, *J. Agric. Food Chem. 38*: 695 (1990).
34. Y. Guang-Qi, in *Selenium in Biology and Medicine*, Part A (G. F. Combs, Jr., J. E. Spallholz, O. A. Levander, and J. E. Oldfield, Eds.), Van Nostrand Reinhold, New York, 1987, pp. 9–31.
35. O. H. Muth and W. H. Allaway, *J. Am. Vet. Med. Assoc. 142*: 1379 (1963).
36. J. A. Erdman, R. C. Severson, J. G. Crock, T. F. Harms, and H. F. Mayland, U.S. Geol. Surv. Open-File Rep. 89-628, 1989.

3
Adsorption, Volatilization, and Speciation of Selenium in Different Types of Soils in China

J. A. Tan, W. Y. Wang, D. C. Wang, and S. F. Hou

Institute of Geography, Chinese Academy of Sciences, Beijing, People's Republic of China

I. INTRODUCTION

At the end of the 1960s, in response to health department requests, we started to engage in ecoenvironmental studies to determine the causes of two endemic diseases of humans, Keshan disease (KD) and Kaschin-Beck disease (KBD). The former is an endemic cardiomyopathy, the latter an endemic osteoarthrosis. Since then, our investigation and sample collection have extended to every province of China, except Taiwan Province, and have covered the main types of geographical environments in China. The distribution of both endemic diseases has been found to relate to some special characteristics of the geographical environment, especially soil characteristics. The two diseases are distributed mainly in a distinct wide belt, usually referred to as the disease belt, running from the northeast to southwest of China and located in the middle transition belt from the southeast coast to the northwest inland region (Figures 1 and 2). The belt is mainly characterized by temperature forest and forest-steppe soils that belong to the brown drab soil (earth) series [1,2].

From the viewpoint of geographical ecology, the geoecosystem, with humans as its core, differs according to characteristics of the geographical environment. The results of analyses of geoecosystem component substances, including rock, water, soils, grains, hair, etc., sampled from

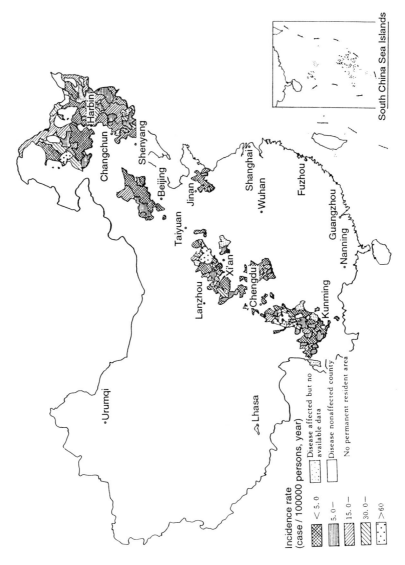

Figure 1 Distribution map of annual average incidence of Keshan disease (acute and subacute) in China. (Adapted from Ref. 10.)

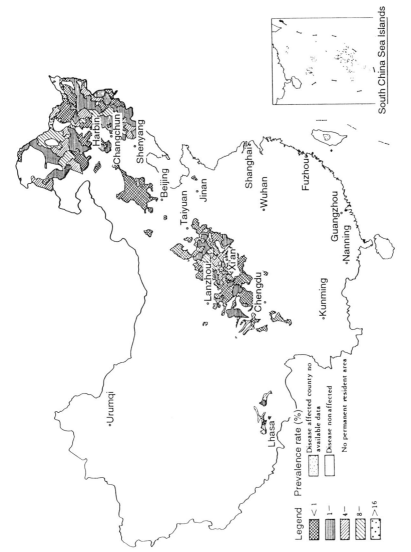

Figure 2 Distribution of Kaschin-Beck disease in China, 1970–1982. (Adapted from Ref. 10.)

locations throughout China prove that the two diseases are always located in low-selenium ecoenvironments that geographically form a low-selenium (low-Se) belt coinciding with the distribution of the two diseases. This finding of the low-Se belt showed us that the two diseases were closely related to Se deficiency in the geoecological environment [3–7]. The direct association of Se in natural geoecosystems to human health was first identified chemicogeographically in China.

Based on our studies, the following geographical features of Se have been found [7–9].

1. The low-Se geoecosystem occurs mainly in and near the temperate forest and forest-steppe landscape as an axis in China, and the relatively high Se concentrations in the geoecosystem usually appear in typical humid tropical and subtropical landscapes and typical temperate desert and steppe landscapes.
2. In juvenile soil landscapes, Se from parent material is a very important factor controlling the level of this element in the geoecosystem.
3. In some mountain districts or elevated areas, the distribution of the low-Se geoecosystem is also associated with a certain vertical geographical landscape zone; for example, in western Sichuan and Yunan provinces and eastern or southeastern Tibet, the low-Se geoecosystem usually occurs at altitudes above approximately 1300–2000 m above sea level and is associated with vertical mountain forest, forest steppe, and meadow steppe landscapes.
4. There are relatively high Se concentrations in the geoecosystem in some large accumulation plains, such as the Songliao, Weihe, and Hua Bei plains, compared with the washing areas within the same type of geographical zone.

These four distributional laws of the element resulting from our research have basically dominated the spatial distribution patterns and structures of Se in the geoecosystems in China, which, on a large scale, exhibit a low-Se belt running from northwest to southwest. Flanking it on both sides, to the southeast and northwest, are two relatively high Se belts [7–10] (Figure 3).

The average Se concentrations in geoecosystem substances from various landscapes are listed in Table 1. Because of these substantive data derived from practical investigations and sample analyses throughout the country, we were able to determine a series of threshold values of Se to divide the low-Se or Se-deficient geoecosystem from the relatively high Se or adequate Se geoecosystem. The threshold values are 0.15 µg/g for total soil Se, 0.003 µg/g for soil-water soluble Se, 0.025 µg/g for food grain Se, and 0.20 µg/g for human hair Se.

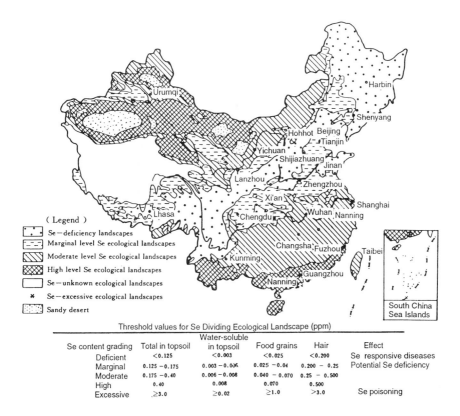

Figure 3 Selenium ecological landscape map of China. (Adapted from Ref. 10.)

Soil Se, which usually plays an important role in determining the Se level in food grains, vegetables, and even drinking water, is the basis of the Se cycle in the geoecosystem. Therefore, the geographical distribution of soil Se determines to a large extent that of grain Se, which, through the ecological food chain, influences human health in some specific areas, resulting in Se-responsive diseases such as Keshan disease, Kaschin-Beck disease, and even endemic selenosis. To gain an understanding of the laws of spatial distribution, the ecocycle of Se, the formation of the low-Se belt, and their influences on human health, we carried out some studies on adsorption, volatilization, and speciation of Se in the main types of soil in China.

Table 1 Mean Se Content (μg/g) in Ecosystematic Materials for Various Landscapes[a]

Landscape	Type	Maize	Wheat	Rice	Hair	Topsoil
Adequate Se	1. Desert	0.052 (66)	0.128 (135)	0.091 (19)	0.470 (387)	0.260 (55)
	2. Desert-steppe and steppe	0.029 (3)	0.057 (15)	0.040 (5)	0.462 (23)	0.173 (24)
Low Se	3. Temperate forest and forest steppe	0.016 (58)	0.017 (25)	0.021 (6)	0.108 (223)	0.118 (25)
	4. Warm temperate forest and forest steppe	0.016 (19)	0.017 (98)	—	0.151 (679)	0.080 (19)
	5. Purplish soil	0.016 (44)	0.023 (50)	0.018 (67)	0.129 (492)	0.076 (21)
	6. Vertical belt temperate forest and weadow steppe	0.008 (11)	0.010 (20)	0.016 (8)	0.158 (101)	0.100 (15)
Adequate Se	7. Northern subtropical yellow-brown earth	0.038 (3)	0.039 (13)	0.040 (25)	0.383 (31)	0.238 (48)
	8. Middle subtropical red-yellow earth	0.044 (8)	0.055 (36)	0.060 (105)	0.333 (143)	0.270 (21)
	9. Southern subtropical lateritic red earth	0.071 (4)	0.061 (22)	0.077 (106)	0.493 (30)	—
	10. Tropical laterite	0.090 (1)	—	0.101 (16)	0.591 (41)	0.320

[a]Number of samples shown in parentheses.
Source: Ref. 9.

II. ADSORPTION OF SELENIUM ON SOILS

Soils are considered an important source or storehouse of Se for plant, crops, and forages and thus important to animal and human health. Therefore, soils play an important role in Se cycling in the geoecosystem. Owing to adsorption processes, soils have the ability to retain Se, avoiding its loss by leaching. Because of differences in the physical, chemical, and biological properties of various soils, apparently the pathways and intensity of Se cycling are different in different geoecosystems. Among the soil components involved in adsorption of Se, clay minerals and metal oxides are considered to be the key factors.

We first conducted studies on adsorption of Se on soils in China [11; unpublished research report, 1985]. These studies had two main purposes:

(a) to research adsorption of Se by clay minerals and metal oxides and (b) to research adsorption of Se by major types of soil in China. In our study, kaolinite, vermiculite, and montmorillonite were used to study the adsorption of Se by clay minerals; Fe_2O_3, Al_2O_3, and MnO_2 were used for studying adsorption of Se on metal oxides. The major soil types were laterite (sampling from Guangdong Province), red earth (Hunan Province), dark brown earth (Jilin Province), drab soil (Shaanxi Province), black soil (Heilongjiang Province), chestnut (Nei Monggol autonomous region), gray desert soil (Xinjiang autonomous region), and purple soil (Sichuan Province). The Se form used in the experiment was Na_2SeO_3, determined by fluorescence spectrophotometry with 2,3-diaminoaphthalene (DAN). The experiment was conducted under pH 6–7, at a temperature of $20 \pm 3°C$, oscillated on an oscillator for 4 h, and with an equilibration time of 48 h. The experimental material quantity is 40 mg for clay minerals and metal oxides and 0.8 g for soils, and the ratio of material to equilibration solution is 1:5.

A. Adsorption of Selenium by Clay Minerals and Metal Oxides

The experimental data were analyzed according to the Langmuir equation. The research results showed that the ability of metal oxides to adsorb Se is much higher than that of clay minerals. The saturation Se-adsorbing capacities of the metal oxides are ranked as follows: Fe_2O_3, 1773 µg/g > MnO_2, 1318 µg > Al_2O_3, 1143.5 µg/g. However, when the experiment was conducted under a low Se concentration, this sequence changed. For example, at the point of half-saturated adsorption, the sequence of adsorbability of the studied metal oxides was Fe_2O_3 > Al_2O_3 > MnO_2; their corresponding adsorption intensities, a, are respectively 5.62 µg/L, 88.85 µg/L, and 147.40 µg/L. The sequence for saturation adsorbing capacity of Se on clay minerals was vermiculite, 295.2 µg/g > kaolinite, 265.0 µg/g > montmorillonite, 119.2 µg/g. Like adsorption of oxides, when the experiment was conducted under low Se concentrations, at half-saturated adsorption, the sequence of adsorbability of clay minerals to Se became kaolinite > vermiculite > montmorillonite; the corresponding adsorption intensities were 192.4 µg/L, 257.3 µg/L, and 444.1 µg/L, respectively. Peng and Xu [12] conducted an experiment on adsorption of Se on clay minerals, metal oxides, and humic acid. The results of the adsorbability sequence were manganese dioxide > ferric oxide > kaolinite > humic acid > silicon dioxide. These research results are beneficial for explaining the geographical distribution of Se in soils in China and in other geoecosystem substances such as plants, crops, animals, and even human hair.

B. Adsorption of Selenium by Main Types of Soils

1. Adsorbability of the Major Types of Soils in China

The experimental results are shown in Table 2. We found that the southern China soils rich in metal oxides and clay minerals developed under high temperature and abundant rainwater have the strongest adsorbability. The soils such as dark brown earth, drab soil, and black soil developed under temperate forest and forest-steppe landscapes have moderate adsorbability, as does the chernozen, which developed under the humid steppe landscape. However, chestnut and gray desert soil developed under arid landscapers, and juvenile purple soil, formed directly from purple sandstone/mudstone of Jurassic and Cretaceous strata, are the lowest in Se adsorbability.

2. Correlation Between Absorbability, Total Se, and Water-Soluble Se in Soils

The soil's ability to adsorb Se is positively correlated with its total Se content; the correlation coefficient was 0.68 ($P<0.001$). This result might indicate that soil adsorbability is an important factor for governing the Se content in soil. A good positive correlation between total soil Se and water-soluble Se in soil was also found; the correlation coefficient was 0.399, $P<0.01$. The rate of water-soluble Se to total Se ranges from 0.46 to 5.27 (Table 2). However, adsorbability and soil water-soluble Se

Table 2 Adsorbability of Total Se and Water-Soluble Se in Main Soils of China (Average of Profile)

	Laterite	Red earth	Drab soil	Dark brown soil	Black soil	Chernozem	Chestnut	Gray desert soil	Purple soil
Q_m (µg/g)	157.1	159.0	54.97	43.03	63.94	54.08	32.91	13.40	16.80
Total Se (µg/g)	0.56	0.39	0.069	0.17	0.23	0.14	0.11	0.22	0.038
Water-soluble Se (ng/g)	2.6	5.6	1.7	4.5	3.5	5.2	5.8	5.6	2.0
W/T rate (%)	0.46	1.44	2.46	1.52	4.73	5.27	2.55	2.55	5.26

Q_m: saturated absorbing capacity.
W/T: percent rate of water-soluble Se to total Se in soil.

have an insignificant negative correlation ($r = -0.163$) between them. This implies that the factors controlling the soil water-soluble Se content are complicated.

3. Variation in the Soil Profile's Se-Adsorbing Capacity, Total Se Content, and Water-Soluble Se Content

Regarding Se in the soil profile, the trend was as follows. For the southern soil (laterite and red earth) and arid area soil (chestnut and gray desert soil), the ability to adsorb Se usually increased with the increase in soil profile depth, whereas for the temperature forest and forest-steppe soil (dark brown earth and black soil) and juvenile soil (purple soil), the ability to adsorb Se was just the opposite, decreasing with increases in soil profile depth. However, the profile distribution of water-soluble Se was generally consistent in all soil types on which experiments were conducted, with exception of chestnut, in which Se concentration declined as soil profile depth increased. Variations in total Se in the soil profile exhibited a similar trend: the upper horizons of soil usually had higher concentrations, which declined with increases in soil profile depth. This distribution of total Se and water-soluble Se in the soil profile means that the biota and rainfall may have some measurable influence on the Se cycle in soils.

4. Soil Adsorbability and Geographical Distribution of Soil Se in China

Our ongoing research on the soil's ability to adsorb Se enabled us to understand the geographical distribution and ecocycle of Se in soil and in our country's entire geoecosystem. For example, since the southern China soils, such as laterite and red earth, have the highest adsorbability, they can retain more Se that will not be leached by the abundant rainfall, which means that these soils can provide adequate Se for the ecocycle to protect animal and human health from the dangers of suffering from endemic Se-responsive deficiency diseases. The soils in the middle transition belt of China, such as dark brown earth, drab soil, black soil, and purple soil, are of lower adsorbability, and strong leaching can occur. The geoecosystems influenced by these soils have low levels of Se in the ecocycle. Keshan disease and Kaschin-Beck disease, which have been identified as being associated with Se deficiency in the geoecosystem, occur only in this transition belt in China. Although the adsorbability of soil types in arid areas is not particularly strong, the leaching process in them is weak, and with high pH there still are relatively high levels of Se in these soils, as either total Se or water-soluble Se. Therefore, in the arid areas with these soil types, relatively high Se concentrations are maintained in the ecocycle.

Besides soil adsorbability, other factors such as Se volatilization and speciation in soil govern the cycle of Se in the soil and in the geoecosystem as a whole. These factors are discussed next.

III. VOLATILIZATION OF SELENIUM FROM SOILS

In the biogeochemical cycle of Se, the volatilization of Se is an important factor. As a result of biological activity, volatile Se compounds can be released from terrestrial and aquatic ecosystems into the atmosphere [13]. The major compound released from soils, lakes, sediments, and sewage sludge is dimethylselenide. Similarly, the dimethylselenide can be released from plants by enzymatic activity.

Dimethylselenide released from soil is dependent on the microbial activity of fungi, yeasts, and bacteria and on the physical and chemical features of the soil, such as temperature, moisture, pH, and organic matter. The volatilization of Se from soil is primarily a microbial process. According to preliminary estimates, the total amount of Se evolved from soil into the atmosphere is almost the same as the amount released by anthropogenic activity, which shows the importance of Se volatilization in the environment.

Human and animal health are highly impacted by the Se status in their environment. Studies on Keshan disease and Kaschin-Beck disease in China have shown that these diseases are related to low Se content in the environment, especially the low flux of Se in the soil–plant–human system. To determine the effects that Se volatilization has on the formation of low-Se environments and the relationship of volatilization to the incidence of both diseases, it was necessary to study volatilization of Se from different soils in China. The research findings will increase not only understanding of the factors that influence the formation of low-Se areas in China but also the general understanding of the cycle of Se in the geoecosystem.

A. Volatilization of Different Soils and the Influence of Key Factors on Se Volatilization

Soils were collected from different environmental areas, both from environments where Keshan disease and Kaschin-Beck disease are endemic and from non-disease areas. Four English soils with different Se contents were collected for comparison. Selenium-75 as sodium selenite was spiked to the soil. The total experimental period was 15 days. Because most volatile Se as methylselenide can be easily trapped by nitric acid,

10 mL of concentrated nitric acid was used. A well-crystal scintillation counter was used to count the radioactivity daily.

The results showed that there is a positive correlation between the percentage of ^{75}Se evolved from soil and the pH of the soil ($r=0.852$, $P<0.01$). Because microbial activity is related to the volatilization of Se, acid and alkali conditions affect the process. Most bacteria develop well in soil under neutral and slightly acidic conditions. When the soil is too acidic, the microbiological activity will be decreased, and as a result only small amounts of Se can be volatilized from the soil. However, the volatilization of Se was found to be negatively correlated to clay particles ($r = -0.705$, $P < 0.05$ for <0.001 mm particles; $r = -0.745$, $P < 0.05$ for 0.001–0.005 mm particles). Different soil textures affect microbial activity; in sandy soil with good ventilation conditions, there is strong microbial activity. On the other hand, clayey soils are poorly aerated and show reduced microbial activity.

B. Differences in Volatilization of Se from Soils Sampled from Disease and Non-Disease Areas in China

The volatilization of Se from different soil types was determined to be in the following order: drab soil (rate 0.092%) > black soil (0.024%) > purple soil (0.017%) > brown earth (0.012%) > red earth (0.001%) and laterite (0.001%). These results are derived from the complex effects of pH, clay particles, and organic matter content on volatilization of Se and are in agreement with the conclusions of Hamdy and Gissel-Nielsen [20]. Drab soil with its high silt and sand contents is slightly alkaline or neutral in pH, which is suitable for microbial activity, and the highest amount of Se is volatilized from this soil. Conversely, red soil and laterite with low pH and high clay and organic matter content (32.5% and 58.0% of clay particles smaller than 0.001 mm, respectively) exhibit only small amounts of Se volatilization (Table 3).

The volatilization of Se from soils where Keshan disease and Kaschin-Beck disease are endemic is higher than in non-disease areas. The total Se in soil in disease areas is at low levels as determined by our earlier research.

The obvious zonal differentiation shows that biological and climatic factors have a decisive effect on the distribution of Se in soil. Figures 4 and 5 show that more Se evolves from soil in disease areas. The volatilization of Se from soil depends on microbial activity. Table 4 shows the biomass of microorganisms in different soils. The biomass is higher in black soil,

Table 3 Selenium-75 Evolved from Different Soils

Soil type	Evolved ^{75}Se (%)	Total Se (μg/g)
Drab soil	0.092	0.12
Black soil	0.024	0.17
Purplish soil	0.017	0.05
Brown earth	0.012	0.13
Red earth	0.001	0.34
Laterite	0.001	0.53

drab-luto soil, dark brown earth, and brown earth than in red earth or laterite [23]. The different rates of volatilization of Se in similar soil types from this study are also listed in Table 4. Those soils with higher microbial biomass are from the low-Se area of China. These results imply that high volatilization of Se in soil from disease areas could be one of the principal causes of low Se in these soils.

It is interesting that the same trend for the volatilization of Se was found in both Chinese soils and English soils. Figure 6 shows that more Se

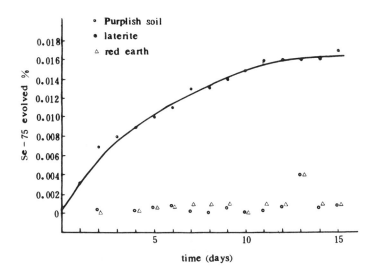

Figure 4 Selenium-75 volatilization from different soils in disease and non-disease areas. (From Ref. 24.)

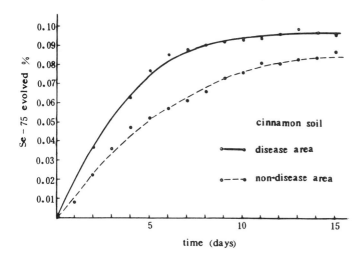

Figure 5 Selenium-75 volatilization from drab soil in disease and non-disease areas. (From Ref. 24.)

evolved from soil with low Se content in Brecon (southern Wales) than from soils with high Se content in both northern Wales and Derbyshire, England.

As discussed above, its volatilization from soil has an important, even critical effect on the content of Se in soils, and more Se is evolved from soil in disease areas than in non-disease areas. Volatilization is an important factor in the cycle of Se in the geoecosystem and affects the formation of low-Se soil. Therefore, it is necessary to calculate the volatilization of Se in low-Se soil under field conditions. If the total Se concentration in soil is 0.05 μg/g, when topsoil is a 30 cm layer per mu (Chinese area unit, about 1/6 acre), the total Se content is 13 g (soil density 1.3). We assume that the total output of crops from each mu is 500 kg/yr and the Se level, 0.04 Se μg/g in the crop, is the minimum necessary for maintaining the health of humans and animals. The total amount of Se taken up from soil by crops, including straw, will be 0.06 g, only 0.46% of the total amount Se in the soil. While the volatilization of Se from 1 mu of land, according to the average value evolved, 0.35 × 10^{-3} g Se/day, the total amount of Se evolved from soil is 0.07 g/yr (assuming a frost-free season of up to 200

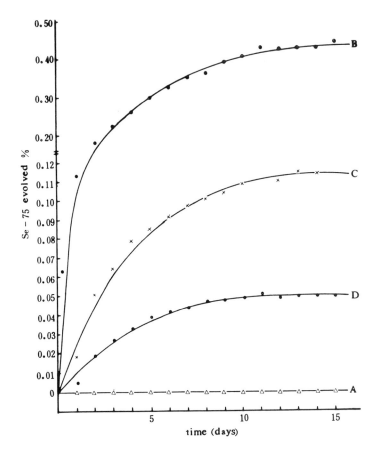

Figure 6 Selenium-75 volatilization from soils with different selenium contents. A, Peaty soil (high Se); B, brown earth (low Se); C, brown earth (normal Se); D, noncalcareous pelosols (high Se). (From Ref. 24.)

days and no microbial activity in soil after the season), which slightly exceeds the amount absorbed by crops from the soil.

IV. SPECIATION OF SELENIUM IN SOILS

The speciation of Se in soil provides important data for clarifying its ecocycle as well as for identifying causes of the low-Se belt in China. The studies on existing forms of soil Se usually deal with two aspects: valence states and combination (binding) forms. Selenium exists in the following

Table 4 Selenium Volatilization and Biomass (X 10^4/g Dry Soil) of Different Soils in China

Soil type	Location	Bacteria	Actino-myces	Fungi	Percent ^{75}Se evolved in this study
Dark brown earth	Huma County, Heilongjiang	2327	612	13	—
Brown earth	Shenyang, Liaoning	1284	39	36	0.012
Black soil	Harbin, Heilongjiang	2111	1024	19	0.024
Drab luto	Northwestern China	2172	241	11	0.092
Yellow brown earth	Nanjing, Jiangsu	1406	271	6	—
Red earth	Hangzhou, Zhejiang	1103	123	4	0.001
Laterite	Xuwen, Guangdong	507	39	11	0.001

Source: Ref. 24.

oxidation states: SeO_3^{2-} ($HSeO_3^-$), SeO_4^{2-} ($HSeO_4^-$), Se^{2-} (HSe^-), and elemental Se. It exists in inorganic, organic, and organic-inorganic forms. As soil is a sophisticated multielement system affected by many unstable factors, there is no commonly accepted research method suitable for studying Se speciation. Past studies were focused mainly on theoretical exploration using simulation experiments. A detailed review was presented by Elrashidi et al. [25]. However, only limited data were available on real soils.

In China, as mentioned above, there exists a low-Se belt in which there have been many problems involving human and animal health. Systematic research on the existing speciation of Se in soil will be important for a theoretical and practical understanding of the causes of the formation of the low-Se belt and its specific effects on human and animal health. Therefore, studies of soil Se speciation in China have been conducted mainly for this purpose. Early studies in China were concentrated on soil water-soluble Se. In recent years, the importance and significance of speciation of Se in the ecoenvironment have been explored in a preliminary fashion.

A. Soil Water-Soluble Selenium

Soil water-soluble Se is one of the major Se species that can be used by organisms. It affects human and animal health by governing the Se content in drinking water and grains. Cheng et al. [26] studied the soluble Se content and related factors using a boiling water extraction method on soils in northeastern China. The results indicated that the water-soluble Se content for dark brown soil, brown earth, and black soil is very low, with an average of 0.2 μg/kg (range 0.1–0.6 μg/kg), less than 1% of the soil's total Se content. In these areas, Keshan disease and Kaschin-Beck disease of humans and white muscle disease of animals were prevalent. But for chernozem, black soil, and drab soil, water-soluble Se concentrations are relatively higher, with an average of 1.88 μg/kg (0.72–7.46 μg/kg), about 2–59.7% of the soil's total Se content. Tan et al. [7] studied the geographical differences of soluble Se in soils from different areas in China. The results revealed that soil water-soluble Se for the red-yellow earth series developed in tropical and subtropical regions in southeastern China was about 5 ± 3 μg/kg, but for the brown-drab earth series developed in the temperate zone in central China (where the low-Se belt is located), the value is 3 ± 2 μg/kg, while for the dry steppe and desert soil series developed in the northwest the Se concentration increases to 7 ± 3 μg/kg. Report presented by Chen et al. [27] shows a strong linear relationship between soil total Se, pH, and soil water-soluble Se. The typical soil soluble Se content was about 2 μg/kg, but for seleniferous and Se-rich soils, the concentration was as high as 42.9 μg/kg. The soluble Se content of 95% of the samples tested was below 5 μg/kg. Selenium-responsive diseases prevailed in areas that had the lowest soluble Se content in their soils.

With an HPLC-FLD system, Wang et al. [unpublished report, 1991] studied the oxidation state of water-soluble Se and its forms in soils of KBD-affected and nonaffected areas. The results show coexistence of organic and inorganic Se in soil water extractions, and the ratio of SeO_3^{2-} to SeO_4^{2-} for inorganic Se varied among soils. The distributions of different Se species in soil had no significant relationship to the pH and Eh of the extracted solution. The total content of water-soluble Se showed significant differences between soils of disease and non-disease areas.

B. Selenium Extracted by Different Extractants from Soils in China

We have established a procedure for sequential fractionation of soil Se based on the Jackson method [28], which was used for fractionation of soil phosphorus, and have studied the content of combined Se, its distribution and regulation, and the factors controlling its chemical behavior in soils in various ecoenvironmental landscapes in China [29,30]. The result showed linear relationships between Se and phosphorus in different extract solutions, demonstrating that the procedure established above is suitable for the sequential fractionation of soil combined Se, and further implies that there is a similar mechanism in soil for preserving and binding Se and phosphorus. The experimental results are shown as Tables 5 and 6. Seven states of Se in soil were extracted in the study.

Selenium state 1 is extracted with neutral (1 N NH$_4$Cl) solution and includes water-soluble Se, loose combined Se, and exchangeable Se. For laterite and red earth, which were developed under humid tropical and subtropical landscapes in the southeastern coastal areas of China, Se state 1 is not higher than 15 µg/kg, about 0.44–3.0% of the extracted Se and 0.34–2.6% of the total soil Se. But in northwestern China, for sierozem, which was developed on loess parent material under warm-temperature semiarid landscapes, and for gray desert soil, which was developed under temperate arid desert landscapes, Se state 1 is present as 8 µg/kg and 22 µg/kg, respectively, which accounts for 5.67% and 8.40% of the extracted Se, or about 3.81% and 6.34% of the total Se content of the two types of soils. In contrast, one of the typical soils distributed within the low-Se belt in central China, drab soil, developed as a dry forest soil in warm-temperate areas lying between the above two types of natural geographical landscapes, has a Se state 1 content of only 2 µg/kg, 2.63% of the total extracted Se or 2.13% of the soil's total Se. In general, Se state 1 is higher in soils in the northwest than in the southeast. This distribution property is in accordance with the common finding that soil water-soluble Se is higher in alkaline soil than in acid soil. Selenium state 1 is the lowest for soils in the low-Se belt.

Selenium state 2 is the Se form that could be extracted with NH$_4$F (0.5 N) solution after the first extraction. It accounted for 10–20% of the total extractable Se or 6–20% of the total Se for different soils. In acid soils, Se state 2 was associated mainly with iron and aluminum oxides and salts, especially with aluminum compounds. A few parts of Se state 2 were bound to clay minerals. In alkaline soils, in addition to the complexes listed above, the Se state 2 was controlled chiefly by calcium and magnesium salts. Much of the Se state 2 was distributed in the surface layers of the soil's solid phase and had strong positive correlations with total Se in soil ($r = 0.7255$).

Selenium state 3 is the Se that could be extracted with NaOH solution (0.1 N). It is mainly associated with iron compounds and distributed in subsurface layers of the solid phase in soils. The Se state 3 content of different soils is sequenced as follows: laterite > red earth > drab soil > sierozen > gray desert soil, which is the same as the distribution as the iron in soils.

Table 5 Concentrations of Different Selenium Species of in Topsoils (µg/g)

			Selenium state						
			1	2	3	4	5	6	7
		Total	Extractant						
Soil	pH	Se	NH$_4$Cl	NH$_4$F	NaOH	H$_2$SO$_4$	C-A[a]	NH$_4$F	resid.
Laterite	4.75	0.573	0.002	0.047	0.258	0.002	0.048	0.101	0.115
Red earth	4.45	0.460	0.012	0.078	0.194	0.013	0.033	0.062	0.068
Drab soil	7.88	0.094	0.002	0.012	0.024	0.021	0.015	0.002	0.018
Sierozem	7.90	0.210	0.008	0.012	0.018	0.025	0.043	0.055	0.069
Gray desert soil	8.58	0.347	0.022	0.067	0.013	0.083	0.026	0.051	0.087

[a]C-A: sodium citrate–sodium dithionite.
Source: Ref. 28.

Table 6 Percentage of Different States of Se to Total Extractable Se and to Total Se in Soils (µg/g)

Soil		Sum, states 1–6	State						Residual Se
			1	2	3	4	5	6	
Laterite	a	0.458[c]	0.44	10.26	56.33	0.44	10.43	22.05	
	b	79.93	0.34	8.20	45.03	0.34	8.38	17.63	20.07
Red earth	a	0.392[c]	3.06	19.90	49.49	3.32	8.42	15.82	
	b	85.22	2.61	16.96	42.17	2.83	7.17	13.48	14.78
Drab soil	a	0.076[c]	2.63	15.79	31.58	27.63	19.74	2.63	
	b	80.85	2.13	12.77	25.53	22.34	15.96	2.13	19.15
Sierozem	a	0.161[c]	5.67	7.45	11.18	17.73	30.49	39.01	
	b	67.14	3.81	5.71	8.57	11.90	20.47	26.19	23.33
Gray desert soil	a	0.262[c]	8.40	25.57	4.96	31.68	9.92	19.67	
	b	75.10	6.34	19.31	3.75	23.92	7.49	14.70	25.07

[a]Percentage of different state Se to total extracted Se amount, which includes states 1–6.
[b]Percentage of different state Se to total Se amount.
[c]Total extractable Se amount.
Source: Ref. 28.

Selenium state 4 is the part of Se that is extracted with H_2SO_4 solution (0.5 N). It is associated mainly with calcium compounds or minerals. The Se state 4 content sequence is laterite < red earth < drab soil < sierozem < gray desert soil, which has the same distribution as Ca in soils.

Selenium state 5 and Se state 6 are extracted with sodium citrate–sodium dithionite solution and NH_4F (0.5 N) solution, respectively. Selenium state 5 is associated chiefly with iron compounds, and Se state 6 has been associated with plumbic compounds. They are located in the inner parts of Fe and Al oxides or minerals, aggregates, concretions, and secondary minerals but not in crystal lattices. The total of Se state 5 and state 6 comprise about 30% of the total extractable Se and about 20% of the total Se in soils in southeastern China, but their distribution is higher in northwestern China, with the largest proportion in the sierozem.

Selenium state 7 is the residual Se in soil that could not extracted by the six extractants used for the other states. It accounts for 20% of total Se in soils and has a significant positive correlation with total Se in soils ($r = 0.9212$, $P < 0.05$).

Selenium states 1–4 are unenveloped Se, which is available to plants. Among them, Se states 2–4 account for 60–70% of total extractable Se (36.36% of Se in sierozem) and 50–60% of total Se in soils (26.38% of Se in sierozem). They also have significant positive relations with total Se in soils ($r = 0.9212$, $P < 0.05$). It is concluded from these results that the Se state 1–4 contents in typical soils in China are laterite (0.309 µg/g) > red earth (0.297 µg/g) > gray desert soil (0.185 µg/g) > sierozem (0.063 µg/g) > drab soil (0.063 µg/g). It could be concluded that the formation of the low-Se geoecosystem in China has been influenced by low Se states 1–4. Therefore, the Se state 1–4 contents in drab soil in disease-affected areas are the lowest. In addition to these results, Wang et al. [31] studied different forms of Se in certain cultivated soils with sequential extraction procedures and concluded that Se in cultivated soils is primarily in the organic state.

It is concluded that although there has been some study of Se speciation in soils in China, difficulties exist in this research area because of the lack of a mature study method, and it is difficult to compare various data from different authors. Therefore, the development of a well-recognized, accepted and unified research method is an urgent need.

REFERENCES

1. EGAS (The Group of Endemic Disease and Environment, Institute of Geography, Academic Sinica), *Acta Geogr. Sin. 34*: 85 (1979) (in Chinese).

2. J. A. Tan, *Nat. Geogr. J. India* 28: 15 (1982).
3. EGAS, *Acta Geogr. Sin.* 34: 369 (1981) (in Chinese).
4. EGAS, *Acta Geogr. Sin*, 37: 132 (1982) (in Chinese).
5. EGAS, *Geogr. Res.* 3: 40 (1984) (in Chinese).
6. EGAS, *Bull. Chin. Acad. Sci.* 3: 54 (1988).
7. J. A. Tan, D. X. Zheng, S. F. Hou, W. Y. Zhu, R. B. Li, Z. Y. Zhu, and W. Y. Wang, in *Selenium in Biology and Medicine*, Part B (G. F. Combs, J. E. Spallholz, O. A. Levader, and J. E. Oldfield, Eds.), Van Nostrand Reinhold, New York, 1987, p. 859.
8. J. A. Tan, W. Y. Wang, Z. Y. Zhu, L. Z. Wang, and Y. L. Lu, *Chin. J. Geochem.* 7: 273 (1988).
9. J. A. Tan, in *Environmental Life Elements and Health* (J. A. Tan, P. J. Peterson, R. B. Li, and W. Y. Wang, Eds.), Science Press, Beijing, 1991, p. 145.
10. J. A. Tan (Editor-in-Chief), *The Atlas of Endemic Disease and Their Environments in People's Republic of China*, Science Press, Beijing, 1989, p. 39.
11. D. X. Zheng, R. B. Li, and J. A. Tan, *Sci. Geogr. Sin.* 6: 22 (1986) (in Chinese).
12. A. Peng and L. Q. Xu, in *Proceedings of Symposium on Heavy Metals in Environment*. Science Press, Beijing, 1988, p. 89.
13. P. J. Peterson, *Reactions and Interactions in Biogeochemical Cycles*, Univ. Warwick, Coventry, U.K., 1980, pp. 1–25.
14. A. T. Francis, J. M. Duxbury, and M. Alexander, *Appl. Microbiol.* 28: 248 (1974).
15. O. E. Olson, E. E. Cary, and W. H. Allaway, *Agron. J.* 68: 839 (1976).
16. Y. K. Chau, P. T. S. Wong, B. A. Silverberg, P. L. Luxon, and G. A. Bengert, *Science* 192: 1130 (1976).
17. J. W. Doran and M. Alexander, *Soil Sci. Soc. Am. J.* 41: 70 (1977).
18. B. G. Lewis, C. M. Johnson, and T. C. Boyer, *Biochim. Biophys. Acta* 237: 603 (1971).
19. L. Barkes and R. W. Fleming, *Bull. Environ. Contam. Toxicol.* 12: 308 (1974).
20. A. A. Hamdy and G. Gissel-Nielsen, *Z. Pflanzenernahr.* 6: 671 (1976).
21. R. Zieve and P. J. Peterson, *Sci. Total Environ.* 19: 277 (1981).
22. EGAS, *Acta Nutrimenta Sin.* 4(3): 209 (1982).
23. Y. X. Hao, *Microorganism in Soil*, Science Press, Beijing, (1982), (in Chinese).
24. W. Y. Wang and P. J. Peterson, in *Environmental Life Elements and Health* (J. A. Tan et al., Eds.), Science Press, Beijing, 1990, p. 79.
25. M. A. Elrashidi, D. C. Adriano, S. M. Workman, and W. L. Lingsay, *Soil Sci.* 144: 141 (1987).
26. B. R. Cheng, J. Shanjian, Y. Shurong, H. Rongzhen, and S. Shijun, *Acta Pedol. Sin.* 17: 55 (1980) (in Chinese).
27. D. Z. Chen, S. X. Ren, and J. Y. Li, *Acta Pedol. Sin.* 21: 247 (1984) (in Chinese).
28. M. L. Jackson, *Soil Chemical Analysis*, Prentice-Hall, Englewood Cliffs, N.J., 1958, pp. 134–182.
29. S. F. Hou, D. Z. Li, L. Z. Wang, W. Y. Wang, and J. A. Tan, *Geogr. Res.* 9: 7 (1990) (in Chinese).

30. S. F. Hou, L. Z. Wang, D. Z. Li, W. Y. Wang, and J. A. Tan, *Acta Geogr. Sin.* 47: 31 (1992) (in Chinese).
31. Z. J. Wang, L. H. Zhao, L. Zhang, J. F. Sun, and A. Peng, *Environ. Sci. (China)* 3: 43 (1991) (in Chinese).

4

Kesterson Reservoir—
Past, Present, and Future:
An Ecological Risk Assessment

Harry M. Ohlendorf and Gary M. Santolo

CH2M HILL, Inc.
Sacramento, California

I. INTRODUCTION

Kesterson Reservoir and the San Luis Drain were constructed by the U.S. Bureau of Reclamation (hereinafter shortened to USBR or Reclamation) between 1968 and 1975 [1]. The drain extends from the Five Points area in Fresno County to Kesterson Reservoir in Merced County (Figure 1). It was originally intended to extend into the Sacramento-San Joaquin Delta, but construction was halted in 1975 because of funding limitations and concern over the potential environmental impacts of discharging agricultural drainwater into the delta. During Kesterson Reservoir's early operation, flow was mainly from surface water, but after 1981 most of the flow into Kesterson Reservoir was subsurface drainage water from irrigated agricultural lands in Fresno County.

Aquatic birds nesting at Kesterson Reservoir in 1983 were found to have high rates of embryo deformities and mortalities [2, 3]. Beginning in 1984, adult birds were also found dead in unusually high numbers. Through a series of field and laboratory studies, these effects were attributed to the exceptionally high concentrations of selenium in the birds' diets and their tissues. This bioaccumulation of selenium occurred because of the high concentrations of selenium in the subsurface agricultural drainage carried to Kesterson by the San Luis Drain. (All references to "Kesterson" or "the reservoir" refer specifically to

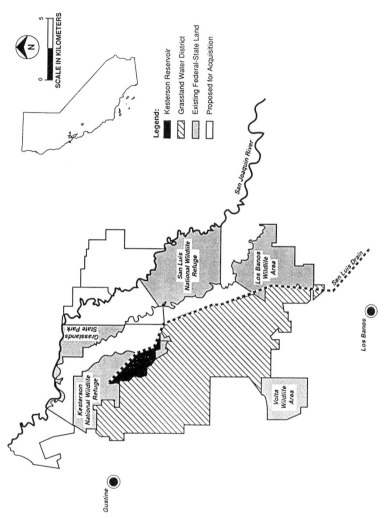

Figure 1 Location of Kesterson Reservoir and nearby areas in northern Merced County, California. (Modified from Refs. 29 and 30.)

Kesterson Reservoir and do not include the adjacent National Wildlife Refuge.)

Reclamation undertook studies and control actions to alleviate the hazard of selenium exposure to aquatic birds [1, 4]. Most notably, Reclamation halted discharge of agricultural drainage to Kesterson in 1986 and subsequently, in 1988, dewatered the Reservoir and filled all areas to at least 15 cm above the expected average seasonal rise of groundwater. These actions effectively transformed the Reservoir into an area with three types of terrestrial habitat, as described in Section III. The decision to dewater and fill the reservoir was based in part on data from Kesterson biomonitoring that indicated that terrestrial animals were not as sensitive to the elevated concentrations of selenium as aquatic birds.

Monitoring has been conducted since 1987 to measure selenium concentrations among plants and animals at the site, document habitat and faunal changes, and determine whether adverse effects are occurring in the ecosystem [5–9]. The findings of that monitoring program, in conjunction with monitoring of soils and groundwater [10–13], served as the primary data for an ecological risk assessment prepared recently for Reclamation [14]. That report follows the guidelines for conducting ecological risk assessments for Superfund sites [15], although Kesterson Reservoir is not designated as such a site. Those guidelines are intended to ensure completeness and consistency in the reporting of assessment results. Because of space limitations, this chapter is a condensed version of the recent assessment.

II. OBJECTIVES AND SCOPE OF THE INVESTIGATION

The objectives of this assessment were to

- Review available pertinent information about selenium from Kesterson research and monitoring and from other terrestrial ecosystems
- Estimate the most likely levels of selenium in various biota (plants and animals) at the site, based on the current levels in biota, knowledge of the selenium inventory at the site, and projections of future biologically available selenium levels for soils and water provided by Lawrence Berkeley Laboratory (LBL)
- Assess the risks of adverse effects to animals caused by the site's selenium inventory, based on observed and estimated future levels of biologically available selenium in various media
- Assess the significance of the site's selenium toxicosis risks

- Identify contingency plans and evaluate their effectiveness in eliminating or reducing any potentially significant risks of selenium toxicosis
- Recommend research and monitoring that would provide information needed to improve management efficiency for the site or other projects

This assessment was conducted by analyzing and comparing data available from the Kesterson Reservoir Biological Monitoring Program through November 1991 with other published and unpublished reports. Most of the focus was on Kesterson Reservoir, but selected reports of biological transfer of selenium through components of other terrestrial ecosystems were used where applicable to Kesterson. For example, uptake of selenium by plants growing on fly ash disposal landfills was not considered applicable to Kesterson because the chemical forms of selenium in the soils probably differ [3,16,17]. In contrast, the biological transfer of selenium from those plants to animals feeding on them was considered to be similar to the biological transfer in the terrestrial ecosystem at Kesterson.

A substantial amount of information is available concerning the hazards of waterborne selenium to wildlife, particularly aquatic birds [3,18,19]. Criteria available from those sources were used to evaluate potential risks at Kesterson if portions of the reservoir were to become flooded during wet years.

Sampling of certain terrestrial plants and animals had been conducted at Kesterson Reservoir before lower elevation portions were filled in 1988 [1,4–6,20–22], and those data were evaluated for use in the assessment. However, the most consistent and useful results came from sampling conducted in all portions of the reservoir after filling was completed [7–9]. Biological monitoring was conducted in accordance with monitoring plans developed in consultation with the U.S. Fish and Wildlife Service (USFWS) and the Central Valley Regional Water Quality Control Board. The objectives of the biological monitoring program were defined as follows by the Central Valley Regional Water Quality Control Board:

- Assess the impact of Kesterson Reservoir on local and migratory wildlife
- Provide a basis for adjusting Kesterson Reservoir management
- Verify the effectiveness of cleanup actions at Kesterson Reservoir
- Provide a basis for modifying future biological monitoring

Sampling and data collection occurred each year in all portions of Kesterson Reservoir with comparisons among trisections, habitats, and years. The biological monitoring and site management plans were revised annually in response to monitoring results (e.g., adding or dropping certain species) and potentially hazardous conditions (e.g., occurrence of ephemeral pools). Scientific names for species mentioned in the text are listed in Table 1.

Selenium concentrations in soil, groundwater, and surface water (which occurred seasonally in pools) also were measured during 1989–1992 [7–10,12,13]. The present ecological assessment is based partly on the results of that monitoring and partly on projections of biologically available selenium in those media being made by LBL as part of this overall assessment of the site [11].

There are general relationships between water-extractable selenium concentrations in soils and selenium concentrations found in plants. However, these relationships apparently are influenced by many variables, including plant species, rainfall, temperature, soil pH and sulfate concentrations, plant growth rates, root depth, and distribution of selenium in the soil profile [17,23,24]. The general soil–plant relationship and modeling were used to characterize selenium relationships and to predict concentrations that can be expected in the future. These models are based on data from Kesterson Reservoir observations and professional judgments concerning probable plant succession, animal diets and home ranges, and selenium biogeochemistry.

Birds are the wildlife species most sensitive to chronic selenium toxicosis because of reproductive effects [3] and are the main focus of this assessment, along with mammals. Small mammals have been sampled most consistently for monitoring selenium in wildlife because they are less mobile than birds, more resident throughout the year (nonmigratory), potentially affected by dietary selenium, and widespread throughout Kesterson Reservoir. Small mammals (e.g., rodents and shrews) are good indicators of local conditions because of their small home ranges.

Previous reports have shown that selenium is bioaccumulated by frogs and snakes at Kesterson [6–8,25]. Although amphibians (toads) and reptiles (snakes and lizards) still occur there, those species were not considered in the risk assessment because of their low populations and the paucity of data about levels of selenium that affect tissues or diet [3]. Amphibian eggs and tadpoles may be affected by waterborne selenium, but they are not likely to be more sensitive than aquatic birds, which have been a primary concern. Unlike the aquatic birds, terrestrial species (meadowlarks and barn swallows) inhabiting Kesterson have shown no signs of reproduction problems.

The Wildlife Habitat Relationships (WHR) database was used, along with professional judgment, to determine wildlife species that may occur as the plant community at Kesterson Reservoir matures. The WHR system is an information system created through multiagency cooperation and maintained by the California Department of Fish and Game [26]. The WHR database consists of many components used to assess terrestrial

Table 1 Common and Scientific Names of Some Plants and Animals Mentioned in Text

Common name	Scientific name
Fungi	
Mushroom	*Agaricus* sp.
Plants	
Alkali goldenbush	*Haplopappus acradenius* ssp. *bracteosus*
Alkali heath	*Frankenia grandifolia* var. *campestris*
Alkali sacaton	*Sporobolus airoides*
Bassia	*Bassia hyssopifolia*
Cattail	*Typha* sp.
Clover	*Trifolium, Medicago,* and *Melilotus* spp.
Iodine bush	*Allenrolfea occidentalis*
Prickly lettuce	*Lactuca serriola*
Saltbush (perennial)	*Atriplex lentiformis*
Salt grass	*Distichlis spicata* var. *nana*
Thistle	*Cirsium* sp.
Invertebrates	
Daphnia	*Daphnia* sp.
Midge larvae	Chrionomidae
Water boatmen	Corixidae
Sowbugs	Isopoda
Fish	
Mosquitofish	*Gambusia affinis*
Birds	
American coot	*Fulica americana*
American avocet	*Recurvirostra americana*
Barn swallow	*Hirundo rustica*
Black-necked stilt	*Himantopus mexicanus*
Killdeer	*Charadrius vociferus*
Mallard	*Anas platyrhynchos*
Northern harrier	*Circus cyaneus*
Red-tailed hawk	*Buteo jamaicensis*
Screech owl	*Otus asio*
Tricolored blackbird	*Agelaius tricolor*
Western meadowlark	*Sturnella neglecta*
Mammals	
Black-tailed hare	*Lepus californicus*
California ground squirrel	*Spermophilus beecheyi*
California vole	*Microtus californicus aestuarinus*
Coyote	*Canis latrans*
Deer mouse	*Peromyscus maniculatus gambelii*
Desert cottontail	*Sylvilagus audubonii*
Heermann's kangaroo rat	*Dipodomys heermanni tularensis*

Table 1 (Continued)

Common name	Scientific name
House mouse	*Mus musculus*
Ornate shrew	*Sorex ornatus californicus*
Raccoon	*Procyon lotor*
San Joaquin kit fox	*Vulpes macrotis mutica*
Southwestern pocket gopher	*Thomomys bottae*
Western harvest mouse	*Reithrodontomys megalotis longicaudus*

vertebrate species occurrences, habitat requirements, life history information, and relative abundance. The emphasis on a need for predictive ability has resulted in a systematic and thorough treatment of the available information. The data for a specific location are accessed by selecting parameters such as habitat, county, and legal status, with a variety of elements or special requirements for each species. The WHR computerized database was used to classify and assess potential effects of habitat changes at Kesterson Reservoir and to provide comparison procedures for species composition and habitat descriptions.

III. STUDY AREA

Kesterson Reservoir is located in the Grasslands area of northern Merced County within the San Joaquin Valley of California (Figure 1). The reservoir consisted of 12 shallow ponds (average 1–1.5 m deep) totaling about 500 ha that were designed to serve as evaporation and holding basins for subsurface agricultural drainage waters from the western San Joaquin Valley [1,27]. Shallow saline groundwater collected by tile drainage systems in Fresno County farmland was transported to Kesterson Reservoir by the San Luis Drain.

The Kesterson ponds and San Luis Drain were constructed between 1968 and 1975 as an initial phase in the long-range development of agricultural drainage facilities for the San Joaquin Valley [1,27]. Kesterson was intended to serve as a 2100-ha regulating reservoir for storage and evaporation of agricultural drainwater during part of the year. At other times, the water would be discharged back into the San Luis Drain, which would carry it to the Sacramento-San Joaquin River Delta (about 125 km farther northwest) when river flows were considered sufficiently high to dilute the contaminants present in the drainwater.

Kesterson and other regulating reservoirs along the San Luis Drain were intended to provide beneficial uses of agricultural drainwater through their management as wetland habitat for wildlife [28]. Under a cooperative agreement between Reclamation and USFWS, Kesterson became part of the National Wildlife Refuge System [1,27]. The undeveloped portion of the refuge (1900 ha) is native grassland and seasonal wetland habitat typical of this area. However, Kesterson Reservoir has been removed from the National Wildlife Refuge System and is being managed now solely by Reclamation [M. Delamore, USBR, personal communication].

Construction of the San Luis Drain and associated facilities was halted in 1975 by funding limitations and concern about potential environmental impacts of drainwater discharge into the delta [1,27]. During 1972–1978, flow into Kesterson Reservoir was mainly surface flow via the San Luis Drain. By 1978, water deliveries included some subsurface drainwater, and after 1981 almost all of the flows into Kesterson Reservoir were subsurface agricultural drainwater. Water from the drain entered the southern end of the reservoir at Ponds 1 and 2 and flowed northward to Pond 12 (Figure 2). Water from the drain also could flow directly into Pond 12 through an emergency spillway intended to prevent excess water from overthrowing into an adjacent slough (Mud Slough).

Following the discovery of selenium contamination in Kesterson Reservoir, Reclamation halted discharge of agricultural drainage to Kesterson, dewatered the reservoir, and filled lower elevation portions to prevent groundwater from rising to the soil surface [1,4]. The drying and filling of Kesterson Reservoir converted the site to upland habitat consisting of three habitat types. *Grassland* habitat includes the higher elevation, upland area that existed at the reservoir before it was dried and filled; *Filled* habitat includes formerly low-lying areas filled with soil to prevent the occurrence of seasonal wetlands; and *Open* habitat includes former cattail areas that were not filled but were disked to prevent use by tricolored blackbirds. Grassland habitat covers about 30% of the reservoir; filled habitat, about 60%, and open habitat, 10% (Figure 2).

Because the ponds will no longer be flooded, and because the levees separating the ponds were removed as part of the filling operation, monitoring is currently being stratified based on "trisections" rather than the 12 former pond areas [4] (see Figure 2). The trisections were chosen on the basis of geographic distribution and operational history of the former ponds, as follows.

- Trisection 1 consists of the southernmost ponds (Ponds 1–4). These ponds are south of Gun Club Road and received the largest amounts of drainwater during 1978–1986. The ponds in this trisection have the

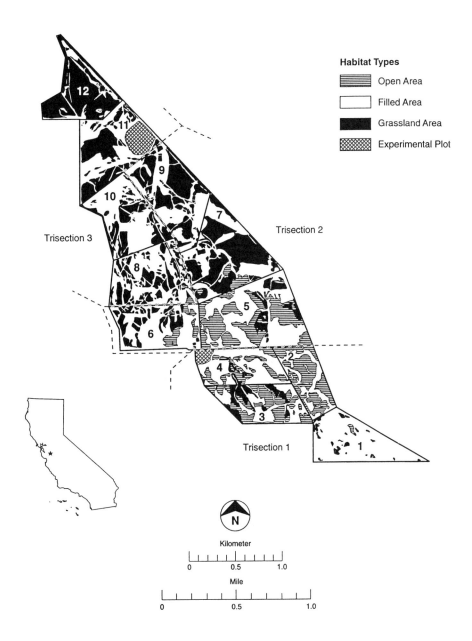

Figure 2 Kesterson Reservoir site habitat types, 1991.

highest soil selenium levels. They contained mainly open water and cattail areas in the past and currently contain mostly Open and Filled habitat.
- Trisection 2 consists of Ponds 5–7 and 9. These ponds are in the central area of Kesterson north of Gun Club Road and also received substantial amounts of drainwater in the past. The ponds in this trisection generally have lower soil selenium levels than those in Trisection 1. They previously contained open water and cattail areas and now contain some Grassland habitat.
- Trisection 3 consists of the northern ponds (Ponds 8 and 10–12), which received the least amount of drainwater. They have the lowest soil selenium levels and contain large areas of Grassland habitat.

Land adjacent and near Kesterson Reservoir consists mainly of native grasslands, seasonal and permanent wetlands, riparian, and agricultural cropland habitats along the San Joaquin River and its tributaries; Mud and Salt Sloughs drain the area west of the San Joaquin River [29,30]. Some portions of the region are currently owned by federal and state governments, others have been proposed for acquisition, and the remainder is expected to remain in private ownership as the northern division of the Grassland Water District. Although most animals collected for the Kesterson Biological Monitoring Program are resident within the reservoir, other species (including predatory birds and mammals, seasonally resident granivorous birds, and others) are more mobile and use Kesterson only as a portion of their home ranges.

Habitats at Kesterson Reservoir are not particularly unique, but no long-term studies have been conducted to document secondary succession to climax communities in San Joaquin Valley lands that may revert to native habitat. Monitoring the succession that occurs at Kesterson Reservoir and selenium concentrations in its ecosystem components may provide useful information for predicting wildlife values of other lands in the western San Joaquin Valley that are taken out of production because of poor drainage or high selenium levels. It is expected that some currently farmed lands within the San Luis Unit will be abandoned for those reasons, but their future wildlife values cannot be predicted readily [31].

A. Existing Terrestrial Habitats and Expected Changes

As part of the biological monitoring program at Kesterson Reservoir, three habitat types have been characterized. These three habitat types, in turn, include varying mixtures of three other habitat types described in the WHR database [26,32], as follows.

- *Grassland* areas are mostly Perennial Grassland habitat (WHR descriptor) and are dominated by dense mats of salt grass. Some areas of Alkali Scrub habitat (WHR descriptor), dominated by iodine bush, occur within the Grassland habitat type.
- *Filled* areas are former lower elevation wetlands that were filled with soil and are currently dominated by annual plant species characteristic of Annual Grassland habitats (WHR descriptor). Some areas of Alkali Scrub habitat (WHR descriptor), dominated by perennial saltbush, occur within the Filled areas.
- *Open* areas are former cattail stands that were drained and disked and are now sparsely vegetated with annual species such as bassia, prickly lettuce, and clover.

Although the monitoring program was established on the basis of habitat conditions that resulted from the filling of Kesterson, the WHR descriptor habitat types are more useful in predicting ecological changes.

As grasslands change along an ecological gradient from tallgrass to shortgrass to desert scrub, the small mammal fauna typically shifts from domination by herbivores to a more equal mixture of omnivores and granivores [33]. This general pattern has been observed in the mammalian fauna at Kesterson, and it also is affected by the density of vegetative cover. Ten species of small mammals and primarily herbivorous medium-sized mammals have been observed commonly during one or more years since 1988; these include ornate shrew, California ground squirrel, southwestern pocket gopher, Heermann's kangaroo rat, deer mouse, western harvest mouse, California vole, house mouse, black-tailed hare, and desert cottontail.

Areas with high-density plant cover tend to have a high total biomass, low-diversity vertebrate community composed of subsurface and litter-dwelling herbivores with a relatively high reproductive potential. In areas where cover is dense, vegetation is generally widely available to all herbivores, which must consume relatively large quantities of low-energy food. Perennial Grassland habitats at Kesterson Reservoir support a limited number of small mammal species. Of the seven small mammal and three larger mammal species (ground squirrel, hare, and cottontail) commonly occurring within the reservoir, only two, the ornate shrew and California vole, occur relatively often in this habitat throughout the reservoir. Other species captured from Perennial Grassland habitat are the harvest mouse, house mouse, and deer mouse; however, they are generally restricted to ecotones or areas dominated by alkali sacaton. Two of the larger mammals, the black-tailed hare and desert cottontail, are common throughout the reservoir but relatively uncommon in Perennial Grassland habitats.

California ground squirrels are common only locally in the reservoir and are relatively uncommon in Perennial Grassland habitats.

Areas with low-density plant cover are high in animal species diversity and in total biomass and are composed of surface-dwelling, relatively long-lived granivores and omnivores having relatively opportunistic reproductive habits. In areas where cover is sparse, the quantity and quality of vegetation is generally insufficient to support large numbers of herbivores. Annual plants at low-cover sites typically deposit a much higher proportion of their energy in seeds than perennial plants associated with high-density cover. The increase in the ratio of energy in seeds to energy in vegetation tends to favor granivores and omnivores. The Annual Grassland habitats support most small mammal species found within the reservoir. The ornate shrew is the only species not captured in this habitat. Before 1991, California voles were present but uncommon in Annual Grassland habitat in Pond 1 and were restricted to locations with dense vegetative cover. Two of the three omnivores, house mice and deer mice, are abundant in this habitat. The third species, western harvest mouse, is uncommon in the Annual Grassland habitat. Heermann's kangaroo rats are locally common and are rapidly becoming abundant in some locations.

Existing flora and fauna of Kesterson Reservoir are described in more detail in the various monitoring reports, but especially in Ref. 9. For the purposes of this assessment, site vegetation was divided into four groups: grasses, transitional annuals, persistent annuals, and perennials (Figure 3). These groups were selected on the basis of their expected occurrence among the various habitat types and their selenium concentrations. Annual broadleaf plants were divided into the transitional species expected to decline in frequency after about 10 years and the species expected to persist and become more common as succession continues.

B. Aquatic Habitats and Birds

Studies conducted at Kesterson during 1983–1985 showed that high levels of selenium were bioaccumulated by aquatic plants, invertebrates, fish, and birds [1,3]. The most pronounced effects on wildlife species were found in birds that fed regularly in the Kesterson ponds. Overall, at least 39% of the 578 nests of aquatic birds at Kesterson Reservoir monitored to hatching during 1983–1985, or from which a later-stage embryo was examined, contained at least one dead or deformed embryo or chick. Embryos or chicks with developmental abnormalities were found in 110

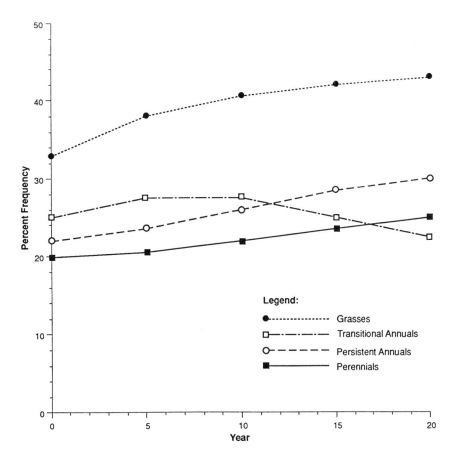

Figure 3 Expected succession among four groups of terrestrial plants at Kesterson Reservoir.

nests (19%). The abnormalities were similar to those in chickens and mallards fed diets containing 7–10 µg Se/g [3,34].

Surface ponding of rainwater will occur in some portions of Kesterson Reservoir during particularly wet years, and selenium concentrations in these surface pools may be high enough to be of concern for wildlife [10,12,13,35]. The greatest concern is that these pools may form during late winter or spring and be present during the nesting season of aquatic birds (such as waterfowl and shorebirds) that could feed on plants and invertebrates in the pools if they persisted long enough to support those food chain components.

Ephemeral pools that form under current conditions (i.e., postfilling)—and those expected to occur in the future—result from the accumulation of rainfall rather than from rising groundwater [10,13]. The filling of topographically lower portions of Kesterson effectively eliminated the probability of groundwater rising above the soil surface.

During 1989, mass balance models and numerical simulation studies were used to investigate the hydrologic factors leading to water table fluctuations in and around Kesterson [10,13]. These studies and considerable water level data indicate the flooding of nearby duck clubs is responsible for most of the rise in water table elevation. Rainfall is a significant source of water table rise only when it exceeds normal values by more than 50% or when it is intense, such as occurred during February and March 1992. Model predictions indicated that in years with twice the average annual rainfall of 240 mm, much of the Reservoir could be covered with standing water during winter. However, selenium and salt concentrations in these pools are expected to be low compared to pre-1989 levels because the pools result from rainwater ponding rather than from soil waters rising from below. Nevertheless, surface water selenium concentrations in some pools exceeded the recommended safe levels of 5 µg/L or less [36,37].

During the winter of 1991–1992, total rainfall at Kesterson Reservoir was about 320 mm [12]. This total is not unusual for the area, and it was similar to the rainfall in the winters of 1982–1986 (330–360 mm). The largest single storm during the 1991–1992 winter (77 mm rainfall) occurred on February 12. This and subsequent storms during February (total of 150 mm) and March (80 mm) produced enough intense rainfall during a short period that many rainwater pools formed in the reservoir. Some of these pools disappeared within a week or two after the heavier rainfall occurred, but others persisted until mid-April. The pools that formed on some fill areas and the large one within the University of California (U.C.), Riverside, experimental plot in Pond 11 were among the more persistent ones.

Selenium concentrations in surface pools that have been monitored since early 1989 have averaged less than 10 µg/L, but concentrations up to 162 µg/L have been measured in some pools [10,12,13,35]. Pools formed by rising groundwater would contain selenium at concentrations about one to two orders of magnitude higher than those observed in pools formed from rainwater.

LBL sampled surface waters from many of the rainwater pools during February 15–21, 1992 [12]. The sampling included collections from 11 of the former ponds (excluding Pond 7) and all habitat types within the reservoir. A number of these pools were monitored until mid-April to determine changes in conductivity and selenium concentrations. The sampling during February 15–21 also included collections from the Fremont Canal (south of Pond 6), Mud Slough (at Pond 10), and pools present immediately outside the Kesterson Reservoir exterior levee (but within the perimeter fences).

Among the 94 samples from rainwater pools during February 15–21, 1992, the highest concentration (162 µg/L) occurred in a small pool at an ongoing LBL surface water monitoring site in Pond 5 [12]. The four other pools in which waterborne selenium concentrations exceeded 50 µg/L included Pond 2, southeast corner, 141 µg/L; Pond 6, LBL 1-ft excavation test plot P6S12, 120 µg/L; Pond 4, east of U.C. Riverside plots, LBL soil site P4X, 74.8 µg/L; and Pond 2, 2 VS Plot D, north end, 54.3 µg/L. Thus, the highest waterborne selenium concentrations tended to occur in some of the experimental plots and in a small portion of the Open habitat type in Pond 2.

The geometric mean selenium concentration for the surface water pools (including all 94 samples) was 2.73 µg/L, indicating that the highest values were very atypical [12]. Overall, the distribution of waterborne selenium concentrations for pools within the Reservoir included 38% greater than 5 µg/L, 24% greater than 10 µg/L, and 11% greater than 20 µg/L. Selenium concentrations in Fremont Canal (24 µg/L) and Mud Slough (7.6 µg/L) were greater than the geometric mean for Kesterson Reservoir pools (2.73 µg/L) or in the pools outside the Reservoir perimeter levee (all 1.6 µg/L or less).

In the 20 pools monitored during mid-February to mid-April 1992 (or until they disappeared, if that occurred earlier), conductivity generally increased, mainly because of evaporation [12]. The relatively low conductivity, even by mid-April (<3.0 dS/m in most pools), indicated that the water was mainly rainfall rather than rising groundwater. Temporal patterns in selenium concentrations were much more variable than those for conductivity. Selenium concentrations in some pools increased (though generally not more than about twofold), whereas the concentrations in other pools decreased or showed no consistent trend.

Selenium concentrations exceeded 20 µg/L during early April in the pool within the U.C. Riverside Pond 11 experimental area (28 µ/L) as well as the Pond 4 experimental area previously used by U.C. Riverside (about 70 µg/L), two pools in Pond 5 (30 and 38 µg/L), and one in Pond 3 (about 22 µg/L). The waterborne selenium concentrations in other pools were generally below 10 µg/L, or the pools had disappeared by that time.

IV. CONTAMINANTS OF CONCERN

Selenium is the only contaminant of particular concern from an ecological perspective at Kesterson Reservoir, and management of the selenium inventory that remains on the site is the focus of this assessment. Selenium contamination was the primary reason for Kesterson's closure and its management as a terrestrial habitat [1,4]. Earlier studies had shown that organochlorine contaminants were not accumulated to significant concentrations by mosquitofish living in the Kesterson ponds or in the San Luis Drain [38]. Although boron accumulated in some of the aquatic plants to levels that were potentially harmful to waterfowl reproduction [39], this was not true of most food chain components [21,40]. Other trace elements generally occurred at concentrations similar to background levels found at reference sites [21,38,40]. Selenium also was identified as the most harmful of the various drainwater contaminants through a series of integrated field and laboratory research studies [34]. Hence, measurement of selenium concentrations in various ecosystem components has been the main focus of the monitoring program [e.g., 5–10,13,35].

Sample collection, handling, and transportation procedures for the biological monitoring program have been described elsewhere [9]. Briefly, the specimens were collected and sorted or dissected using standard procedures to ensure that they were not contaminated by the investigators. After processing, each sample was placed in a chemically cleaned polyethylene bottle or Whirl-Pak bag and double-labeled with its unique code, indicating taxon, collection date, pond number, habitat type, and sampling site number. This information was also transcribed onto preformatted data sheets. The sealed and numbered sample containers were grouped by taxon. Samples were frozen immediately after sorting or dissection and were later shipped to the analytical laboratory by overnight delivery.

Samples of both plant and animal tissues were analyzed at Environmental Trace Substances Research Center (ETSRC) in Columbia, Missouri, for selenium content by hydride generation and atomic absorption on a Varian VGA-76 system. Plant and invertebrate samples collected from Kesterson Reservoir have been analyzed by ETSRC since 1984. Some vertebrate wildlife samples (amphibians, reptiles, birds, and mammals) collected before 1987 were analyzed at ETSRC, whereas others were analyzed at the U.S. Fish and Wildlife Service's Patuxent Wildlife Research Center in Laurel, Maryland. However, all plant and wildlife samples collected since 1987 have been analyzed at ETSCR, and the Patuxent Research Center has monitored quality control of ETSRC analyses performed for USFWS. Therefore, results are considered comparable among years.

Surface soil (0–15 cm depth) samples were collected in conjunction with the biological monitoring program. They were taken from the sampling stations used for plants to determine the relationship between soil selenium and plant selenium. These soil samples were analyzed by LBL for years 1989–1991. Detailed results of selenium analyses are provided in biological monitoring reports for each year since 1987 [5–9]. All data were transformed to common logarithms for analyses because the distributions were log-normal.

V. EXPOSURE CHARACTERIZATION

Generalized exposure pathways for Kesterson Reservoir biota are illustrated in Figure 4. Under normal conditions, all exposed plants and animals are terrestrial species, but during extended periods of heavy rainfall ephemeral pools may form in some portions of the reservoir.

A. Terrestrial Habitats

Plants and animals representative of the various wildlife exposure pathways have been analyzed in one or more of the years since the filling of lower elevation areas in 1988. Those species were generally selected to facilitate comparison of spatial, temporal, and habitat trends for selenium levels in the ecosystem.

For modeling purposes, common plant species were grouped into four categories based on ecological considerations (see Figure 3) and selenium concentrations:

- *Grasses* are generally lower in selenium than other plants in Kesterson Reservoir; grasses are expected to persist over time as dominant plant species in Annual Grassland habitat and as a component of Perennial Grassland and Alkali Scrub habitats.
- *Transitional annuals* tend to have the highest concentrations of selenium; they are fast-growing, early invader species expected to be replaced by slower growing annuals and perennials.
- *Persistent annuals* are also generally higher in selenium than grasses but lower than the transitional annuals; they are slower growing, secondary invader species expected to remain a part of Annual Grassland, Perennial Grassland, and Alkali Scrub climax communities over time.
- *Perennials* are expected to be dominant species of the Alkali Scrub climax community; they tend to have lower selenium concentrations than annual species. The biological monitoring program has sampled alkali heath and perennial saltbush, but the heath is apparently not an

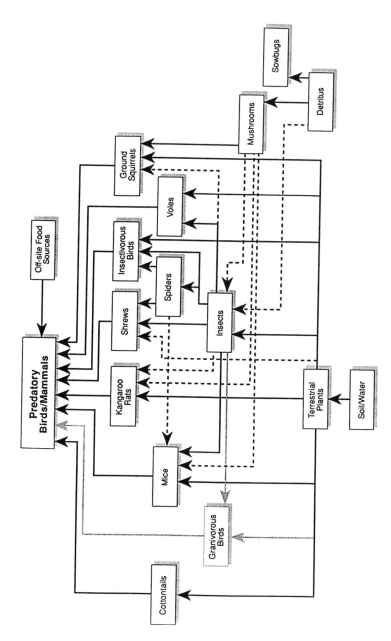

Figure 4 Exposure pathways to selenium for terrestrial wildlife at Kesterson Reservoir. (Note that four types of plants are considered in the model.)

important food of mammals living at Kesterson. Thus, the assessment is based only on the limited data for perennial saltbush.

Transitional annual plant species are among the more important components of the risk assessment because their occurrence in the ecosystem is expected to increase during the next 10 years; they have relatively high selenium concentrations; and they are important foods for wildlife [8,9].

In selecting the groupings of plants, root depth was considered one of the criteria. However, there was no clear relationship between root depth and plant selenium concentration, and for some species the depth of the root system was not known. Thus, the ecological grouping resulted in more homogeneous sets of plants that reflected similar selenium concentrations and expected occurrence through plant succession in the future.

Wildlife exposure was characterized on a reservoir-wide basis because most of the wildlife species of interest move readily from one habitat to another and also could move from one trisection to another. A model was constructed from available information to correlate selenium concentrations in various components of the ecosystem under current conditions and to predict concentrations expected to occur in the next 20 years. The model is built on the pathways for selenium exposure illustrated in Figure 4. It takes into account three different projections for water-extractable selenium in soils at a depth of 15 cm to 1 m provided by Benson et al. [11]. This selenium fraction and soil depth are considered the best indicators of bioavailable selenium for plant uptake on the basis of other studies. The summary results of the model are included as Table 2 and show the expected range of mean selenium concentrations for biota when selenium bioavailability in soils ranges from low to high.

Data from previous monitoring [7–11,13] were synthesized to represent current conditions (Year 0), and selenium concentrations in various ecosystem components were estimated for Years 1–20 (1992–2011). Most of the animals included in the biological monitoring program are resident on the site. In addition, four representative predatory birds and mammals were included in the assessment. Their potential exposures were modeled to reflect the portions of their home ranges that might occur on the reservoir, as follows: northern harrier (modeled as 100% resident on the reservoir), red-tailed hawk (expected to be 50% resident), San Joaquin kit fox (25% resident), and coyote (10% resident). Home ranges for these species were estimated from other studies [41–44].

Relationships between water-extractable soil selenium and selenium concentrations in each of the four plant groups were determined through regression analyses; these relationships for persistent annuals are illustrated

Table 2 Geometric Mean Selenium Concentrations During 1989–1991 and Range of Mean Selenium Concentrations Expected to Occur Under Conditions of Lowest to Highest Bioavailability During 1992–2011 (μg/g, dry weight)

Year	Soil[a]	Grasses	Transitional annuals	Persistent annuals	Peren-nials	Mush-rooms	Insects (TF = 2.14)[b]	Spiders (TF = 1.28)[b]	Cottontails (TF = 2.01)[b]	Mice (TF = 1.03)[b]
1989	0.09	2.86	6.09	3.92	2.64		7.90	11.0		5.15
1990	0.06	2.63	5.86	3.60	2.58	448	6.50	8.8		6.64
1991	0.10	2.97	6.20	4.08	2.67	211	9.80	15.0	15.3[c]	6.46
1989–1991	0.08	2.82	6.05	3.86	2.68	218	8.17	10.4	8.3[d]	6.02
1992	0.06–0.16	2.61–3.31	5.83–6.52	3.56–4.56	2.57–2.74	199–264	7.73–9.1	9.88–11.7	7.79–9.3	5.69–6.75
1993	0.06–0.19	2.61–3.45	5.83–6.65	3.56–4.76	2.57–2.77	199–278	7.73–9.4	9.88–12.0	7.78–9.6	5.69–6.95
1994	0.06–0.21	2.61–3.53	5.83–6.73	3.56–4.88	2.57–2.79	199–286	7.74–9.6	9.89–12.3	7.77–9.8	5.69–7.08
1995	0.06–0.24	2.61–3.65	5.83–6.83	3.56–5.04	2.57–2.82	199–297	7.74–9.8	9.89–12.6	7.76–10.0	5.68–7.24
1996	0.05–0.26	2.61–3.72	5.83–6.89	3.56–5.15	2.57–2.83	199–304	7.74–10.0	9.89–12.7	7.75–10.1	5.68–7.35
1997	0.05–0.28	2.61–3.79	5.83–6.95	3.56–5.24	2.57–2.85	199–311	7.73–10.1	9.88–12.9	7.74–10.3	5.68–7.44
1998	0.06–0.29	2.61–3.82	5.83–6.98	3.56–5.29	2.57–2.85	199–314	7.72–10.1	9.86–13.0	7.72–10.3	5.67–7.48
1999	0.06–0.31	2.61–3.89	5.83–7.03	3.56–5.38	2.57–2.87	199–320	7.70–10.3	9.84–13.1	7.71–10.4	5.66–7.56
2000	0.06–0.32	2.61–3.92	5.83–7.06	3.56–5.42	2.57–2.87	199–323	7.69–10.3	9.82–13.2	7.69–10.5	5.65–7.60
2001	0.06–0.33	2.61–3.95	5.83–7.09	3.56–5.46	2.57–2.88	199–326	7.67–10.3	9.81–13.2	7.67–10.5	5.63–7.63
2002	0.06–0.34	2.61–3.97	5.83–7.11	3.56–5.50	2.57–2.88	199–329	7.65–10.4	9.78–13.3	7.66–10.6	5.62–7.66
2003	0.06–0.35	2.61–4.00	5.83–7.13	3.56–5.54	2.57–2.89	199–332	7.62–10.4	9.74–13.3	7.62–10.6	5.60–7.68
2004	0.06–0.36	2.61–4.03	5.83–7.16	3.56–5.58	2.57–2.90	199–335	7.60–10.4	9.71–13.3	7.59–10.6	5.58–7.70
2005	0.06–0.37	2.61–4.06	5.83–7.18	3.56–5.62	2.57–2.90	199–337	7.56–10.4	9.66–13.4	7.55–10.7	5.55–7.72
2006	0.06–0.37	2.61–4.06	5.83–7.18	3.56–5.62	2.57–2.90	199–337	7.53–10.4	9.62–13.3	7.52–10.6	5.53–7.70
2007	0.06–0.38	2.61–4.08	5.83–7.20	3.56–5.66	2.57–2.91	199–340	7.50–10.4	9.58–13.4	7.49–10.7	5.51–7.72
2008	0.06–0.38	2.61–4.08	5.83–7.20	3.56–5.66	2.57–2.91	199–340	7.47–10.4	9.55–13.3	7.46–10.7	5.49–7.71
2009	0.06–0.39	2.61–4.11	5.83–7.22	3.56–5.70	2.57–2.91	199–343	7.45–10.4	9.52–13.3	7.44–10.7	5.47–7.73
2010	0.05–0.39	2.49–4.11	5.71–7.22	3.40–5.70	2.54–2.91	188–343	7.42–10.4	9.48–13.3	7.40–10.7	5.45–7.71
2011	0.05–0.39	2.49–4.11	5.71–7.22	3.40–5.70	2.54–2.91	188–343	7.16–10.4	9.14–13.3	7.12–10.6	5.25–7.69
							7.12–10.3	9.10–13.2	7.09–10.6	5.23–7.67

KESTERSON RESERVOIR—PAST, PRESENT, AND FUTURE

Year	Kangaroo rats (TF = 1.06)	Shrews (TF = 2.41)	Meadowlarks (TF = 0.98)	California voles (TF = 2.92)	Ground squirrels (TF = 1.04)	Kit foxes[e] (TF = 0.65)	Coyotes[e] (TF = 2.89)	Red-tailed hawks[e] (TF = 1.38)	Northern harriers (TF = 0.83)
1989		21.00	23.0[c]	14.0			3.4[c]		7.2[f]
1990		19.00	16.0[c]	9.3					
1991	4.40	15.00	6.8[c]		19.0[c]	1.7[g]	5.4[c]	5.3[c]	5.6[f]
1989–1991	4.39	20.28	7.8[d]	11.9	10.6[d]		4.4[c]	5.3[c]	6.6[f]
0	4.10–5.05	19.2–22.7	7.37–8.72	11.2–13.5	9.03–11.4	1.56–1.85	4.19–4.81	4.99–5.91	6.25–7.45
1992	4.09–5.23	19.2–23.4	7.37–8.99	11.2–14.0	9.02–11.9	1.56–1.91	4.18–4.94	4.99–6.08	6.24–7.68
1993	4.09–5.34	19.2–23.8	7.37–9.15	11.2–14.3	9.02–12.2	1.56–1.95	4.18–5.01	4.99–6.19	6.24–7.83
1994	4.08–5.49	19.2–24.4	7.37–9.38	11.2–14.7	9.02–12.6	1.56–1.99	4.18–5.11	4.98–6.34	6.24–8.02
1995	4.08–5.58	19.2–24.7	7.37–9.52	11.2–14.9	9.02–12.9	1.56–2.02	4.18–5.17	4.98–6.43	6.24–8.14
1996	4.07–5.67	19.2–25.1	7.37–9.64	11.2–15.1	9.02–13.1	1.56–2.05	4.18–5.23	4.98–6.51	6.24–8.25
1997	4.07–5.71	19.2–25.2	7.35–9.69	11.1–15.2	9.01–13.2	1.55–2.06	4.18–5.26	4.97–6.54	6.22–8.29
1998	4.06–5.79	19.1–25.5	7.34–9.80	11.1–15.4	9.00–13.4	1.55–2.09	4.17–5.31	4.96–6.62	6.21–8.39
1999	4.05–5.82	19.1–25.6	7.33–9.84	11.1–15.5	8.99–13.5	1.55–2.10	4.17–5.33	4.95–6.65	6.20–8.43
2000	4.05–5.86	19.0–25.7	7.31–9.88	11.1–15.5	8.98–13.6	1.55–2.11	4.16–5.35	4.95–6.68	6.19–8.47
2001	4.04–5.89	19.0–25.8	7.30–9.92	11.0–15.7	8.97–13.7	1.55–2.12	4.16–5.37	4.94–6.71	6.18–8.50
2002	4.03–5.92	18.9–25.8	7.27–9.95	11.0–15.7	8.95–13.8	1.54–2.13	4.15–5.39	4.92–6.73	6.15–8.53
2003	4.02–5.95	18.9–25.9	7.24–9.97	10.9–15.8	8.94–13.9	1.54–2.14	4.14–5.41	4.90–6.75	6.13–8.56
2004	4.01–5.97	18.8–25.9	7.21–9.98	10.9–15.8	8.92–14.0	1.53–2.14	4.13–5.43	4.88–6.77	6.10–8.57
2005	4.00–5.96	18.7–25.9	7.18–9.98	10.9–15.8	8.90–14.0	1.53–2.14	4.12–5.42	4.87–6.75	6.08–8.55
2006	3.99–5.99	18.6–25.9	7.15–9.98	10.9–15.8	8.88–14.0	1.52–2.15	4.11–5.44	4.85–6.77	6.06–8.58
2007	3.98–5.98	18.6–25.9	7.13–9.96	10.8–15.8	8.87–14.0	1.52–2.15	4.10–5.43	4.83–6.76	6.04–8.56
2008	3.97–6.01	18.5–25.9	7.10–9.98	10.8–15.9	8.85–14.1	1.52–2.15	4.10–5.45	4.82–6.78	6.02–8.58
2009	3.96–6.00	18.4–25.8	7.08–9.95	10.8–15.8	8.84–14.1	1.51–2.15	4.09–5.44	4.80–6.77	5.99–8.56
2010	3.79–5.99	17.8–25.8	6.83–9.93	10.4–15.8	8.44–14.1	1.46–2.15	3.98–5.43	4.64–6.75	5.77–8.54
2011	3.78–5.99	17.7–25.7	6.80–9.90	10.3–15.8	8.42–14.1	1.46–2.15	3.97–5.43	4.62–6.74	5.75–8.52

[a]Water-extractable soil selenium (15 cm to 1 m).
[b]TF = Transfer factor (selenium concentration in consumer divided by concentration in food).
[c]Measured liver concentration.
[d]Whole-body selenium concentrations were calculated from measured liver selenium data by using the regression equation derived from small mammal data (see text). Whole-body selenium values were used for modeling purposes because these animals may be eaten by predators.
[e]Off-site food sources for these wide-ranging predators were modeled as containing 0.5 μg/g selenium.
[f]Assumed from 1989 and 1991 barn owl liver value.
[g]Liver value, 1987 [43].

in Figure 5. The illustrated relationships are based on measured water-extractable selenium concentrations in the top 15 cm of soil and in plants growing nearby. There is a general increase in plant selenium as water-extractable selenium increases in surface soils, although the relationship is highly variable for reasons discussed earlier. The slopes of the regression lines were highly significant ($P<0.0001$) for each of the four plant groups except perennials, probably because of their small sample size. The relationships between soil selenium and the concentrations in detritus or sowbugs are not known, and the model considered values to remain static at previously measured concentrations. The limited data available ($n=6$) on the relationship between water-extractable soil selenium and selenium concentrations in mushrooms were incorporated into the model (along with data on other mushrooms from Kesterson) to simulate concentrations expected reservoir-wide.

Insects and other herbivores were assumed to feed on various kinds of plants in proportion to their expected frequency of occurrence in the reservoir. For insects and other consumers, a fractional *weighting factor* (≤ 1) was assigned to each component of their diet. A *transfer factor* (concentration in consumer divided by concentration in food item) was determined. Because dietary shifts are expected, the weighting factor assigned to each plant group that was represented in the consumers' diet was adjusted in accordance with the expected frequency of the plant group in the future. However, the transfer factors remain constant because no changes in uptake rate are expected.

There was fair to good information on food habits of birds and mammals collected for analysis in the monitoring program during some years, but some food items found in stomachs could not be assigned to specific categories used in the assessment. There has been no sampling of granivorous birds, some of which may become more common components of the Kesterson ecosystem as habitat succession occurs. Diets of the predatory bird and mammal species were estimated on the basis of various reports [43,45], and a weighting factor was assigned to each component of the diet. Off-site food sources were assumed to have background concentrations of selenium (0.5 µg/g) for modeling purposes, and they received a weighting factor that represented the off-site fraction of the predators' overall diets (i.e., 0 for harrier, 0.5 for red-tail, 0.75 for kit fox, and 0.9 for coyote).

The smaller mammals (shrews, voles, mice, and pocket gophers) sampled in the biological monitoring program usually have been analyzed as whole-body samples, but since 1989 livers were analyzed separately for those specimens. The following generalized equation was developed to

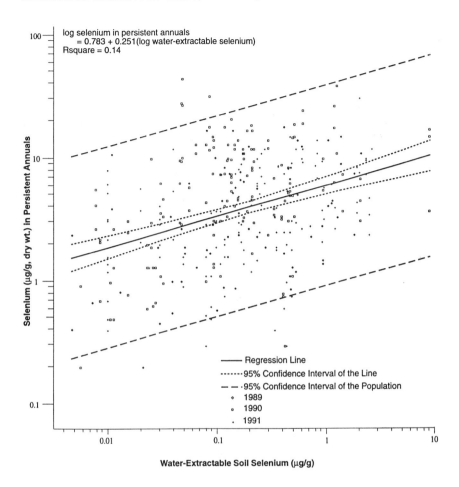

Figure 5 Regression relationship between selenium in persistent annuals and water-extractable selenium in soil (0–15 cm).

describe whole-body selenium concentrations on the basis of liver concentrations:

$$\log \text{Se in whole body} = -0.378 + 1.0925 \, (\log \text{liver Se})$$

The regression equation was then used to estimate whole-body selenium concentrations in larger mammalian prey (ground squirrels, hares, and cottontails) and in meadowlarks, which may be consumed by predatory birds and mammals but for which only the livers were analyzed.

Selenium concentrations in the plants and animals inhabiting Kesterson Reservoir are not expected to change markedly during the next 20 years. The concentrations expected to occur in the year 2011 under conditions of moderate, low, and high bioavailability of soil selenium (based on values from Benson et al. [11]) are presented in Table 3, where they are compared to current levels. If selenium bioavailability remains moderate, average concentrations are expected to be about 10–20% greater than current levels in most biota. With low bioavailability, selenium concentrations in plants and animals are expected to be somewhat lower (10–15%) after 20 years than the current levels. Highest expected concentrations in plants and animals are about 20–30% above those in 1989–1991. Selenium concentrations are expected to be higher in mushrooms than in all other

Table 3 Mean Selenium Concentrations Expected to Occur in Plants and Animals at Kesterson Reservoir in 2011 Under Varying Levels of Bioavailability Compared to 1989–1991 Means (μg/g, dry weight)

Samples	1989–1991 Mean	Expected concentrations		
		Moderate bioavailability	Low bioavailability	High bioavailability
Soil[a]	0.1	0.2	0.1	0.4
Grasses	2.8	3.6	2.5	4.1
Transitional annuals	6.0	6.8	5.7	7.2
Persistent annuals	3.9	4.9	3.4	5.7
Perennials	2.6	2.8	2.5	2.9
Mushrooms	218.0	290.0	188.0	343.0
Insects	8.2	9.3	7.1	10.3
Spiders	10.4	11.9	9.1	13.2
Cottontails	8.3	9.5	7.1	10.6
Mice	6.0	6.9	5.2	7.7
Kangaroo rat	4.4	5.3	3.8	6.0
Shrew	20.3	23.1	17.7	25.7
Insectivorous birds	7.8	8.9	6.8	9.9
California vole	11.9	14.0	10.3	15.8
Ground squirrel	10.6	12.2	8.4	14.1
Kit fox[b]	1.7	1.9	1.5	2.1
Coyote[b]	4.4	4.9	4.0	5.4
Red-tailed hawk[b]	5.3	6.0	4.6	6.7
Northern harrier[b]	6.6	7.6	5.7	8.5

[a]Water-extractable 15 cm to 1 m depth.
[b]Liver values; others are whole-body values.

organisms. Because plants serve as the primary source of selenium uptake and there are many similarities in the dietary exposure pathways for animals at Kesterson (Figure 4), the degree of change in selenium concentrations among the various animals is similar (Tables 2 and 3).

B. Aquatic Habitats

During years with annual rainfall in excess of 350 mm (about 50% more than normal), surface water pools up to 5 cm deep are expected to occur over large areas within the reservoir [10]. In the event of a 500-mm annual rainfall (100-year event), pools 10–25 cm deep may occur over most of the reservoir. The water in the pools will be derived from rainwater ponding at the soil surface. The most extensive ponding is anticipated to occur from early February to mid-March, as it did during 1992. Selenium and salts present at the soil surface will dissolve in these pools, potentially creating selenium concentrations ranging from less than detectable to several hundred micrograms per liter. Estimates of a reservoir-wide average selenium concentration were made in 1990. For 20-cm-deep pools covering the entire reservoir, estimates of the average selenium concentrations range from about 10 to 30 µg/L. Measured data from the 1991–1992 winter yield an average selenium concentration of 12.9 µg/L and a geometric mean concentration of 2.73 µg/L [12]. Predicted and measured concentrations are in reasonable agreement, although measured values tend to be on the low side of the estimated range. Given the limited data from which these estimates were developed, it was not possible to narrow the expected range of average concentrations. These concentrations are in the range of values sometimes measured during 1989 and 1990 in surface waters near Kesterson, including Mud Slough, Fremont Canal, and the San Luis Canal [10].

Soil selenium concentrations in the top 15 cm are strongly influenced by rainfall infiltration [10]. Selenate, in particular, is readily leached down deeper into the soil profile with winter rains, which tends to decrease the inventory of soluble selenium available for dissolution in rainwater pools that form at Kesterson. About 5% of the total selenium inventory in Kesterson surface soils (top 15 cm) is water-extractable, indicating that only a limited amount of the selenium is currently mobile and available for plant uptake or dissolution into rainwater pools. No significant changes in total or water-extractable selenium concentrations were observed between 1989 and 1991. However, on a reservoir-wide basis there was a significant increase ($P<0.05$) in the ratio of water-extractable selenium to total selenium in soils [11].

When pools of rainwater formed in portions of the reservoir during 1989–1991, they usually disappeared within 2 or 3 weeks [7–9], but in 1992 some pools persisted until mid-April. Typically, persistence was much shorter after 1989, probably because of the increased vegetative cover in most areas, the deep-soil ripping done by Reclamation to improve water percolation, and the additional fill placed in areas where rainwater pools had formed. The pools that were observed during 1989–1991 were too ephemeral to develop significant populations of aquatic plants or invertebrates, and they received no significant wildlife use. However, during 1992, aquatic invertebrates were present in some pools in sufficient numbers for aquatic birds to feed there.

Aquatic invertebrates were collected from six of the pools on March 31, 1992, by CH2M HILL biologists and analyzed for selenium. Other pools were examined, but invertebrate population densities were too low to warrant collection. The selenium concentrations in these invertebrates are shown in Table 4, along with waterborne selenium concentrations measured by LBL on or near March 31. Relationships between waterborne selenium and the concentrations found in invertebrates were highly variable, as reflected by the three daphnia samples. Among those three pools the waterborne selenium ranged only from 6.5 to 12.0 µg/L, but the concentrations in daphnia ranged from 7.1 µg/g (where waterborne selenium was highest) to 48 µg/g. The geometric mean for invertebrates

Table 4 Selenium Concentrations in Aquatic Invertebrates (µg/g, dry weight) and Water (µg/L) from Surface Water Pools at Kesterson Reservoir, March 31, 1992

	Invertebrates		Water
Location[a]	Type	µg/g	(µg/L)
1	Daphnia	21.0	6.5
3	Water boatmen	6.7	13.0
6A	Midge larvae	6.0	3.0
6B	Daphnia	7.1	12.0
10	Daphnia	48.0	8.3
10	Water boatmen	16.0	(8.3)[b]
11	Water boatmen	11.0	11.0
11	Water boatmen	11.0	(11.0)[b]
Geometric mean		12.4	8.07

[a]Pond number; two pools were sampled in Pond 6.
[b]Waterborne concentration for pool included only once in computing geometric mean.

(12.4 µg/g) was about 6 to 10 times the normal background for these organisms.

Bird counts throughout the reservoir on four days during late March 1992 indicated a daily average of about 200 migrant shorebirds (including various species that do not nest in this region). Four American avocets, 32 black-necked stilts, 10 killdeer, and 17 ducks were observed [46]. The avocets, stilts, and ducks were birds that could potentially nest within this region. Other observations, however, indicated that the numbers of birds foraging in the Kesterson National Wildlife Refuge wetlands east of the San Luis Drain were typically greater than those observed within the reservoir.

The numbers of aquatic birds decreased during April and May 1992 [46]. Daily average shorebird counts for four days in April included 6 migrant shorebirds, 3 avocets, 10 stilts, 5 killdeer, and 14 ducks; the daily average for May, based on 11 days, included no migrant shorebirds, 2 killdeer, and 6 ducks. These counts reflect decreasing use of the reservoir as surface water pools disappeared, which occurred before the beginning of the nesting season for most shorebirds and waterfowl.

VI. RISK CHARACTERIZATION

A. Terrestrial Habitats

The model used in this assessment produced estimates of selenium concentrations that could be expected to occur in plants and animals living at Kesterson. The ranges of values presented in Table 2 are based on projections for low and high bioavailability of soil selenium provided by LBL [11]. Monte Carlo simulation [47] for Year 10 (2001) was used to describe the uncertainties associated with the estimated selenium concentrations in terrestrial plants and in the overall diets of several bird and mammal species. The simulations were performed mainly because of the large variability in the soil–plant selenium relationships as shown in Figure 5.

Each Monte Carlo simulation included 100 runs using randomly selected values for water-extractable soil selenium. These values were taken from the range between low and high projected water-extractable soil selenium concentrations, assuming a uniform distribution within that range. Data from the monitoring program show that the soil–plant selenium relationship is log-normal. Therefore, for each randomly selected soil selenium concentration, a corresponding plant selenium concentration was randomly selected from within the log-normal distribution of expected values. The range of those expected values for plants was based on the regression relationship for soil selenium to plant selenium and the variability in that relationship. These simulations for persistent annual plants are illustrated in the upper panel of Figure 6. The vertical line for each panel of the figure shows the geometric mean selenium concentration found during 1989–1991 [9] (Table 2).

Shaded areas in each panel of Figure 6 show the expected range of reservoir-wide mean selenium concentrations that can be expected. This range of expected mean concentrations is taken from the values for Year 10 based on estimates for low and high bioavailability of soil selenium (Table 2). The cumulative probability line illustrates the simulation results for the probability of selenium concentrations in individual plant samples being lower than the X-axis values. Because of uncertainty in future bioavailability of soil selenium and in the soil–plant relationships for selenium, future concentrations could be different from those produced by the simulations. However, these simulations provide the best estimate of future conditions.

Expected selenium concentrations in mushrooms, insects, and the diets of selected terrestrial bird and mammal species are illustrated in lower portions of Figure 6. The best estimate is that mean selenium concentrations for the diets of these animals should be generally similar to current levels. The cumulative probability line shows that selenium concentrations in some samples will be considerably higher than the 1989–1991 mean (as they were during those years), but there is a relatively narrow range of expected mean concentrations for the diets of most animals.

Mean selenium concentrations in the diets of mice are expected to be lower than 10 µg/g (Figure 6). In the laboratory, rats were fed diets containing wheat with biologically incorporated selenium grown in a seleniferous area of South Dakota [48]. Diets with 8 µg/g or more selenium caused mortality in rats, and the 6.4 µg/g diet depressed weight gain and caused changes in organ weights. However, no effects of selenium on body condition or liver weights of small mammals were observed at Kesterson during 1984 when mean selenium concentrations were higher than those expected in the next 20 years [20].

Selenium concentrations in small mammals (Table 2) are expected to remain below those measured in 1984 at Kesterson [20]. No clear effect of selenium on small mammal reproduction was found, although whole-body analyses by species for all ponds at Kesterson averaged about 11 µg/g. That level of selenium in prey species was sufficient to suggest concern for possible effects on predators, such as raptorial birds, raccoons, coyotes, and the endangered San Joaquin kit fox. In domestic dogs, dietary levels of natural selenium above 7 µg/g inhibited growth and 20 µg/g

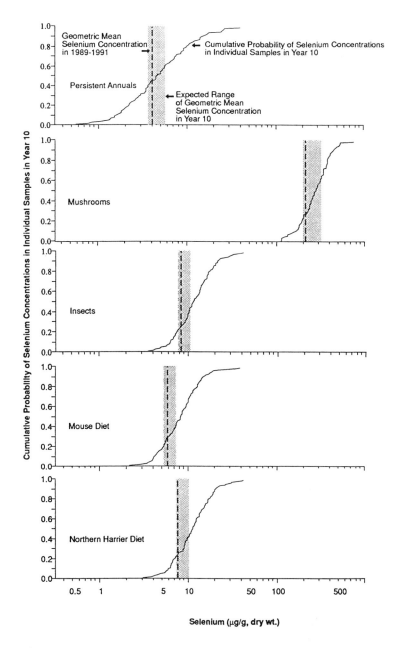

Figure 6 Range and probability of possible selenium concentrations in representative plants and in animal diets at Kesterson Reservoir in Year 10.

caused severe nervous disorders, nodular cirrhosis of the liver, and ascites [49]. Consequently, follow-up studies of screech owls, raccoons, coyotes, and kit foxes were conducted.

Screech owls fed 10 µg/g selenium in a laboratory study had reproductive success equivalent to controls [34; S. N. Wiemeyer, USFWS, personal communication]. No embryo mortality or deformities were observed, and hatching weight was similar to the controls. However, two of 12 screech owls (one from each of two pairs) fed 30 µg/g selenium died after 3 weeks on the study. All of the four remaining pairs laid eggs, but only one pair hatched a single young, which died within 1 day. Thus, the dietary selenium threshold for screech owls appears to be between 10 and 30 µg/g. Mean selenium concentrations in the common small mammals (mice, kangaroo rats) at Kesterson are expected to remain below 8 µg/g, even at the highest rates of bioavailability (Table 3). Although concentrations in voles and shrews may increase to 16 and 26 µg/g, respectively, such levels are not markedly higher than those found in these species since 1988 (i.e., after filling). They also are comparable to or lower than those found in voles and shrews in 1984 [20]. Based on dietary composition, home ranges, and selenium concentrations in prey species, no significant risks are expected for the two predatory bird species (red-tailed hawk and northern harrier) used in the model (Figure 6; Table 2).

Selenium concentrations in the liver, hair, feces, and blood of Kesterson raccoons during 1986 averaged 10–30 times higher than in raccoons from the nearby Volta Wildlife Area [50]. Selenium concentrations in livers of Kesterson raccoons were less than those in five of nine other mammal species sampled in 1984, and there was no evidence that selenium had negative effects on raccoons inhabiting Kesterson.

The results of a 1988 study on the endangered kit fox [43] showed that it is unlikely that elevated levels of selenium in the small mammal prey at Kesterson have had a negative effect on kit foxes, owing to their minimum use of Kesterson. The results of modeling indicate that kit foxes should not be exposed to significant risk in the future, even at the highest expected rates of selenium bioavailability (Table 2). The biological monitoring program continues to show little or no use of Kesterson by kit foxes [7–9]. The reservoir forms only a small portion of a typical kit fox home range (about 25%), and utilization of the site is low enough to minimize risks of exposure to this species. No other threatened or endangered animal species are known to occur on the site, and no significant risks to threatened or endangered species are expected to occur in the foreseeable future.

In addition to kit foxes, Paveglio and Clifton [43] studied coyotes collected from Kesterson. Two of 11 coyotes contained selenium levels within the range associated with chronic selenium toxicosis in domestic dogs. However, these elevated levels of selenium were probably due to the coyotes' consumption of American coots at the reservoir. (Those birds were frequently found dead or moribund during field studies there in 1984–1988, but they were part of the aquatic food chain eliminated from Kesterson by the filling operation.) Although coyotes continue to forage at Kesterson and have had a den there each year [7–9], the reservoir represents only a small portion of the typical range for a coyote. The single coyote sampled in 1991 was a pup salvaged from the Kesterson den. No significant risk of impacts to coyotes is expected (Table 2).

Selenium concentrations in insectivorous birds (e.g., meadowlarks and barn swallows) are not expected to increase to biologically significant levels. The model assumes that meadowlarks remain on the reservoir throughout the nesting season, but they apparently forage extensively off-site, and the population has declined as habitat conditions (vegetation) have become less favorable for them. No effects have been observed in meadowlarks or swallows that nest at Kesterson, and selenium concentrations in these species have declined since 1988 [7–9] (Table 2).

The greatest risk of selenium exposure to terrestrial wildlife appears to be the possible consumption of mushrooms, which contain the highest levels of selenium currently found at the site and the highest levels expected in the future (Tables 2 and 3). Wildlife receptors could be exposed by eating the mushrooms directly (as observed in ground squirrels and some small mammals) or by eating other consumers of mushrooms, such as insects. However, the consumption of mushrooms by animals appears to be very limited. The species most likely to be affected would be those that have small home ranges and feed on very localized food resources, such as some of the small mammal species. More mobile animals, such as birds, are less likely to feed consistently in small areas or on a narrow range of foods that could be affected by the seleniferous mushrooms.

Any effects of wildlife exposure to selenium at Kesterson would be limited to the exposed individuals or their offspring, and those should normally be minimal, as discussed above. No effects are expected to be transferred among individual animals through contact or hereditary means, and selenium contamination is not expected to move at significant levels beyond the immediate vicinity of Kesterson Reservoir through biotic or physical transport.

There is no evidence of significant community or ecosystem effects, either now or expected in the future. Changes in animal population levels (e.g., voles and shrews) have been attributed to changing habitat conditions rather than selenium exposure [9]. There also is no evidence of cumulative effects of selenium on exposed animals.

B. Aquatic Habitats

Potential exposure of aquatic birds (such as waterfowl and shorebirds) to seleniferous rainwater pools may become significant if rainfall is adequate to form extensive, persistent pools. Since the low-lying areas were filled with soil, no pools have persisted long enough to develop significant aquatic flora and fauna during most years. However, this could happen if rainfall is greater than normal or if much of the rainfall occurs during a short period in late winter or early spring. Earlier modeling studies [10] concluded that average selenium concentrations in the rainwater pools could be in the range of 10–30 µg/L, based on limited data. If this occurred and the pools persisted for several weeks, they probably would become attractive to waterfowl and shorebirds. This would be of greatest concern during late winter and spring when birds are preparing to nest. During the winter of 1991–1992, the geometric mean concentration of waterborne selenium in mid-February (February 15–21, 1992) was 2.73 µg/L (an arithmetic average of 12.9 µg/L) in 94 samples from pools located throughout the reservoir. By March 26, only a small fraction of those original pools remained and the geometric mean concentration from 20 pools was 11.3 µg/L. In these remaining pools, there was no consistent pattern of increasing selenium concentrations. Some individual pools (e.g., the Pond 11 U.C. Riverside experimental area), however, did show increasing selenium concentrations during the two-month monitoring period from mid-February to mid-April 1992.

Waterborne selenium concentrations of 10 µg/L have been associated with impaired hatchability in shorebirds, and concentrations of 10–20 µg/L have been associated with teratogenic effects [19]. Selenium uptake and loss occur rapidly in mallards [51]. When fed a diet containing 10 µg/g selenium, concentrations in the mallards' livers reached an equilibrium in about 8 days, although the rate for muscle equilibrium was much slower (81 days). Half-times for loss were 18.7 days for liver and 30.1 days for breast muscle. These observations indicate that birds could accumulate enough selenium within a week or two of feeding on high-selenium invertebrates in pools containing 10–30 µg Se/L that their reproductive success could be impaired.

An experimental study was conducted to determine if migratory birds overwintering in a selenium-contaminated area could accumulate, retain, and carry enough selenium burden to the breeding area to impair reproduction [52]. To address this question, adult mallards were fed 15 µg/g selenium, as selenomethionine, for 21 weeks during winter. Had such a diet been continued throughout the breeding season, reproduction would have been completely eliminated judging by results from earlier studies [53]. However, at the onset of the breeding season, selenium was removed from the diet to simulate migration to a clean breeding area. Females that began egg laying just 2–3 weeks after the switch to the control diet had normal reproduction compared to birds that had always been fed a control diet. Since birds would normally take longer than 2–3 weeks to migrate to their breeding grounds and begin egg laying, it is unlikely there would be a harmful carryover effect of selenium contamination from the wintering area. The reason for the quick return to normal reproduction was a rapid loss of selenium from the bodies of the females and, consequently, little selenium incorporated into their eggs. The dietary selenium concentration in this study was similar to the geometric mean (12.4 µg/g) found in the aquatic invertebrates collected from surface water pools at Kesterson on March 31, 1992.

Although many trace elements have been found at concentrations above background levels in the water or biota at evaporation basins in the San Joaquin Valley, the observed adverse effects on bird reproduction have been limited to those apparently caused by selenium [3,19,54]. There is a reasonably clear relationship between selenium concentrations in water, invertebrates, and waterbird eggs and also between those selenium concentrations and reproductive effects in exposed bird populations. Selenium concentrations in eggs of birds nesting at uncontaminated reference sites throughout the United States average less than 3 µg/g dry weight. These background concentrations of selenium in eggs were associated with normal rates of egg hatchability and teratogenesis (deformities) in the populations. The threshold for mean egg selenium associated with impaired hatchability was 8 µg/g. In the Tulare Basin, average egg selenium concentrations of 8 µg/g or more in avocets and stilts were associated with impaired hatchability in these populations. The threshold for mean egg selenium associated with increased teratogenic effects in waterbird populations was found to be in the range of 13–24 µg/g; the midpoint of this range (18.5 µg/g) was used in further estimates. More detailed estimates for dietary and waterborne selenium concentrations associated with the three risk characterization thresholds are presented elsewhere [3,19,54].

The relationships found in these studies between selenium concentrations in birds' eggs, their diet, and waterborne selenium in evaporation basins are illustrated in Figure 7. The regression lines in this figure reflect

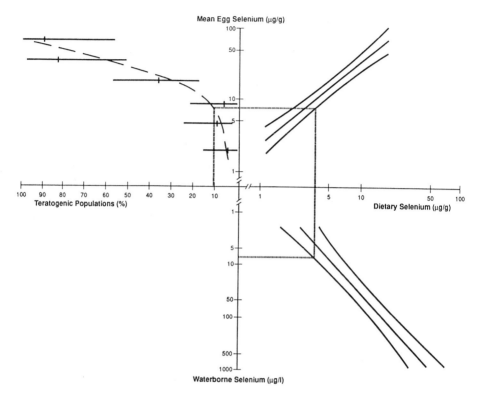

Figure 7 Hatchability effect risk characterization for aquatic birds. Mean egg selenium of 8 μg/g dry weight is associated with impaired hatchability but low probability of teratogenicity in avocet and stilt populations, an average of less than 3.5 μg/g selenium in the diet, and less than 7.8 μg waterborne selenium per liter. (Modified from Ref. 19.)

the process of bioaccumulation of selenium in the food chain. The figure illustrates (in a counterclockwise direction, starting lower right) the relationships among (a) selenium concentrations found in aquatic invertebrates (i.e., "dietary selenium" for aquatic birds) and water from evaporation basins [J. Shelton et al., DWR, unpublished data], (b) average selenium concentrations found in mallard eggs when the ducks were fed various concentrations of dietary selenium [53], and (c) the frequency of detecting embryo abnormalities (teratogenesis) among bird populations for which mean egg selenium concentrations were determined [19]. The dashed line (or "box") in the center of the figure connects mean egg selenium concentration (at the threshold effect level of 8 µg/g) to the maximum dietary and waterborne selenium concentrations expected to produce that mean egg selenium concentration. (The dashed line intercepts the lower 95% confidence intervals for those two regression equations.)

The upper left quadrant of Figure 7 shows the relationship between mean egg selenium and teratogenicity in aquatic bird populations. In that portion of the figure, the vertical line is the calculated percentage of the populations, within certain ranges of mean egg selenium, in which embryo deformities were detected [19]. The horizontal line shows the 95% confidence interval for this value. The number of individuals in those populations showing evidence of teratogenicity is usually not known.

The species of birds (waterfowl and shorebirds) most likely to be affected by seleniferous rainwater pools are protected by the Migratory Bird Treaty Act. The numbers of birds likely to use the pools can be estimated only in a very general way, because the size and duration of the pools (and the aquatic food chains that develop in them) can be predicted only in general terms. The attractiveness of the pools to aquatic birds would increase in proportion to their size and persistence. Avian food would probably not be abundant in the pools unless they persisted for several weeks. If they did, there probably would be large areas of wetland habitat on nearby portions of Kesterson National Wildlife Refuge, duck clubs, or other areas in the Grasslands. Regional flooding would diminish the attractiveness of Kesterson Reservoir pools, and the numbers of birds they would attract would be small. Thus, the exposed birds would represent a small fraction of the regional waterfowl and shorebird populations. Such a scenario was borne out in 1992 when large pools (i.e., the U.C. Riverside experimental plot in Pond 11) persisted and produced avian food, but more ducks and shorebirds used adjacent areas in Kesterson National Wildlife Refuge. The surface water pools in the reservoir had essentially disappeared by mid-April.

In summary, waterborne selenium concentrations that may occur in surface water pools at Kesterson could result in significant bioaccumulation of selenium by aquatic birds. If the surface water pools persist throughout the nesting season, the selenium concentrations could be high enough to reduce egg hatchability or cause teratogenesis in bird embryos. The likelihood of those effects is difficult to predict because of the uncertainty in estimating waterborne selenium in ephemeral pools at Kesterson and in assessing biological effects on the basis of waterborne selenium concentrations. During February 1992, geometric mean waterborne selenium concentrations in surface pools were 2.73 µg/L, including samples from several experimental areas that tended to have among the highest concentrations. By March 26, 1992, the geometric mean selenium concentration in the pools being monitored was 11.3 µg/L. Invertebrates collected on March 31, 1992, from six pools had a geometric mean selenium concentration of 12.4 µg/g when the mean waterborne selenium concentration was 8.07 µg/L in the pools from which they were collected. In the Tulare Basin, waterborne selenium concentrations of 10 µg/L have been associated with average egg concentrations of 8 µg/g and with hatchability effects; waterborne concentrations of 10 to 20 µg/L have been associated with average egg selenium concentrations up to 25 µg/g and teratogenic effects. Thus, waterborne selenium concentrations in the ranges that have been predicted for rainfall ponding within Kesterson are of concern for aquatic birds that may nest there. This risk of migratory birds being affected suggests a need for a management plan that could be implemented in the event of significant rainwater ponding in the reservoir.

VII. CONCLUSIONS AND LIMITATIONS

The objectives of this risk assessment were met generally through review and interpretation of available information about selenium in the Kesterson ecosystem. The most likely levels of biologically available selenium were estimated in various biota at the site. Expected selenium levels were estimated for conditions of moderate, low, and high rates of bioavailability over a period of 20 years (1992–2011). At the highest bioavailability rate, expected mean selenium concentrations in most plants and animals were estimated to be about 20–30% greater than current levels, which have not been associated with biological effects in terrestrial animals living at Kesterson. No significant effects are expected to occur among those animals as a result of selenium exposure, although there are a number of limitations and uncertainties of these predictions, as described below.

The two greatest potential sources for significant ecological effects resulting from the site's selenium inventory appear to be those related to (a) mushrooms (because of selenium exposure to their consumers) in the terrestrial habitats and (b) persistent rainwater pools that may occur over large areas if annual rainfall is more than 50% greater than normal. If these rainwater pools are present for several weeks in late winter or early spring, they could support significant populations of aquatic plants and invertebrates that could be consumed by aquatic birds, which appear to be more sensitive to the adverse effects of selenium than terrestrial birds or mammals. Although surface water pools in the reservoir disappeared before the nesting season for most aquatic birds during spring 1992, management plans for controlling exposure of aquatic birds appear to be the greatest current need for site management. Our recommendations for future research and monitoring are stated in Section VIII.

There are a number of limitations and uncertainties associated with the model used in this assessment.

1. Although there is a significant positive relationship between water-extractable selenium in the soil and selenium concentrations found in plants, many factors cause variability in the relationship. The factors that are probably important ones at Kesterson are the plant species, rainfall, soil sulfate concentration, plant growth rate, root depth, distribution of selenium in the soil profile, and timing of the sample collection.
2. The data currently available were obtained largely during three years of below-average rainfall. There were no consistent trends in selenium concentrations found in most types of samples (water-extractable selenium in soils or selenium in plants and animals). The high degree of spatial, seasonal, and annual fluctuation in water-extractable selenium in soils is reflected in concentrations found in plants and animals, and this variability limits the conclusions that can be drawn from available data.
3. Because of the drought, no systematic sampling of the terrestrial habitat has occurred during wet years since Kesterson filling was completed in 1988. Changes in weather patterns could greatly affect the bioavailability of the selenium inventory at Kesterson.
4. The trophic relationships among various ecosystem components (e.g., specific plant species, or other foods such as detritus or mushrooms, upon which various insects feed and the range of diets for all vertebrate consumers) are not known in detail. Because of the high selenium concentrations found in mushrooms and the limited information about their ecology and wildlife consumers at Kesterson, the significance of mushrooms in the cycling of selenium is not known.
5. The rate of ecological succession that will occur at Kesterson during the next 20 years cannot be estimated precisely because such ecosystems have not been studied previously. Furthermore, the model assumes that secondary succession will not be disrupted by fire or other factors.
6. The available database is very limited for selenium bioaccumulation in perennial plants (e.g., saltbush, alkali goldenbush, and thistle) eaten by mammals at Kesterson. Because these plants are expected to become more important components of the ecosystem, additional sampling is needed to determine relationships between soil and plant selenium concentrations.

7. The probability and extent of ponding expected to occur during wet years and the waterborne selenium concentrations expected in the rainwater pools were estimated through modeling with a limited database. These estimates give an approximation of conditions that can be expected, but they are not precise. In particular, new influences, such as creating wetlands to the east and south of Kesterson Reservoir, must be evaluated. Moreover, we have yet to observe the extent and duration of ponding that occurs in a year that is much wetter than average (>350 mm rainfall). The average waterborne selenium concentrations found in rainwater pools during the winter of 1991–1992 fell within the expected range, but individual pools varied widely. During the period in which water was in the pools, selenium concentrations increased in some but decreased in others.

VIII. RECOMMENDATIONS

Unless a more immediate threat related to mushrooms is identified, no management actions should be required to mitigate or control mushrooms. The following management actions are recommended to further reduce the probability of adverse effects from selenium bioaccumulation in surface water pools.

1. Experimental plots no longer being used for ongoing research or monitoring programs should be ripped and filled with soil to reduce the amount of water they cause to pool.
2. Any surface water pools that are present during late winter and early spring should be monitored weekly for bird use. If birds are observed feeding in these pools, the frequency of monitoring should be increased to daily, and invertebrates and water should be collected for analysis. These analyses will assist in documenting bioavailability of selenium in the exposure pathway for aquatic birds and in evaluating the need for further management actions related to the pools. (For example, there may be a need for additional fill material in some limited areas if pools form there consistently and have high concentrations of waterborne selenium).
3. If pools persist beyond mid-April and are being used as feeding areas by species that may nest there, hazing of birds should be implemented selectively to reduce the amount of time they feed in pools having high concentrations of selenium in invertebrates. It should be noted that if pools are persisting this late in the season within the reservoir, there is a high probability that alternative habitats will be available nearby for the birds. Also, hazing should not be used until it is really needed (late April or May) to avoid the birds' becoming habituated to the

disturbance, thereby reducing its effectiveness. During the nonbreeding season, the aquatic birds found at Kesterson would be sufficiently mobile that reproductive or health effects would not be likely to occur.
4. During very wet years, one possibility for management of rainwater pools that should be considered is to allow surface water to flow off the site. Surface water could be sampled to verify that selenium concentrations do not exceed the ambient water quality criterion of 5 µg/L [36] before water is allowed to drain from the site. With low selenium concentrations in water and high rates of dilution as the water enters Mud Slough (which should occur during times when water is pooling in Kesterson), drainage from the site would probably cause less risk to aquatic birds than having extensive pools persist for long periods onsite. Findings by LBL indicate that waterborne selenium concentrations in some pools increase as water remains in them. The potential for this increase to occur in extensive pools could be reduced by allowing water to flow off the site while concentrations are low.

More information on several subjects would improve the estimation of ecological risks associated with the selenium still present in the ecosystem at Kesterson Reservoir. This information can be obtained through continued monitoring and research programs. Information that would be the most valuable toward improving the ecological risk assessment and recommendations for acquisition of such information include the following.

- Better definition of the factors influencing bioavailability of selenium in Kesterson soils. Although there is a general relationship between water-extractable selenium in the soil and selenium concentrations found in plants (and animals), the concentrations are not highly correlated and factors causing the variability are not well understood. Moreover, the depth distribution and speciation of selenium will continue to evolve and approach thermodynamic equilibrium with the surrounding soil environment. Soils and representative plants should be collected each year, and soil chemistry parameters should be measured to define better the factors influencing selenium bioavailability. If no unexpected changes in selenium concentrations in the sampled species occur within three more years of sampling, plant sampling could be reduced in either species or frequency (perhaps to every other year instead of every year). Likewise, if soil chemistry parameters do not provide meaningful data within that time period and the model predictions of selenium concentrations are verified, the soil chemistry analyses could be dropped from the monitoring program.

- More detailed information about selenium trends in terrestrial habitats. Monitoring has shown no consistent trend in selenium concentrations found in plants and animals during the three years since Kesterson Reservoir filling was completed in 1988, although those have all been dry years. Additional data for some of the representative plants, invertebrate animals, and small mammals would help to establish trends. The monitoring program could be modified to focus on critical food chain samples by collecting selected plants, invertebrates, and small mammals each year for three more years. These analyses would determine trends and refine the accuracy of predictions developed during this assessment.
- Information about selenium bioavailability in wet years. When wetter than normal conditions occur again, additional monitoring in the terrestrial habitats would help determine whether selenium bioaccumulation by plants and animals is significantly affected. If rainwater pools form, surface soil and water could be sampled to improve the ability to estimate waterborne selenium concentrations from total or water-extractable selenium in surface soils. This information would help in developing plans for management of rainwater pools that may be extensive if annual rainfall is greater than 350 mm (50% more than normal). Rainwater pools also could be monitored for aquatic plants and invertebrates if the pools persist more than 2 weeks, and biological samples could be collected when available.
- Updated predictions of extent of flooding and waterborne selenium concentrations in surface water pools during very wet years. In addition to the monitoring of pools when they occur, updated modeling predictions could further assess management alternatives for pools if flooding is extensive. Model predictions of the extent of flooding and concentrations of the waterborne selenium were made previously using a limited database [10,13]. By incorporating the recent data (since 1989) into the model, it should be possible to improve estimates of potential risks to aquatic birds during very wet years. This information would be useful in deciding the best approaches for management of surface water pools.
- More information about selenium levels in detritus, detritivores (e.g., sowbugs), and mushrooms and their relationship to animal consumers. Organic detritus has not been sampled under terrestrial conditions since the Kesterson Reservoir filling, and relationships of this selenium reservoir to the concentrations in detritivores are not known. More important, the relationships between mushrooms and the sources of selenium they contain are not known, nor is the fate of selenium that invertebrate and vertebrate consumers have accumulated. Mushrooms may become

an important source of selenium mobilization into the biota if they become more abundant and are eaten by more animals. Detritus, detritivores, and mushrooms could be included in future monitoring program collections to monitor these components of the terrestrial ecosystem. In addition, animal trapping could be conducted in the vicinity of mushrooms that show evidence of wildlife consumption.
- Rate of ecological succession that occurs at the site. There have been no studies of other areas comparable to Kesterson in terms of ecological disturbance, and further monitoring of plant succession and animal populations using the site would contribute to better predictions of wildlife values as well as ecological risk associated with the site. Some of the findings may be applicable to other lands that are currently farmed but may be abandoned because of drainage problems. Continued monitoring is recommended, included every-other-year surveys of vegetation and birds along the established transects currently being used. During those surveys, qualitative observations of mammals and other wildlife also would be recorded. It is recommended that small mammal trapping be conducted annually for the next 3 years (and thereafter perhaps every other year) for the biological monitoring program and for qualitative surveys of species present.
- More information about selenium levels in those plants and animals that become more important components of the Kesterson ecosystem. The greatest information voids now include data on the perennial plants that have been found in the stomach contents of small mammals and are expected to become more common.

IX. SUMMARY

Following the discovery of selenium contamination in Kesterson Reservoir, the U.S. Bureau of Reclamation halted discharge of agricultural drainage to Kesterson in 1986, dewatered the reservoir, and filled lower elevation portions to prevent groundwater from rising to the soil surface in 1988. Monitoring has been conducted since 1987 to measure selenium concentrations among plants and animals at the site, document habitat and faunal changes, and determine whether adverse effects are occurring in the ecosystem. The primary data for this risk assessment, recently prepared for Reclamation and condensed in this chapter, are the findings of that monitoring program in conjunction with monitoring of soils and groundwater, especially the water-extractable selenium.

Soil selenium concentrations in the top 15 cm are strongly influenced by rainfall infiltration. With winter rains, selenate, in particular, is leached readily deeper down into the soil profile, which tends to decrease the inventory of soluble selenium available for dissolution in rainwater pools that form at Kesterson. About 5% of the total selenium inventory in Kesterson surface soils (top 15 cm) is water-extractable, indicating that only a limited amount of the selenium is currently mobile and available for plant uptake or dissolution into rainwater pools. No significant changes in total or water-extractable selenium concentrations were observed between 1989 and 1991. However, on a reservoir-wide basis, there was a significant increase ($P<0.05$) in the ratio of water-extractable selenium to total selenium in soils.

The drying and filling of Kesterson Reservoir converted the site to upland habitat consisting of three habitat types. Grassland habitat includes the higher elevation, upland areas that existed at the Reservoir before it was dried and filled; Filled habitat includes formerly low-lying areas filled with soil to prevent the occurrence of seasonal wetlands; and Open habitat includes former cattail areas that were not filled but were disked to prevent use by tricolored blackbirds. Grassland habitat covers about 30% of the reservoir; Filled habitat, about 60%, and Open habitat, 10%.

For modeling purposes, common plant species were grouped into four categories based on ecological considerations and selenium concentrations. These four plant groups are components of the various plant communities that will occur at Kesterson Reservoir over time at varying degrees of dominance. Wildlife exposure was characterized on a reservoir-wide basis because most of the wildlife species of interest can move readily from one habitat to another and also from one area to another. A model was constructed from available information to correlate selenium concentrations in various components of the ecosystem under current conditions and to predict concentrations that can be expected to occur in the next 20 years.

Selenium concentrations in the plants and animals inhabiting Kesterson Reservoir are not expected to change markedly during the next 20 years. If selenium bioavailability remains moderate, average concentrations are expected to be about 20–20% greater than current levels in most biota. With low bioavailability, selenium concentrations in plants and animals are expected to be 10–15% lower than the current levels after 20 years. Highest expected concentrations are about 20–30% above those in 1989–1991. Selenium concentrations are expected to be higher in mushrooms than in all other organisms.

The greatest risk of selenium exposure to terrestrial wildlife appears to be the possible consumption of mushrooms, which contain the highest levels of selenium currently found at the site. Wildlife receptors could be exposed by eating the mushrooms directly (as observed in ground squirrels and some small mammals) or by eating other consumers of mushrooms, such as insects; however, available data indicate that consumption of mushrooms by wildlife at Kesterson is very limited. The species most likely to be affected would be those that have small home ranges and feed on very localized food resources, such as some of the small mammal species. More mobile animals, such as birds, are less likely to feed consistently in small areas or on a narrow range of foods that could be affected by the seleniferous mushrooms.

Selenium concentrations in small mammals are expected to remain below those measured during a study in 1984 when the reservoir was partially flooded with drainwater. In that study, there was no clear effect of selenium on small mammal reproduction, although whole-body analyses by species for all ponds at Kesterson averaged about 11 µg/g. That level of selenium in prey species was sufficient to suggest concern for possible effects on predators, such as raptorial birds, raccoons, coyotes, and the endangered San Joaquin kit fox. Follow-up studies of screech owls, raccoons, coyotes, and kit foxes suggested that adverse effects of selenium should not be expected in those or similarly exposed predators at Kesterson.

Selenium concentrations in insectivorous birds (e.g., meadowlarks and barn swallows) are not expected to increase to biologically significant levels. No effects have been observed in meadowlarks or swallows that nest at Kesterson; their selenium concentrations have declined since 1988.

When pools of rainwater formed in portions of the reservoir during 1989–1991, they usually disappeared within 2–3 weeks. In most areas, the pools' persistence was much shorter after 1989, probably because of increased vegetative cover, the deep-soil ripping that was done by Reclamation to improve water percolation, and the additional fill that was placed in areas where rainwater pools previously formed. The pools observed during 1989–1991 were too ephemeral to develop significant populations of aquatic plants or invertebrates, and they received no significant wildlife use. In 1992, some pools did persist until mid-April, long enough for aquatic invertebrates to be present in sufficient numbers for aquatic birds to feed there. Potential exposure of aquatic birds (such as waterfowl and shorebirds) to seleniferous rainwater pools could become significant if rainfall is adequate to form extensive surface water pools that persist throughout the nesting season. However, the likelihood of reproductive effects in birds is difficult to predict because of uncertainty in estimating waterborne selenium concentrations in ephemeral pools at

Kesterson and in assessing their biological effects. Other studies have concluded that average selenium concentrations in the rainwater pools could be in the range of 10–30 µg/L, based on limited data. If this occurred and the pools persisted for several weeks, they probably would become attractive to waterfowl and shorebirds, which would be of greatest concern during late winter and spring when birds are preparing to nest. Waterborne selenium concentrations of 10 µg/L have been associated with impaired hatchability in shorebirds, and concentrations of 10–20 µg/L have been associated with teratogenic effects.

The species of birds most likely to be affected by seleniferous rainwater pools are protected by the Migratory Bird Treaty Act. The numbers of birds likely to use the pools can be estimated only in a very general way, because the size and duration of the pools (and the aquatic food chains that develop in them) can be predicted only in general terms. Attractiveness of the pools to aquatic birds would increase in proportion to their size and persistence. Avian food would probably not be abundant in the pools unless the pools persisted for several weeks. If they did, there probably would be large areas of wetland habitat on nearby portions of Kesterson National Wildlife Refuge, duck clubs, or other areas in the Grasslands. This regional flooding would diminish the attractiveness of Kesterson Reservoir pools, and the numbers of birds they would attract would be small. Thus, the exposed birds would represent a small fraction of the regional waterfowl and shorebird populations.

During March 1992, the surface water pools at Kesterson were used by aquatic birds, most of which were migrant shorebirds that would not nest in the region. Bird use decreased during April, and essentially all of the aquatic birds had disappeared by May. Bird use of wetland areas nearby on Kesterson National Wildlife Refuge appeared to be greater than within the reservoir, although comparative censuses were not conducted.

The potential for migratory birds to be affected suggests a need for a management plan that could be implemented in the event of significant rainwater ponding in the reservoir. Persistent, widespread rainwater pools that may form in very wet years (especially those years wetter than 1992 when rainfall is more than 50% above normal) could represent a significant threat to aquatic migratory birds protected by the Migratory Bird Treaty Act.

No active site management of the terrestrial habitat is recommended unless further information is obtained that indicates a significant threat of selenium toxicosis to wildlife. Although mushrooms accumulate high concentrations of selenium, no specific threat to wildlife has been identified. No significant changes in selenium exposure are expected in the near future, and no impacts have been identified in terrestrial wildlife at Kesterson. Therefore, no active site management appears necessary to limit wildlife exposure or to reduce the selenium inventory more rapidly than is occurring now.

No further actions appear warranted for management of rainwater pools during dry years, because these pools contain only moderate concentrations of selenium and disappear soon after the rainfall ends. However, if rainfall increases or is higher than normal during early spring, there will be a need for management of rainwater pools that cover more extensive areas or that persist long enough for significant aquatic plant or invertebrate populations to become established. These pools would be most significant during the late winter and spring because aquatic birds would be attracted to them during or shortly before their nesting season.

The following management actions for pools are recommended:

1. Experimental plots no longer being used for ongoing research or monitoring programs should be ripped and filled with soil to reduce the amount of water they cause to pool.
2. Any surface water pools present during late winter and early spring should be monitored weekly for bird use. If birds are observed feeding in these pools, the frequency of monitoring should be increased to daily, and invertebrates and water should be collected for analysis.
3. If pools persist beyond mid-April and are being used as feeding areas by species that may nest there, hazing of birds should be implemented selectively to reduce the amount of time they feed in pools having high concentrations of selenium in invertebrates.
4. Consideration should be given to allowing surface water to flow off the site during very wet years because it would probably cause lower risk to aquatic birds than having the water remain in surface pools onsite.

More information is needed on the following subjects to improve the estimation of ecological risks associated with the selenium still present at Kesterson Reservoir:

- Better definition of the factors influencing bioavailability of selenium in Kesterson soils during wet years as well as dry or normal years
- More detailed information about selenium trends in terrestrial habitats
- Updated predictions of extent of flooding and waterborne selenium concentrations in surface water pools during very wet years
- Current information about selenium levels in detritus and detritivores (e.g., sowbugs) and mushrooms from representative sites within Kesterson and their relationship to animal consumers
- Rate of ecological succession that occurs at Kesterson

- Continued focus of the selenium monitoring program on those plants and animals that become more important components of the Kesterson ecosystem

Recommendations are given for meeting these information needs. Briefly, biological monitoring should be continued to assess whether adverse effects are likely to occur in wildlife. The monitoring program could be modified to fulfill the information needs identified during this assessment. The modified program would increase the level of effort regarding certain needs, but the number of analyses for most plants and animals would be reduced. For example, the level of effort for monitoring plants should increase for perennials as more species occur there, whereas two representative species for each of the three other categories (grasses, transitional annuals, and persistent annuals) would be adequate. In addition to plants, the monitoring program should include soils, invertebrates, and small mammals to determine whether model predictions are accurate. The biological monitoring plan should remain flexible to include collection of new species of plants and animals that become important components of the Kesterson ecosystem.

ACKNOWLEDGMENTS

The U.S. Bureau of Reclamation provided funding under contract number 9-CS-20-00440 for the biological monitoring program and for the recent ecological risk assessment. We appreciate the advice and guidance of Noel Williams in overall aspects of the monitoring program and risk assessment, particularly in the modeling and simulation used in the assessment. Sally Benson provided unpublished data and insights concerning mobility and bioavailability of selenium in the Kesterson ecosystem. Marjorie Castleberry and Cheryl Johnson provided valuable technical assistance through all the field, laboratory, and analytical phases of the project.

REFERENCES

1. U.S. Bureau of Reclamation, Mid-Pacific Region, in cooperation with U.S. Fish and Wildlife Service and U.S. Army Corps of Engineers, Final Environmental Impact Statement, Kesterson Reservoir, 1986.
2. H. M. Ohlendorf, D. J. Hoffman, M. K. Saiki, and T. W. Aldrich, *Sci. Total Environ.* 52: 49 (1986).
3. H.M. Ohlendorf, Bioaccumulation and effects of selenium in wildlife, in *Selenium in Agriculture and the Environment* (L. W. Jacobs, Ed.), SSSA Spec. Publ. 23, American Society of Agronomy and Soil Science Society of America, Madison, Wisc., 1989, pp. 133–177.

4. U.S. Department of the Interior, Submission to California State Water Resources Control Board in Response to Order No. WQ-88-7, Effectiveness of Filling Ephemeral Pools at Kesterson Reservoir, Kesterson Program Upland Habitat Assessment, and Kesterson Reservoir Final Cleanup Plan, 1989.
5. U.S. Bureau of Reclamation, Kesterson Program, Biological Monitoring Winter and Spring 1987, USBR Mid-Pacific Region, Sacramento, Calif., 1987.
6. U.S. Bureau of Reclamation, Kesterson Program, Biological Monitoring Fall 1987, Winter, Spring, and Summer 1988, and Preliminary Upland Habitat Assessment, Mid-Pacific Region, Sacramento, Calif., 1988.
7. U.S. Bureau of Reclamation, Kesterson Program, Kesterson Reservoir Biological Report, Mid-Pacific Region, Sacramento, Calif., 1989.
8. U.S. Bureau of Reclamation, Kesterson Program, Kesterson Reservoir Biological Report and Monitoring Plan, Mid-Pacific Region, Sacramento, Calif., 1990.
9. U.S. Bureau of Reclamation, Kesterson Reservoir Biological Monitoring Report and 1992 Biological Monitoring Plan, Mid-Pacific Region, Sacramento, Calif., 1991.
10. S. M. Benson, T. K. Tokunaga, P. Zawislanski, A. W. Yee, J. S. Daggett, J. M. Oldfather, L. Tsao, and P. W. Johannis, Hydrological and Geochemical Investigations of Selenium Behavior at Kesterson Reservoir, Univ. California, Lawrence Berkeley Laboratory, Earth Sciences Division, LBL-29689, 1990.
11. S. M. Benson, T. K. Tokunaga, and P. Zawislanski, Anticipated Soil Selenium Concentrations at Kesterson Reservoir, Univ. California, Lawrence Berkeley Laboratory, Earth Sciences Division, LBL-33080, 1992.
12. S. M. Benson, T. Tokunaga, P. Zawislanski, C. Wahl, P. Johannis, M. Zavarin, A. Yee, L. Tsao, D. Phillip, and S. Ita, 1991–1992 Investigation of the Geochemical and Hydrological Behavior of Selenium at Kesterson Reservoir, Univ. California, Lawrence Berkeley Laboratory, Earth Sciences Division, LBL-33532, 1993.
13. Lawrence Berkeley Laboratory, Hydrological, Geochemical, and Ecological Characterization of Kesterson Reservoir: Annual Report, October 1, 1988–September 30, 1989, Univ. California, Berkeley, Earth Sciences Division, LBL-27993, 1990.
14. U.S. Bureau of Reclamation, Ecological Risk Assessment for Kesterson Reservoir, Mid-Pacific Region, Sacramento, Calif., 1992.
15. U.S. Environmental Protection Agency, Risk Assessment Guidance for Superfund, Vol. II, *Environmental Evaluation Manual*, Interim Final, EPA/540/1-89/001, 1989.
16. H. F. Mayland, L. F. James, K. E. Panter, and J. L. Sonderegger, Selenium in seleniferous environments, in *Selenium in Agriculture and the Environment* (L. W. Jacobs, Ed.), SSSA Spec. Publ. 23, American Society of Agronomy and Soil Science Society of America, Madison, Wisc., 1989, pp. 15–50.
17. A. Kabata-Pendias and H. Pendias, *Trace Elements in Soils and Plants*, 2nd ed. CRC Press, Boca Raton, Fl., 1992.
18. R. Eisler, Selenium Hazards to Fish, Wildlife, and Invertebrates: A Synoptic Review, U.S. Fish and Wildlife Service Biol. Rep. 85 (1.5), 1985.

19. J. P. Skorupa and H. M. Ohlendorf, Contaminants in drainage water and avian risk thresholds, in *The Economics and Management of Water and Drainage in Agriculture* (A. Dinar and D. Zilberman, Eds.), Kluwer, Boston, Mass., 1991, pp. 345–368.
20. D. R. Clark, Jr., *Sci. Total Environ.* 66: 147 (1987).
21. R. L. Hothem and H. M. Ohlendorf, *Arch. Environ. Contam. Toxicol.* 18: 773 (1989).
22. C. A. Schuler, R. G. Anthony, and H. M. Ohlendorf, *Arch. Environ. Contam. Toxicol.* 19: 845 (1990).
23. Z. Z. Huang and L. Wu, *Ecotoxicol. Environ. Safety* 22: 251 (1991).
24. L. Wu and Z. Z. Huang, *Ecotoxicol. Environ. Safety* 22: 267 (1991).
25. H. M. Ohlendorf, R. L. Hothem, and T. W. Aldrich, *Copeia* 1988: 704 (1988).
26. California Department of Fish and Game, *Wildlife Habitat Relationships*, Sacramento, Calif., 1989.
27. U.S. Bureau of Reclamation, San Luis Unit, Central Valley Project California, Inf. Bull. 1, 2, 3, and 4, Mid-Pacific Region, Sacramento, Calif., 1984.
28. U.S. Bureau of Reclamation, San Luis Drain: Status of Study Plans for Completion of Report of Discharge, prepared for the California State Water Resources Control Board, Mid-Pacific Region, Sacramento, Calif., 1982.
29. H. M. Ohlendorf, R. L. Hothem, T. W. Aldrich, and A. J. Krynitsky, *Sci. Total Environ.* 66: 169 (1987).
30. U.S. Department of the Interior and State of California, San Joaquin Basin Action Plan/Kesterson Mitigation Plan, Merced County, Calif., U.S. Dept. of the Interior, Bureau of Reclamation, and Fish and Wildlife Service; State of California Resources Agency and Department of Fish and Game, 1989.
31. U.S. Bureau of Reclamation, San Luis Unit Drainage Program, Central Valley Project, California, Draft Environmental Impact Statement, Mid-Pacific Region, Sacramento, Calif., 1991.
32. K. E. Mayer and W. E. Laudenslayer, Jr. (Eds.), *A Guide to Wildlife Habitats of California*, Dept. of Forestry and Fire Protection, Sacramento, Calif., 1988.
33. W. E. Grant and E. C. Birney, *J. Mammal.* 60: 23 (1979).
34. U.S. Fish and Wildlife Service, Summary Report: Effects of Irrigation Drainwater Contaminants on Wildlife, Patuxent Wildlife Research Center, Laurel, Md., 1990.
35. U.S. Bureau of Reclamation, Revised Monitoring and Reporting Program No. 87-149 for Kesterson Reservoir, Merced County, Annual Report, Oct. 15, 1991, Mid-Pacific Region, Sacramento, Calif., 1991.
36. U.S. Environmental Protection Agency, Ambient Water Quality Criteria for Selenium—1987, Office of Water Regulations and Standards, Criteria and Standards Division, Washington, D.C., EPA 440/5-87-006, 1987.
37. J. A. Peterson and A. V. Nebeker, *Arch. Environ. Contam. Toxicol.* 23: 154 (1992).
38. M. K. Saiki, Concentrations of selenium in aquatic food-chain organisms and fish exposed to agricultural tile drainage water, Proc. Second Selenium Symp., Bay Institute of San Francisco, Tiburon, Calif., 1986, pp. 25–33.
39. G. J. Smith and V. P. Anders, *Environ. Toxicol. Chem.* 8: 943 (1989).

40. C. A. Schuler, Impacts of agricultural drainwater and contaminants on wetlands at Kesterson Reservoir, California, Master's Thesis, Oregon State Univ., Corvallis, 1987.
41. J. Verner and A. S. Boss, California Wildlife and Their Habitats: Western Sierra Nevada, Gen. Tech. Rep. PSW-37, U.S. Forest Service, Berkeley, Calif., 1980.
42. A. Y. Cooperrider, R. J. Boyd, and H. R. Stuart, Inventory and Monitoring of Wildlife Habitat, U.S. Bureau of Land Management, Washington, D.C., 1986.
43. F. L. Paveglio and S. D. Clifton, Selenium Accumulation and Ecology of the San Joaquin Kit Fox in the Kesterson National Wildlife Refuge Area, U.S. Fish and Wildlife Service, Los Banos, Calif., 1988.
44. D. C. Zeiner, W. F. Laudenslayer, Jr., K. E. Mayer, and M. White, *California's Wildlife*, Vol. 2, *Birds*, Department of Fish and Game, Sacramento, Calif., 1990.
45. C. R. Preston, *Condor 92*: 107 (1990).
46. U.S. Bureau of Reclamation, Kesterson Reservoir Monthly Summaries of Wildlife Monitoring, Mid-Pacific Region, Sacramento, Calif., 1992.
47. W. H. Press, B. P. Flannery, S. A. Teukolsky, and W. T. Vetterling, *Numerical Recipes—The Art of Scientific Computing*, Cambridge Univ. Press, New York, 1986.
48. A. W. Halverson, I. S. Palmer, and P. L. Guss, *Toxicol. Appl. Pharmacol. 9*: 477 (1966).
49. M. Rhian and A. L. Moxon, *J. Pharmacol. Exp. Ther. 78*: 249 (1943).
50. D. R. Clark, Jr., P. A. Ogasawara, G. J. Smith, and H. M. Ohlendorf, *Arch. Environ. Contam. Toxicol. 18*: 787 (1989).
51. G. H. Heinz, G. W. Pendleton, A. J. Krynitsky, and L. G. Gold, *Arch. Environ. Contam. Toxicol. 19*: 374 (1990).
52. G. H. Heinz and M. A. Fitzgerald, *Environ. Pollut. 81*: 117 (1993).
53. G. H. Heinz, D. J. Hoffman, and L. G. Gold, *J. Wildl. Manage. 53*: 418 (1989).
54. H. M. Ohlendorf, J. P. Skorupa, M. K. Saiki, and D. A. Barnum, Food-chain transfer of trace elements to wildlife, Proc. Am. Soc. Civil Engineers Nat. Conf. Irrigation and Drainage Engineering, *Management of Irrigation and Drainage Systems: Integrated Perspecticves* (R. G. Allen and C. M. U. Neale, Eds.), Park City, Utah, July 21–23, 1993, pp. 596–603.

5
Field Investigations of Selenium Speciation, Transformation, and Transport in Soils from Kesterson Reservoir and Lahontan Valley

Tetsu K. Tokunaga, Peter T. Zawislanski,
Paul W. Johannis, and Sally Benson

*Lawrence Berkeley Laboratory
Berkeley, California*

Douglas S. Lipton

*Levine-Fricke, Inc.
Emeryville, California*

I. INTRODUCTION

The volume of literature concerning selenium (Se) in the soil environment has grown considerably over the past 50 years. Reviews [1–5] provide comprehensive surveys of this literature. In recent years, results from studies concerned with Se contamination in sediments, soils, and groundwater at Kesterson Reservoir (Merced County, California) have also become available [6–12]. In this chapter, data are presented that (a) suggest uncertainties with respect to characterization of adsorbed Se using sequential extraction methods, (b) illustrate the distribution of various Se fractions in field soil profiles, (c) demonstrate reoxidation and movement of recently reduced field soil Se inventories, and (d) show the relative mobilities of selenate (Se(VI)) and selenite (Se(IV)) under oxidizing field conditions. Data presented in the following sections come primarily from a larger set of studies conducted at Kesterson Reservoir and also in the Lahontan Valley (Churchill County, Nevada). These results are not intended to represent "typical" soils from the western United States but rather were selected to illustrate various characteristics of Se behavior in the soil environment.

II. PHOSPHATE-EXTRACTED SELENIUM

Numerous procedures have been developed for determinations of Se oxidation states and phases in soils [13–18]. These studies have differentiated between various water-soluble, surface (adsorbed), and solid-phase species of Se using a number of sequential extraction methods. However, other studies [19–21] point out limitations associated with presumed specificity of some sequential extraction methods. Detailed explanations of why adsorbed and precipitated species cannot be distinguished in the macroscopic, batch-type procedures employed in sequential extractions are presented in Ref. 22. Acknowledging the operationally defined nature of certain aspects of soil Se fractionation and the fact that many uncertainties still exist in characterizing the various oxidation states and phases in which Se occurs in soils, a limited set of Se fractionation data is presented in this section. The data raise questions concerning the characterization of adsorbed Se(IV) by equilibration with an excess concentration of a competing oxyanion (phosphate). In later sections, evidence of sustained lack of redox equilibria with respect to Se speciation and of reoxidation of more refractory portions of the Se inventories will be discussed.

Investigations of selenite–phosphate exchange and equilibration of soils with various phosphate solutions to determine specifically adsorbed anion compositions have been conducted by numerous researchers [18,23–25]. The majority of studies have relied heavily on interpretations of supernatant solution chemistry in batch or sequential extractions. Direct measurements of binding mechanisms associated with Se adsorption have become possible through the use of various spectroscopic methods. For example, a bidentate, inner-sphere complexation adsorption model for Se(IV) on goethite surfaces is supported by X-ray absorption spectroscopic studies [26,27]. Much weaker, outer-sphere surface complexes with Se(VI) were demonstrated in the same studies.

Data on Se in Kesterson Reservoir and Lahontan Valley soils have come solely through macroscopic measurements that do not provide the detailed information accessible through spectroscopy. Nevertheless, useful information may be obtained with cautious interpretation. Results from one study on a particular phosphate extraction method are presented in this section. In Section II, depth profiles of the phosphate-extractable Se fractions in several soils are presented.

In order to develop a better understanding of the nature of phosphate extracts, several studies of the general method were conducted. Examples of the operationally defined nature of phosphate extracts are provided in Figure 1. In this figure, the phosphate-extractable Se(IV) and Se(VI) are plotted as functions of the Na_2HPO_4 concentration used in the extraction (1:10 soil to solution mass ratio, 24 h shaker equilibration). The choice of Na_2HPO_4 for the extraction solution rather than the more commonly used K_2HPO_4 was based on the fact that the soils of interest are sodic. The extractable Se(IV) concentrations are shown to increase with increased phosphate concentrations, without a unique plateau. Substantial releases of silica have been observed when phosphate concentrations exceed 10 mM. Such results suggest that Se(IV) is released by desorption as well as dissolution of solid phases containing Se(IV) (G. Sposito, Univ. California, Berkeley, personal communication, 1988). These two phenomena cannot be distinguished in data obtained with macroscopic methods [22]. Molecular scale methods such as those of the X-ray absorption spectroscopies (XAS) can provide data relevant for discerning bonding configurations in surface and near-surface environments [e.g., 28,29]. These methods have been used to provide perhaps the only definitive evidence of Se(IV) and Se(VI) adsorption [26,27]. Although Se concentrations in even highly contaminated soils are still several orders of magnitude below

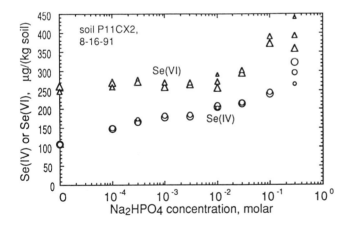

Figure 1 The dependence of phosphate-extractable Se(IV) and Se(VI) on the concentration of Na_2HPO_4 used in the extraction. Note the lack of a distinct plateau in extractable Se(IV) and the additional releases of Se(VI) at high phosphate concentrations. The duplicate to triplicate samples are shown individually.

current detection limits for most synchrotron-based XAS determinations of Se species (ca. 0.1 g/kg^1), this gap is closing with the development of more powerful synchrotron facilities.

Selenium(VI) occurs exclusively as SeO$_4^{2-}$ in the neutral to slightly alkaline extracts shown in Figure 1. The existence of a stable region spanning phosphate concentrations of up to 10 mM over which the extractable Se(VI) concentrations remain at 270±20 µg/kg soil suggests that essentially no adsorption of this species occurs in these soils. A study of Se(VI) in several San Joaquin Valley, California, soils [30] also demonstrated lack of selenate adsorption over a wide range of pH (5.5–9).

A second feature of interest in Figure 1 concerns the release of additional Se(VI) in extracts with Na$_2$HPO$_4$ concentrations in excess of 10 mM. Since the additional yield of Se(VI) occurs over the phosphate concentration range in which substantial solid-phase dissolution is indicated, these data are suggestive of the presence of occluded or coprecipitated Se(VI). However, the possibility of Se(IV) oxidation resulting in the observed Se(VI) increases cannot be ruled out.

In view of the data presented in this section, phosphate-extractable soil Se inventories should be viewed as procedure-specific quantities. Although direct measurement of Se adsorption is possible through XAS when Se concentrations are very high, current detection limits preclude applications of these methods for the in situ speciation of Se even in contaminated soils such as those at Kesterson Reservoir.

III. DISTRIBUTIONS OF VARIOUS SELENIUM FRACTIONS IN SOIL PROFILES

Water-extractable, phosphate-extractable, and total soil Se inventories are described in this section. Water-extractable and phosphate-extractable forms of Se are equated commonly with the water-soluble and adsorbed fractions, respectively. In addition, Se associated with soil carbonate and soil organic matter are described for two of the profiles. Details of the various extraction procedures employed are provided in Refs. 9, 11, and 18. All Se analyses were conducted using hydride generation atomic absorption spectrometry [15] or X-ray fluorescence spectrometry (for a limited number of total Se determinations) [31].

A. Site Descriptions and Results

One soil profile from the Lahontan Valley (located in the Fallon Indian Reservation) and two from Kesterson Reservoir are described in this section. The Lahontan Valley soil profile has developed in the vicinity of a low terrace associated with the pluvial Lake Lahontan. Soils of this area are mapped [32] under the East Fork clay loam series (Fluvaquentic Haploxeroll), although some profile characteristics are more similar to those of the Ragtown clay loam (slightly saline, Typic Torriorthent). The site is located in a field cropped for alfalfa for over 20 years. Detailed site description is provided in Ref. 33. Soil samples were collected in December 1989 when the water table was about 1.7 m below the soil surface. Field observations and laboratory measurements of very high soil salinity (0.021–0.13 S/m in 1:5 soil–water extracts) and high pH (7.8–9.3) indicated that this site is a localized, ineffectively leached portion of an otherwise agriculturally productive area. High clay contents (24–73%) observed in this profile appear to restrict leaching relative to soils in the rest of the field, thus permitting the saline-sodic conditions to persist.

Depth profiles of water-soluble, phosphate-extractable, and total Se in this Lahontan Valley soil are shown in Figure 2a. Water-soluble Se data were obtained from 1:5 (soil/water mass ratio) batch extractions, whereas phosphate extracts were performed using 1:20 (soil to 1 mM Na_2HPO_4 mass ratio) batch extractions. It should be noted that the phosphate-extracted Se profile shown here actually represents the equivalent of the sum of water-extractable and phosphate-extractable Se. Total soil Se concentrations for the subset of samples analyzed by X-ray fluorescence spectrometry are also shown in Figure 2a. The profile of ratios of Se(IV) to Se(VI) in the H_2O extracts of the Lahontan Valley is shown in Figure 2b.

Kesterson Reservoir, located along the southern portion of the Kesterson National Wildlife Refuge, served as a set of terminal evaporation ponds for saline, seleniferous agricultural drain waters from 1981 to 1986. During this period, about 9000 kg of Se and 3×10^8 kg of salts were discharged into 12 evaporation ponds covering a 520-ha area. Drainwater Se concentrations were commonly in the range of 250–350 µg/L. This source of Se was predominantly Se(VI) (ca. 95%), with minor proportions of Se(IV) (ca. 5%). Selenium poisoning of wildlife [34] prompted termination of drainage water discharges at Kesterson Reservoir in 1986.

Over 95% of the soils in Kesterson Reservoir are mapped as Turlock sandy loam (Albic Natraqualf) [35]. The remaining areas have been mapped under the Triangle series (Aquic Chromoxerert). Soils of the general area have developed in the western San Joaquin Valley basin rim, on alluvium of primarily Sierran granitic origin. The semiarid Mediterranean climate combined with the presence of a shallow water table have resulted in saline and saline-sodic soils. Despite coarse-scale uniformity in

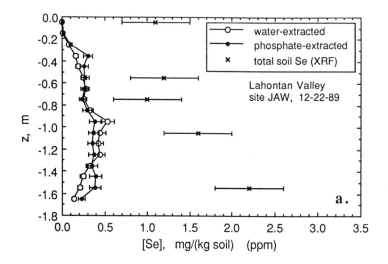

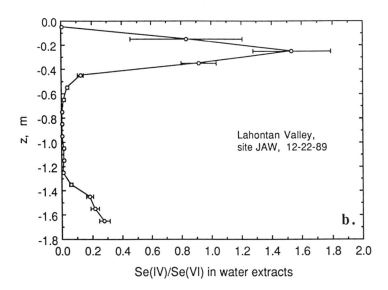

Figure 2 Selenium distributions in the Lahontan Valley soil profile. (a) Depth distributions of water-extracted, phosphate-extracted, and total soil Se. (b) Depth distributions of the selenite/selenate ratio in water extracts.

these soils, natural and site management–associated soil surface differences exist that give rise to important differences with respect to soil Se and salinity distributions in various regions within Kesterson Reservoir. Such differences have been described in Ref. 6. Two of the soil environments are described here. The site 9BE "playa" environment [6,12] has become revegetated with *Bassia hyssopifolia* since termination of regular surface water ponding. In such playa environments, flooding cycles prevented emergence of both terrestrial and rooted aquatic vegetation. The P11C salt grass site, located in an area vegetated continuously with *Distichlis spicata* [6,9], was subject to only intermittent flooding with seleniferous drain waters during the same period.

Sequential extraction depth profiles of the two sites from Kesterson Reservoir are shown in Figures 3a and 4a. The relative amounts of Se in various fractions (at each depth) are shown in Figures 3b and 4b for the two sites. Water-extracted and phosphate-extracted fractions were obtained in a manner similar to that for the Lahontan Valley soil, with the exceptions that 0.25 M KCl and 0.10 M K_2HPO_4 (pH 8) were used in the respective extraction solutions according to Ref. 18. The carbonate-associated and soil organic matter–associated Se fractions were extracted with 1 M sodium acetate (pH 5) and ca. 5% NaOCl (pH 9.5), respectively, as described in Ref. 18.

B. Discussion

Selenium in soil waters from the sites described here occurs primarily as Se(VI), with Se(IV) accounting for most of the remaining soluble inventory. Given the low pK_{a2} of H_2SeO_4 (pK_{a2} = 1.91, cited in Ref. 36), selenate occurs exclusively as SeO_4^{2-} in these neutral to alkaline soil solutions. Selenite, with a pK_{a2} of 7.3, occurs as both $HSeO_3^-$ and SeO_3^{2-} in these waters. Previous experience with a range of soil/water mass ratios in extracts of seleniferous vadose zone soils has indicated some sensitivity of water-extractable Se to the soil/water mass ratio. Unpublished results of water extracts of aerated Kesterson soils show increases of up to 10% in extracted Se when the soil/water ratio was increased from ca. 0.5 up to 20. Increases in extracted Se were due primarily to increases in extracted Se(IV). The observed dependence of water-extractable Se on the soil/water ratio results from Se(IV) desorption and/or dissolution. Extracts of soils with lower pH are expected to exhibit greater sensitivity to the soil/water mass ratio since adsorption-desorption effects become more important. The presence of salt crusts can make water-extractable Se concentrations more procedure-specific because of the limited solubility of

Se salts and other salts in which Se may be occluded or coprecipitated [37]. Determinations of water-extractable Se in reducing soils are complicated by the strong redox dependence of Se adsorption and solubility. In the following paragraphs, the terms "water-extractable" and "water-soluble" will be used interchangeably, while recognizing the aforementioned reasons for the possible lack of a unique water-soluble Se content.

In the Lahontan Valley soil, a very shallow surface zone of apparent Se depletion with respect to water- and phosphate-extractable fractions is evident (Figure 2a). Such a pattern probably resulted from irrigation and the flushing of the more soluble Se fractions to a greater depth. Lower soil salinity in this zone supports this interpretation. The depth profile of the Se(IV):Se(VI) ratios in the Lahontan Valley soil water extracts (Figure 2b) is in qualitative agreement with expectations drawn from the redox dependence of these species [e.g., 4,36], although another explanation will also be suggested. Maxima in the Se(IV):Se(VI) ratio would be expected in regions of lower redox potentials. The maximum at the bottom of the profile is associated with the water table where the lowest oxygen availability would be expected. The maximum in the upper profile is located at the top of the first clay horizon in the profile. Periodic perching of irrigation waters in this zone probably promotes a periodic reducing environment. However, lack of correlation between redox potentials and Se(IV):Se(VI) ratios has been documented previously [38]. In an earlier work, poor correlations between redox potential measurements and other redox couples (e.g., $Fe^{2+}:Fe^{3+}$, $HS^-:SO_4^{2-}$, $NH_4^+:NO_3^-$) were noted in a carefully selected set of groundwater data [39].

Another likely influence on the Se(IV):Se(VI) profile shown in Figure 2b is that of Se(IV) desorption (and maybe dissolution) during the extraction process. As noted earlier, this effect can be significant. In the Lahontan Valley soil, the peaks in the Se(IV):Se(VI) ratio roughly correlated with maxima in the clay content profiles, suggesting the importance of adsorption/desorption.

Profiles of water-soluble Se in the two Kesterson Reservoir soils are shown in Figures 3a and 4a. The shapes of these profiles, with highest concentrations at or near the soil surface, are explainable in terms of the soil water flow and redox conditions during and following the years of Se disposal [6]. During periods of drain water disposal, seleniferous waters were ponded over Kesterson Reservoir soils. Concentrations of selenium in the ponded waters were commonly in the vicinity of 300 µg/L and occurred primarily (≥95%) as Se(VI). Under these conditions, Se was reduced and removed from the percolating waters in the anaerobic zone via precipitation and/or adsorption at, and immediately below, the pond water–soil interface. Consequently, over 90% of the Se inputs were restricted initially to the upper 0.15 m of soil. Incomplete immobilization of Se accounted for its distribution at depth prior to 1986, while reoxidation and leaching from the Se-rich surface have become dominant since termination of drainwater disposal [9,12,40,41]. These more recent processes are described in Section IV.

It is important to note that the water-soluble fraction of the soil Se inventories presented in this section commonly account for only a small fraction of total soil Se. In the Lahontan Valley soil profile, only 10–30% of the soil Se occurs in a water-soluble form, whereas in the Kesterson Reservoir soils this fraction amounts of 2–48% of total soil Se. Since aerobic conditions strongly favoring Se(VI) prevail in each of these soils, these data clearly illustrate the common lack of redox equilibria with respect to Se speciation in both the Lahontan Valley soil and Kesterson Reservoir soils.

Phosphate extraction of the Lahontan Valley soil did not release additional Se relative to the water-soluble inventories, which indicate that insignificant amounts of Se in this profile are desorbable or dissolvable with phosphate. The lack of a measurable adsorbed Se(IV) fraction in this soil is consistent with numerous previous studies of Se(IV) adsorption envelopes [e.g., 25,42–44]. In these studies, Se(IV) adsorption is observed to decrease with increasing pH. Such a pH dependence is explained by a required protonation of the adsorption site prior to ligand exchange [22].

Profiles from two Kesterson soils (Figures 3a and 4a) show significant quantities of phosphate-extractable Se(IV), typically accounting for 10–20% of the total Se inventory. The revegetated playa site, although seasonally flooded (winters of 1986–1987 and 1987–1988) or nearly flooded (with the water table extending up to 0.28 m below the soil surface in 1988–1989), was probably aerated during most of the three years prior to sampling. The salt grass site soils experienced largely aerobic conditions for the three years prior to sampling, although the seasonal rise of the shallow water table extended up to about 0.9 m depth. Significant quantities of phosphate-extractable Se(IV) in these soils may be due to a combination of (1) Se(IV) accumulated during drainwater disposal via Se(VI) reduction, (2) reoxidation to Se(IV) of Se reduced to Se(0) and Se(–II) during deposition of seleniferous drainwaters, and (3) Se(VI) reduced to Se(IV) during the more recent wet seasons. The data presented here cannot distinguish among these possibilities.

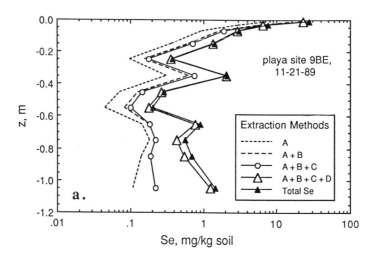

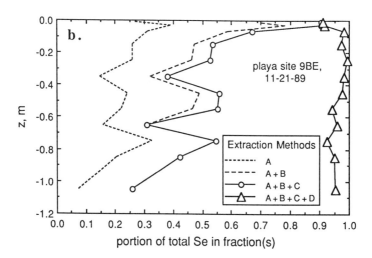

Figure 3 Selenium distributions in the Kesterson Reservoir revegetated playa soil 9BE. (a) Depth profiles of Se in the water-extracted (A), phosphate-extracted (B), carbonate-associated (C), and soil organic matter–associated (D) fractions. The depth distribution of total soil Se concentrations is also shown in this figure. (b) Depth profiles of relative amounts of Se associated with each of the fractions. See text for description of extraction methods.

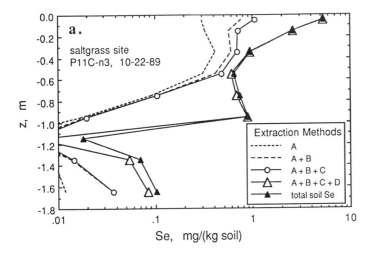

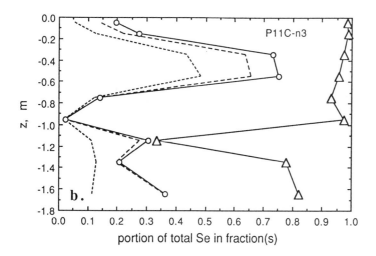

Figure 4 Selenium distributions in the Kesterson Reservoir salt grass site P11C. (a) Depth profiles of Se in the water-extracted (A), phosphate-extracted (B), carbonate-associated (C), and soil organic matter–associated (D) fractions. The depth distribution of total soil Se concentrations is also shown in this figure. (b) Depth profiles of relative amounts of Se associated with each of the fractions. See text for description of extraction methods.

IV. SELENIUM REOXIDATION AND MOVEMENT IN KESTERSON RESERVOIR SOIL PROFILES

A. Background and Methods

Data on Se speciation in the previously described field soils showed a lack of redox equilibria. While Se(VI) accounted for only a fraction of the total soil Se, equilibrium analyses of Se speciation would require that the majority of this element occur as Se(VI) under the prevailing oxidizing conditions [4,36]. The Kesterson Reservoir sites have been maintained as semiarid, upland environments since termination of drainwater discharges. Thus, the deposition of Se in these soils provides an unusual opportunity for monitoring reoxidation of recently reduced Se in the field. In this section, time trends of water-extractable Se from Kesterson Reservoir revegetated playa and salt gras sites are presented. The cored or hand-augered samples were collected in the dry summer through fall months of 1988, 1989, 1990, and 1991. (Samples were not obtained in the saltgrass site in 1990.) Sampling depth intervals ranged from as thin as 0.020 m in the surface crust of the revegetated playa up to 0.15 m in the saltgrass profile. Sampling times coincided with what are believed to be relatively stable periods of the annual Se redox cycle. Samples were collected after periods of below-average winter and spring precipitation and prior to major rainfall events of the following wet season. The previously described water-extraction and hydride-generation atomic absorption spectrometric Se analysis procedures were followed.

B. Results and Discussion of Selenium(VI) Time Trends

Depth profiles of water-extractable Se from the Kesterson Reservoir revegetated playa and salt grass sites are shown in Figures 5a and 6a, respectively. The extracted Se occurred largely as Se(VI). These data, combined with estimates of bulk density profiles, were transformed into the depth-integrated water-extractable Se inventories shown in Figures 5b and 6b. Results presented in this latter manner provide a convenient method for comparing Se inventories. The total mass of water-extractable Se contained in a given profile sampling down to a particular depth is shown directly in these graphs. Thus, year-to-year variations of profile Se inventories are directly comparable at each depth. Changes occurring between specific depth intervals can be determined simply by comparing depth integrals over appropriate ranges.

Two trends common to the revegetated playa and salt grass sites shown here, as well as numerous other Kesterson Reservoir sites, are the general

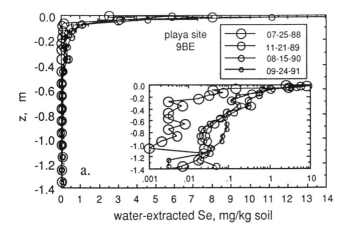

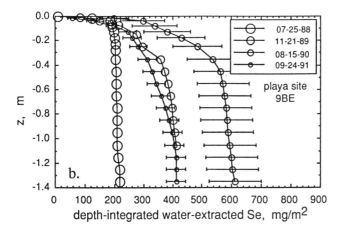

Figure 5 Time trends in water-extracted Se in the Kesterson Reservoir revegetated playa site 9BE. (a) Depth profiles of water-extracted Se. (b) Depth-integrated inventories of water-extracted Se.

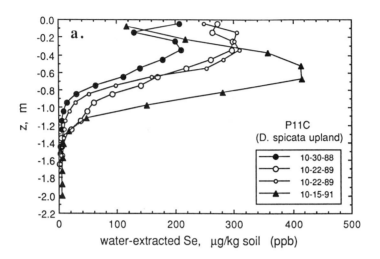
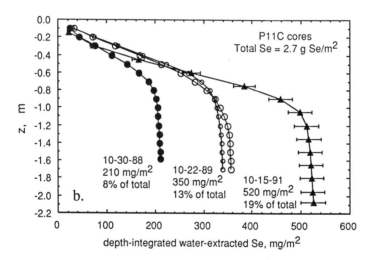

Figure 6 Time trends in water-extracted Se in the Kesterson Reservoir salt grass site P11C. (a) Depth profiles of water-extracted Se. (b) Depth-integrated inventories of water-extracted Se.

increase in the water-extractable Se inventories with time and the shifting of the region of maximum water-extractable Se concentrations downward into the plant root zone. Increases in the water-soluble Se inventories result from reoxidation of reduced Se species. No information concerning relative contributions from the various reduced Se pools is currently available. Declines in the phosphate-extractable fraction over time have been measured at a limited number of sites. A possible additional source of increased water-soluble Se is from an occluded or coprecipitated Se(VI) fraction, suggested by the Figure 1 data. Apparent reoxidation rates of the reduced Se inventory in the range of 0–9% per year have been determined [41]. Note that while the salt grass site has exhibited a year-to-year increase in the water-soluble Se inventory, the revegetated playa site exhibited a reversal of this trend in the 1991 sampling. This reversal appears to have resulted from leaching and reduction during intense late spring rains.

Seasonal leaching and partial reduction of Se(VI) during the winter months is inferred from soil solution sampler data at these and other Kesterson Reservoir sites. Such seasonal declines typically are on the order of tenfold decreases in the soluble Se inventories. Data presented here demonstrate that net oxidation has occurred over most yearly cycles observed to date. Combined influences of seasonal rainfall leaching of Se from the soil surface and transpiratively driven soil water flow into the root zone have resulted in the shifting of the region of maximum soluble Se concentrations deeper into the soil profile. Salinity trends at these and other sites also indicate net concentrations of solutes in the subsurface root zone. Movement of Se deeper than about 1.5 m into the Kesterson Reservoir soil profiles is currently constrained not only by limited recent precipitation relative to evapotranspiration but also by the tendency for Se reduction during leaching through the medium- to fine-textured soils typical of the area. The few Kesterson Reservoir soils that are predominantly sandy tend to coincide with greater Se movement into the shallow groundwater, where Se(VI) reduction and removal from pore waters may occur within a few days [10].

C. Relative Mobilities of Selenate and Selenite in Kesterson Soils

Although numerous adsorption isotherm and adsorption envelop studies comparing the behavior of Se(IV) and Se(VI) have been reported, relatively few have investigated directly the implications of such phenomena on selenium transport in soils. Adsorption and precipitation processes may strongly influence Se mobility since these processes determine the magnitude of local Se storage versus transport. Such considerations are important regardless of whether transport occurs primarily by advection or diffusion. Studies of Se(IV) and Se(VI) transport in soil columns [45–49] and in groundwater [10,50] have reported much higher mobility of Se(VI) relative to Se(IV).

Field data on Se(IV) and Se(VI) movement from seleniferous Kesterson Reservoir soils into initially nonseleniferous soil covers are presented in this section. These data were obtained from a location within Kesterson Reservoir that was covered with 0.53 m of imported, nonseleniferous soil. Approximately 50% of Kesterson Reservoir has been covered in a similar manner to prevent ponding of seleniferous soil waters during winter months in response to the annual rise of the shallow water table [11]. The site was sampled in September 1989, 1.1 years after the emplacement of nonseleniferous fill (soil cover). Profiles of water-extractable Se(IV) and Se(VI) from this site are shown in Figures 7a and 7b. The horizontal line in each of these plots denotes the location of the interface between the original Kesterson Reservoir soil and the introduced fill material (±0.03 m). The data show large increases in water-extractable Se(VI) and no significant increase in Se(IV) in the fill.

The apparently greater upward advance of Se(VI) relative to Se(IV) requires cautious interpretation, as the possibility of significant Se(IV) movement into the fill material followed by oxidation to Se(VI) cannot be ruled out. Additional complications include seasonal Se(VI) reduction and the presence of a large fraction of operationally defined soil organic matter–associated Se in Kesterson Reservoir soils. Future studies that employ both nonradioactive Se and ^{75}Se to distinguish the fate of the initial Se(IV) and Se(VI) inventories simultaneously under various redox environments may provide useful information for short and intermediate time scales. However, the relatively short half-life of ^{75}Se (120 days) may limit the use of this isotope in long-term studies.

V. SUMMARY

Results of our studies on Se in soils from Kesterson Reservoir and the Lahontan Valley were presented to illustrate various aspects of the behavior of Se in soil environments. The persistence of disequilibria with respect to redox conditions was typical in these soils. After 5 years of generally oxidizing conditions, the Kesterson Reservoir soils appear still to be dominated by Se at valences lower that Se^{6+}, with reoxidation rates [to Se(VI)] commonly lower than 10% per year. The significance of annual Se

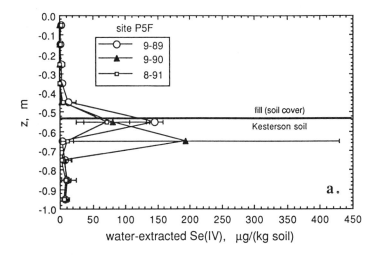

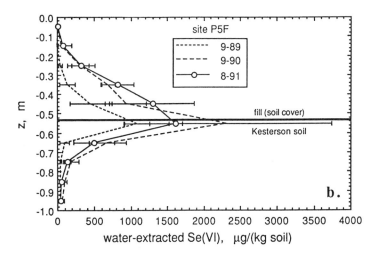

Figure 7 Selenium movement into an initially nonseleniferous soil cover at Kesterson Reservoir fill site P5F. (a) Water-extracted Se(IV) profile. (b) Water-extracted Se(VI) profile.

reduction cycles was also shown in Kesterson Reservoir soils and may help explain the persistence of reduced Se in soil of other arid and semi-arid regions. Some difficulties associated with assigning soil Se to various oxidation states and phases were discussed. The specific problem of distinguishing adsorbed versus precipitated Se(IV) was noted in this context. The potential for obtaining clearer information in this area through spectroscopic methods was noted. Recent X-ray absorption spectroscopy studies of Se(IV) and Se(VI) on model oxide surfaces have contributed substantially to the molecular scale understanding of Se adsorption. Data from various field plots at Kesterson Reservoir illustrate Se reoxidation, Se leaching and reconcentration in the plant root zone, and upward movement of Se into initially nonseleniferous fill soil. The latter set of data are suggestive of the well-recognized, much greater mobility of Se(VI) relative to Se(IV), although the likely simultaneous net oxidation of Se(IV) to Se(VI) in these soils complicates this interpretation. The use of both nonradioactive Se and ^{75}Se in studies comparing Se(IV) and Se(VI) transport and transformations may assist in avoiding the aforementioned ambiguities.

ACKNOWLEDGMENTS

We are grateful to Oleh Weres, Art White, Harold Wollenberg, Andy Yee, Alex Horne, and Rob Long for their substantial efforts and insights provided during the earlier phases of our studies. We thank Joan Oldfather, Leon Tsao, Dan Phillips, and Robert Giauque for providing various chemical analyses. Additional technical assistance by Ray Solbau, Gene Duckart, Kathy Halvorsen, and David Poister is gratefully acknowledged. We also thank the U.S. Bureau of Reclamation staff associated with our studies at Kesterson Reservoir (Susan Hoffman and Mike Delamore, Sacramento; the Los Banos field office) and in the Lahontan Valley (Frank Dimick, Gene Harms, and Robert MacDougal), and the Fallon Indian Reservation members, tribal council, and staff (Merlyn Dixon, Bill Dubois, Jack Allen, Fred Hicks, Sr., Iola Byers, and Floyd Hicks). Discussions with researchers at the University of California (Berkeley, Davis, and Riverside) and the U.S. Geological Survey have also been helpful. Support for this work was provided by the U.S. Bureau of Reclamation (U.S. Department of Interior Interagency agreement 9-AA-20-07250 and 9-AA-20-07250TJ) and by the U.S. Department of Energy (contract DE-AC03-76SF00098).

REFERENCES

1. M. S. Anderson, H. W. Lakin, K. C. Beeson, F. F. Smith, and E. Thacker, *Selenium in Agriculture*, Agricultural Handbook No. 200, Agric. Res. Service, U.S. Dept. Agriculture, U.S. Govt. Printing Office, Washington, D.C., 1961.
2. I. Rosenfeld and O. A. Beath, *Selenium Geobotany, Biochemistry, Toxicity, and Nutrition*, Academic, New York, 1964.
3. S. Sharma and R. Singh, *CRC Crit. Rev. Environ. Control* 13: 23 (1983).
4. C. E. Cowan, *Review of Selenium Thermodynamic Data*, EPRI EA-5655, Battelle Pacific Northwest Laboratories, Richland, Washington, 1988.
5. L. W. Jacobs (Ed.), *Selenium in Agriculture and the Environment*, Soil Sci. Am. Spec. Publ. 23, Am. Soc. Agronomy, Madison, Wisc., 1989.
6. O. Weres, A. R. Jaouni, and L. Tsao, *Appl. Geochem.* 4: 543 (1989).
7. R. H. B. Long, S. M. Benson, T. K. Tokunaga, and A. Yee, *J. Environ. Qual.* 19: 302 (1990).
8. S. M. Benson, A. F. White, S. Halfman, S. Flexser, and M. Alavi, *Water Resour. Res.* 27: 1071 (1991).
9. T. K. Tokunaga, D. S. Lipton, S. M. Benson, A. W. Yee, J. O. Oldfather, E. C. Duckart, P. W. Johannis, and K. E. Halvorsen, *Water Air Soil Pollut.* 57/58: 31 (1991).
10. A. F. White, S. M. Benson, A. W. Yee, H. A. Wollenberg, Jr., and S. Flexser, *Water Resour. Res.* 27: 1085 (1991).
11. T. K. Tokunaga and S. M. Benson, *J. Environ. Qual.* 21: 246 (1992).
12. P. T. Zawislanski, T. K. Tokunaga, S. M. Benson, J. O. Oldfather, and T. N. Narasimhan, *J. Envir. Qual.* 21: 447 (1992).
13. E. E. Cary, G. A. Wieczorek, and W. H. Allaway, *Soil Sci. Soc. Am. Proc.* 31: 21 (1967).
14. H. R. Geering, E. E. Cary, L. H. P. Jones, and W. H. Allaway, *Soil Sci. Soc. Am. Proc.* 32: 35 (1969).
15. O. Weres, G. A. Cutter, A. Yee, R. Neal, H. Moehser, and L. Tsao, Section 3500-Se, in *Standard Methods for the Examination of Water and Wastewater*, 17th ed., 1989.
16. T. T. Chao and R. F. Sanzolone, *Soil Sci. Soc. Am. J.* 53: 385 (1989).
17. M. M. Abrams, R. G. Burau, and R. J. Zasoski, *Soil Sci. Soc. Am. J.* 54: 979 (1990).
18. D. S. Lipton, Associations of selenium in inorganic and organic constituents of soils from a semi-arid region, Ph.D. Thesis, Univ. California, Berkeley, 1991.
19. C. Kheboian and C. F. Bauer, *Anal. Chem.* 59: 1417 (1987).
20. K. A. Gruebel, J. A. Davis, and J. O. Leckie, *Soil Sci. Soc. Am. J.* 52: 390 (1988).
21. P. H. T. Beckett, *Adv. Soil Sci.* 9: 143 (1989).
22. G. Sposito, *The Surface Chemistry of Soils*, Oxford Univ. Press, New York, 1984.
23. S. S. S. Rajan and J. H. Watkinson, *Soil Sci. Soc. Am. J.* 40: 51 (1976).
24. M. Singh, N. Singh, and P. S. Relan, *Soil Sci.* 132: 134 (1976).
25. L. S. Balistrieri and T. T. Chao, *Soil Sci. Soc. Am. J.* 51: 1145 (1987).
26. K. F. Hayes, A. L. Roe, G. E. Brown, Jr., K. O. Hodgsen, J. O. Leckie, and G. A. Parks, *Science* 238: 783 (1987).
27. G. E. Brown, Jr., G. A. Parks, and C. J. Chisholm-Brause, *Chimia* 43: 248 (1989).
28. G. E. Brown, Jr., and G. A. Parks, *Rev. Geophys.* 27: 519 (1989).

29. J. R. Chen, E. C. T. Chao, J. A. Minkin, J. M. Back, K. W. Jones, M. L. Rivers, and S. R. Sutton, *Nucl. Instr. Methods Phys. Res.* **B49**: 533 (1990).
30. R. H. Neal and G. Sposito, *Soil Sci. Soc. Am. J.* **53**: 70 (1989).
31. R. D. Giauque, R. B. Garrett, and L. Y. Goda, *Anal. Chem.* **49**: 62 (1976).
32. W. E. Dollarhide, *Soil Survey of Fallon-Fernley Area, Nevada*, U.S. Dept. Agriculture, Soil Conservation Service, U.S. Govt. Printing Office, Washington, D.C., 1975.
33. T. K. Tokunaga and S. M. Benson, Evaluation of Management Options for Disposal of Salt and Trace Element Laden Agricultural Drainage Water from the Fallon Indian Reservation, Fallon, Nevada, Lawrence Berkeley Laboratory Report LBL-30473, 1991.
34. H. M. Ohlendorf, in *Selenium in Agriculture and the Environment* (L. W. Jacobs, Ed.), Soil Sci. Soc. Am. Spec. Publ. 23, Am. Soc. Agronomy, Madison, Wisc., 1989.
35. P. G. Nazar, *Soil Survey of Merced County, California, Western Part*, U.S. Dept. Agriculture, Soil Conservation Service, U.S. Govt. Printing Office, Washington, D.C., 1989.
36. M. A. Elrashidi, D. C. Adriano, S. M. Workman, and W. L. Lindsay, *Soil Sci.* **144**: 141 (1987).
37. P. T. Zawislanski, in *Hydrological and Geochemical Investigations of Selenium Behavior at Kesterson Reservoir, Annual Report Oct. 1, 1989 through Sept. 30, 1990* (S. M. Benson, T. K. Tokunaga, P. Zawislanski, A. W. Yee, J. S. Daggett, J. M. Oldfather, L. Tsao, and P. W. Johannis, Eds.) Lawrence Berkeley Laboratory, Univ. California, Rep. LBL-29689, 1990.
38. D. D. Runnells and R. D. Lindberg, *Geology* **18**: 212 (1990).
39. R. D. Linsberg and D. D. Runnells, *Science* **225**: 925 (1984).
40. P. T. Zawislanski, Bare soil evaporation at Kesterson Reservoir, Merced County, California: estimation by physical and chemical methods, M.S. Thesis, Univ. California, Berkeley, 1989.
41. S. M. Benson, T. K. Tokunaga, and P. T. Zawislanski, Anticipated Soil Selenium Concentrations at Kesterson Reservoir, Lawrence Berkeley Laboratory Rep. LBL-33080, UC-100, 1992.
42. B. Bar-Yosef and D. Meek, *Soil Sci.* **144**: 11 (1987).
43. R. H. Neal, G. Sposito, K. M. Holtzclaw, and S. J. Traina, *Soil Sci. Soc. Am. J.* **51**: 1161 (1987).
44. N. J. Barrow and B. R. Whelan, *J. Soil Sci.* **40**: 17 (1989).
45. G. Gissel-Nielsen and A. A. Hamdy, *Z. Pflanzenernaehr. Bodenkd.* **140**: 193 (1977).
46. T. Ylaranta, *Ann. Agric. Fenn.* **21**: 103 (1982).
47. L. E. Wangen, A. M. Martinez, and M. M. Jones, The Subsurface Transport of Contaminants from Energy Process Waste Leachates: Final Report, LA-10502, UC-11, Los Alamos National Laboratory, 1985.
48. J. S. Ahlrichs and L. R. Hossner, *J. Environ. Qual.* **16**: 95 (1987).
49. M. H. Alemi, D. A. Goldhamer, and D. R. Nielsen, *J. Environ. Qual.* **20**: 89 (1991).
50. J. H. Hatten, *Geochim. Cosmochim. Acta* **41**: 1665 (1977).

6
Geologic Origin and Pathways of Selenium from the California Coast Ranges to the West-Central San Joaquin Valley

Theresa S. Presser

U.S. Geological Survey
Menlo Park, California

I. INTRODUCTION

Contamination at Kesterson Reservoir led to the assessment of a naturally occurring, potentially toxic, trace element, selenium (Se), from source geologic formations in the surrounding California Coast Ranges. Selenium has been mobilized and concentrated by weathering and evaporation in the process of soil formation and alluvial fan deposition in the arid and semiarid climate of the west-central San Joaquin Valley. This deposition has created a soil salinization problem, and impeding alluvial clay layers have led to waterlogged soils as lands have been reclaimed by flooding to remove the salt and then have been continued to be irrigated. Through current practices of agricultural wastewater management in the valley, subsurface drains are used to collect the saline, shallow groundwater threatening the crop root zone. This wastewater is essentially a soil leachate that was transferred to and stored in wetland areas. This in turn led to extended ecological community exposure to toxic levels of Se and consequent bird deaths and deformities [1].

The contaminant Se was leached from the salinized soils of the extensive alluvial fans developed by Panoche and Cantua Creeks (Figure 1). The highest concentrations of Se (maximum 4.5 µg/g, median 1.8 µg/g) and sulfur in soils [2] and the highest degree of soil salinization occurred in an interfan area between the two creeks. Ephemeral streams and debris flows

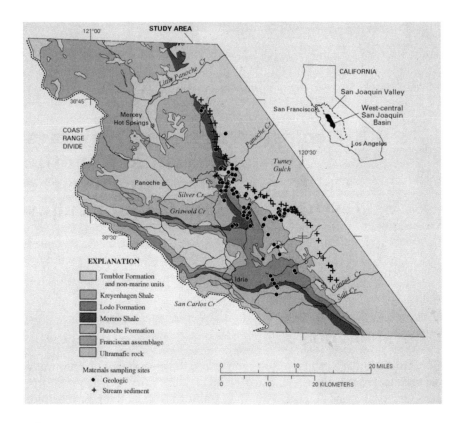

Figure 1 Surficial geology and areal distribution of sampling sites for geologic materials in the study area. (Compiled from Refs. 9 and 16–19 by the Geographic Information Service of the U.S. Geological Survey.) (See color plate.)

that rarely reach the valley axis terminate in this interfan area and develop the most elevated concentrations of Se, as in a closed basin. The highest concentrations of Se in shallow groundwater occur in the same area and along the adjacent southern distal end of the Panoche Creek fan [3]. Selenium concentrations in soils do not follow creek drainages like those of arsenic and mercury, which are deposited at the base of the creeks on the fans and remain there, suggesting that the majority of Se is mobilized in a soluble form. This Se then may be carried by the streams or leached out of the alluvial fans to ultimately reach the San Joaquin River trough, if not intercepted by subsurface drains. Soluble Se and salts continually concentrate at the toes of the alluvial fans through downslope movement and evapotranspiration and move into the shallow groundwater as the valley floor is irrigated. Recharge at fan heads also contributes to Se in shallow groundwater [4]. The shallow groundwater aquifer, once thought to be a series of perched aquifers, is continuous [5] and extends in this area to depths of 120–240 m below land surface [6].

Tracing of sulfate, selenate, and sodium, which were significantly correlated (r = 0.93 for Se and sulfate; r = 0.99 for sodium and sulfate) in the subsurface agricultural drainage [7], eventually led to the consideration of marine, pyritic (FeS_2) sedimentary rocks in the Coast Ranges as the possible geologic source. This type of rock could provide the abundant, soluble sodium and magnesium sulfate minerals creating the soil salinization of the valley. Berner [8] considers pyrite to be the main source of sulfate in marine shales formed from the reduction of sulfate-containing pore waters during diagenesis. Upon weathering, this sulfur is again oxidized to sulfate and can be moved in aqueous systems, appearing as efflorescences or evaporites in arid climates. Preliminary sampling of Panoche, Silver (a major tributary to Panoche Creek), and Cantua Creeks showed sodium-magnesium-sulfate waters with low concentrations of Se (<10 µg/L) [9].

From the geologic record, Se contamination has been proposed to result from the dissolved mineral load ultimately drained from seleniferous Cretaceous marine sedimentary strata underlying and surrounding basins such as the San Joaquin Valley [10]. Extensive volcanic eruptions during Cretaceous time are thought to be the primary source of Se through deposition in Cretaceous seas that had invaded a considerable part of the western states [11]. Selenium was either eroded from igneous rocks or became incorporated in rainfall from volcanic gases and dust in the atmosphere. In either case, Se was incorporated in sediments that have been uplifted over time and exposed to weathering and erosion. Shales commonly contain more Se than other sedimentary rocks [12]. Historically, analytical data from the Cretaceous marine Pierre and Niobrara Shales have shown the highest concentrations of Se (maximum, 103 µg/g; median 6.5 µg/g) in the western United States [13]. Plants grown on soils derived from these shales were found to be toxic to range animals [13]. More recently, reconnaissance of refuge areas in the western United States has shown significant Se contamination [14].

This chapter investigates the water quality of seeps, springs, and streams and characterizes evaporative salts, bedrock (geologic materials), soils, and stream sediments in the areas contributing to the Panoche and Cantua Creek alluvial fans. The principal geologic formations have been identified, and a conceptual model has been developed that explains the predominant mobilization and transport processes. The model accounts for the spatial and temporal variations in Se concentrations.

II. METHODS

Methods of collection, preservation, and elemental analysis of water samples were adapted for saline geochemical environments and are given in Presser and Barnes [7]. The analytical scheme for analysis and speciation of Se in water and salt samples is given by the same authors in Ref. 15. Sampling and analysis methods of geologic materials are given in Presser et al. [9]. All of the analyses for Se used hydride generation atomic absorption spectrophotometry. Care must be taken in pretreatment steps for this method to ensure that all the different forms of Se are available for signal generation [15]. Minerals were identified by X-ray diffraction; analysis was difficult because of the transient nature of some of the hydrated salts.

III. STUDY AREA AND GEOLOGIC SETTING

Alluvial deposits of the west-central San Joaquin Valley are derived primarily from the eastern side of the Coast Ranges. The study area is in the part of the eastern Coast Ranges between Panoche and Cantua Creeks. It encompasses 2600 km^2 (1000 sq m) and 11 drainage basins, from the drainage divide in the west to the alluvial fans in the east and from Salt Creek in the south to Little Panoche Creek in the north. Geologic formations of significant extent in the study area [16–19] are listed in Table 1 and shown in Figure 1.

The Coast Ranges evolved from complex folding and faulting of sedimentary and igneous rocks of Mesozoic and Tertiary age. Major deformation began in Miocene time and continued at intervals through the Pleistocene, when the mountains were raised to their present heights of 1200–1500 m [20].

A simplified structural model of central and northern California Coast Ranges is given by Irwin and Barnes [21]. It shows the Salinian block along the west side of the San Andreas fault and the Franciscan assemblage and Great Valley sequence along the east side. The Great Valley sequence consists of the marine Upper Cretaceous Panoche Formation and the Upper Cretaceous-Paleocene Moreno Shale. The Coast Range thrust fault is the regional contact between the Franciscan assemblage (lower plate) and serpentinite at the base of the Great Valley sequence (upper plate). A thick sequence of Tertiary marine and nonmarine sediments and some

Table 1 Sampled Geologic Units of the California Coast Ranges Study Area

Age	Type	Formation	Water type
Quaternary	Nonmarine sediments	Older alluvium and terrace deposits	Dilute Na-Cl, HCO_3
Pliocene to Pleistocene	Nonmarine sediments	Tulare Formation	Dilute Na-Cl, HCO_3
Miocene (?)	Ores	Mineralization of the New Idria Mercury Mining District	Fe,Mg-SO_4 acid mine drainage
Eocene to Oligocene	Marine sediments	Kreyenhagen Shale (includes Tumey Formation)	Saline Na,Mg-SO_4
Eocene	Marine sediments	Domengine Formation	Saline Na,Mg-SO_4
Paleocene to Eocene	Marine sediments	Lodo Formation	Saline Na,Mg-SO_4
Late Cretaceous to Paleocene	Marine sediments	Moreno Shale	Saline Na,Mg-SO_4
Late Cretaceous	Marine sediments	Panoche Formation	Saline chloride
Late Jurassic to Late Cretaceous	Meta-sedimentary	Franciscan assemblage	Bicarbonate
Jurassic	Intrusive	Ultramafics, mainly serpentinized	Mg ultrabasic

Source: Compiled from Ref. 9.

volcanic rocks locally overlie the region. Tectonic piercement of the Great Valley nappe by the underlying Franciscan rocks occurs at Mt. Diablo and New Idria. In the northern Coast Ranges, most of the nappe has been removed by erosion, so the structurally lower Franciscan rocks are exposed over a wide area. The New Idria piercement structure is a prominent feature in the area of the Coast Ranges implicated to be the source of selenium contamination at Kesterson Reservoir. Acid mine drainage from the New Idria Mining District [22] discharges into Panoche Creek via Silver Creek. The district consists of about 20 quicksilver deposits that rim a fault-bounded core of serpentinite and Franciscan rocks that have been extruded through beds of Upper Cretaceous Panoche Formation. The most productive mercury deposits are found in hydrothermally altered and fractured Panoche Formation along the perimeter of the core.

IV. TYPES OF FLUIDS OF THE COAST RANGES FORMATIONS

Characteristic fluids issue from the principal tectonic blocks just described. Those encountered and sampled in the study area during a survey of creeks, springs, and seeps, which encompassed 62 hydrologic sample sites [9], yielded the following information on dissolved mineral type:

1. Sodium- and magnesium-sulfate waters were derived from Upper Cretaceous and Tertiary marine sedimentary rocks, including the Moreno and Kreyenhagen Shales and the Lodo Formation.
2. Chloride waters were derived from the marine Upper Cretaceous Panoche Formation (part of the Great Valley sequence) and a zone of hydrothermal alteration at Mercey Hot Springs.
3. Bicarbonate waters were derived from the Franciscan assemblage.
4. Ultrabasic, high-magnesium waters were derived from serpentinites.
5. Iron-magnesium-sulfate acid mine drainage was derived from the New Idria Mercury Mining District.

Only the sodium- and magnesium-sulfate waters contained concentrations of Se greater than 3 µg/L. In previous discussions of Coast Ranges geohydrology, sulfate waters have been largely ignored.

To help determine the source of these waters, isotopic compositions of $\delta^{18}O/^{16}O$ and δ D/H for all the spatially and chemically divergent groundwater and surface water samples were analyzed [9]. These data fall on an evaporative trend line with a correlation coefficient of 0.93, indicating a single source for the waters. This isotopic limitation suggests that the solute chemistry of the fluids is controlled and/or modified by solution-reprecipitation processes of surficial and subsurface evaporites rather than by mixing of distinct water sources.

V. EVALUATION OF GEOLOGIC SOURCES OF SELENIUM

A. Concentrations in Bedrock

Cretaceous and Tertiary marine sediments dominate the Coast Ranges study area, which drains to the west-central San Joaquin Valley. These marine shales and sandstones weather into a sulfate regime evidenced by extensive surficial salt efflorescences and evaporites at water and shale surfaces. It is along this eastern flank of the Coast Ranges that a survey of bedrock and stream sediment samples was taken (Figure 1). The Monocline Ridge area is immediately above the interfan area that contains the highest degree of Se contamination. Geologically, it is a massive exposure of the Kreyenhagen Shale shown in yellow in Figure 1.

To determine sources of Se in the study area, 117 bedrock samples were collected in 1986 (Figure 1) [9]. The marine Upper Cretaceous-Paleocene Moreno Shale and the Eocene-Oligocene Kreyenhagen Shale (Figure 1) contain elevated levels of Se, with a maximum of 45 ppm and medians of 6.5 and 8.7 ppm, respectively. Selenium concentrations of these shales are compared with the medians (Figure 2) of other formations in the area (Figure 1): the Upper Cretaceous Panoche Formation, 1.0 ppm; the Paleocene-Eocene Lodo Formation, 0.8 ppm; the Pliocene-Pleistocene Tulare Formation, 1.2 ppm; and Quaternary nonmarine sediments, 0.7 ppm. The Panoche Formation is mainly sandstone, and the Lodo Formation, within this study area, is mostly sandstone. Few samples were collected in the Panoche Formation because of the poor exposure of the shale facies and the presence of nonseleniferous chloride waters. The other geologic units were evaluated only through the characteristic water chemistries associated with them, which were not indicative of probable Se enrichments.

B. Lithology of Seleniferous Shales

1. Upper Cretaceous to Paleocene

MORENO SHALE (MARINE) Foraminiferal and diatomaceous chocolate-brown to maroon, platy, friable shales; lower part contains numerous beds of sandstones with concretions and dikes; upper part is more nearly pure shale and contains a greater proportion of material of organic origin; forms extensive colluvial slopes [23]. Near Cretaceous-Tertiary boundary, planktonic and benthic foraminifera generally indicate a submarine slope and associated basinal depositional facies [24]. Type locality is Moreno Gulch; part of Great Valley sequence. Thickness: 300–900 m.

2. Eocene and Oligocene

KREYENHAGEN SHALE (MARINE) Thin-bedded, chocolate-brown shale and mudstone, grading upward to diatomaceous shale and diatomite; lithic sandstone in lower part; lens of gray lithic sandstone containing prominent oyster beds in upper part along Monocline Ridge [25]. On the basis of Eocene planktonic and benthic foraminifera and calcareous nannoplankton, deposition was in a submarine slope environment [26]. Includes the Tumey Formation because of lithological similarity [27]. Thickness: 450 m.

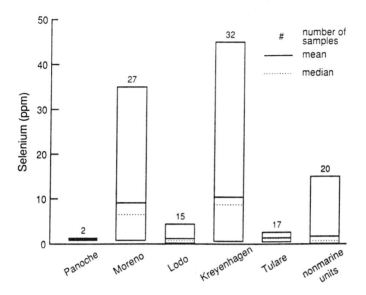

Figure 2 Number of samples, means, medians, and ranges of selenium concentrations in geologic samples by formation. (Modified from Ref. 9.)

VI. MOBILIZATION MECHANISMS

A. Dissolution of Salts and Dissolved Mineral Load

Chemistry of streams and seeps of the drainage basins in the study area is controlled by sodium and magnesium sulfates. The waters are characterized by high dissolved solids (up to 421,400 mg/L). The highest concentration of sodium is 44,000 mg/L; of magnesium, 56,000 mg/L; and of sulfate, 318,000 mg/L.

Thirteen predominantly hydrous, sodium magnesium sulfate minerals (Table 2), visible as extensive surficial efflorescences at water/shale interfaces, were identified in the Coast Ranges study area. The salts contain elevated levels of Se as selenate up to 25 ppm [9]. These sulfate minerals also contain up to 10 waters of hydration and consequently increase in volume by as much as 29% from the dehydrated state. Theoretically, these loose, open-lattice structures of the hydrous minerals could incorporate the selenate (SeO_4^{-2}) anion in the sulfate (SO_4^{-2}) space. These minerals can therefore act as temporary geologic sinks for Se. Samples of the mixed sodium and calcium hydrated mineral hydroglauberite contain up to 14 ppm Se, but samples of gypsum contain <0.5 ppm Se. These results

Table 2 Efflorescent Sulfate Minerals Containing Elevated Selenium in the California Coast Ranges Study Area

Name	Formula
Mirabilite	$Na_2SO_4 \cdot 10\ H_2O$
Thenardite	Na_2SO_4
Bloedite	$MgNa_2(SO_4)_2 \cdot 4\ H_2O$
Konyaite	$MgNa_2(SO_4)_2 \cdot 5\ H_2O$
Loeweite	$MgNa_2(SO_4)_2 \cdot 2^1/_2\ H_2O$
Kieserite	$MgSO_4 \cdot H_2O$
Starkeyite	$MgSO_4 \cdot 4\ H_2O$
Pentahydrite	$MgSO_4 \cdot 5\ H_2O$
Hexahydrite	$MgSO_4 \cdot 6\ H_2O$
Epsomite	$MgSO_4 \cdot 7\ H_2O$
Ferrohexahydrite	$FeSO_4 \cdot 6\ H_2O$
Hydroglauberite	$Na_{10}Ca_3(SO_4)_8 \cdot 6\ H_2O$
Jarosite	$KFe_3(SO_4)_2(OH)_6$

Source: Modified from Ref. 31 with permission.

indicate an exclusion of Se from gypsum and a concentrating mechanism for Se in the more soluble sulfate salts. A majority of halite, as opposed to sulfate salts, was found in the Panoche Formation, which also contains <0.5 ppm Se.

Efflorescent salt crusts on exposed source shales are enriched up to sevenfold in Se compared with their substrates. These deeply weathered shales at higher elevations can contain 95% of the total Se as soluble selenate. In typical hillslopes developed on the Moreno and Kreyenhagen Shales, evaporation from the capillary fringe on summit slopes and salt layering from downgradient throughflow on toeslopes concentrates Se in these slope positions (median values of 12 and 8.2 ppm, respectively) relative to the midslope position (median 4.4 ppm). Therefore, subsurface water movement and fractional crystallization appear to effectively redistribute soluble Se.

B. Weathering of Pyritic Shale Evidence

These sulfate minerals show that the chemistry of the weathering of reduced shale (oxidation of pyrite) is largely a reversal of the chemistry of the early diagenesis of the shale (reduction of sulfate) [8]. Because of similar chemical and physical properties, Se substitutes for sulfur in pyrite (FeS_2) in sedimentary rock at concentrations up to 300 ppm [28]. Up to 3% Se has been found in surveys of Colorado Plateau deposits along with ferroselite ($FeSe_2$), the Se analog of pyrite [29]. The presence of acidic (pH 4) sodium- and magnesium-sulfate seeps gives evidence in the study area for the acid-producing reaction of the weathering of pyrite [30] and, by inference, ferroselite. Five examples of these seeps with elevated levels of Se, up to 420 µg Se/L, were found in widely separated areas of the Moreno Shale [31]. Water chemistries of two of these seeps, at Tumey Gulch and Arroyo Hondo, are given in Table 3. Besides elevated Se, these acid waters contain high levels of dissolved metals; maximum concentrations are 188 mg/L aluminum, 95 mg/L manganese, 13 mg/L zinc, 9 mg/L nickel, and 1.7 mg/L cobalt.

Neutralization and evaporative concentration of these waters and fractional crystallization of sulfate minerals result in a near-surface sodium-sulfate fluid that exceeds the USEPA limit for a toxic Se waste (1000 µg/L). These processes occur in an ephemeral stream at the Tumey Gulch site directly above the interfan zone containing the maximum Se concentrations in soils on the valley floor. The alkaline ephemeral stream (Table 3) from this drainage rarely reaches the valley floor, making this a mainly closed evaporative basin. The alkaline site is geologically similar to the acid site, all being dominated by the Moreno Shale, but the stream water is strongly buffered by bicarbonate water (pH 8.6 and 1415 mg/L HCO_3) derived from the Lodo Formation. As this water is neutralized by adjacent alkaline carbonates, heavy metals are precipitated out (<0.3 mg/L aluminum, 1 mg/L manganese, <0.1 mg/L zinc, 0.6 mg/L nickel, <0.1 mg/L cobalt). The Eh–pH diagram for Se [13], however, shows that neutralization aids the oxidation of Se to soluble selenate, thus mobilizing the Se up to 3500 µg/L. This oxidation would not have taken place in an acid environment, where it is known that insoluble iron-selenite compounds containing 12 ppm Se can exist under an annual rainfall of 100 in. [13]. The alkaline ephemeral stream water is also 2 M in sodium sulfate salt and precipitates mirabilite upon standing. However, analysis of the reaction states of the aqueous solution with respect to various minerals, using the chemical equilibrium computer program SOLMNEQ [32], shows that the water is undersaturated with mirabilite. Theoretically, this could indicate impure crystal growth or slightly inaccurate data on either thermochemical solubility or aqueous species complexation. If this undersaturation represents equilibrium, then similar results for some of the pH 4 series of waters suggest that the Se content of both of these types of waters could be controlled by mirabilite (or an equally soluble magnesium or mixed

Table 3 Selenium Concentrations and Water Chemistry of Acid and Alkaline Seleniferous Waters[a]

		Water type	Water components (mg/L)										
	Se (μg/L)		TDS[b]	Na	SO$_4$	K	Mg	Ca	H[c]	HCO$_3$[d]	Cl	SiO$_2$	B	pH
Acid Waters														
Acid seeps in Moreno Shale														
1 Arroyo Hondo	420	Na-SO$_4$	18,000	3,850	12,500	50	895	395	10		150	61	8.0	4.40
2 Tumey Gulch tributary II	195	Na-SO$_4$	50,000	9,850	34,500	35	3700	470	16		1,100	89	13.	3.85
Alkaline Waters														
Ephemeral stream														
3 Tumey Gulch I	3500	Na-SO$_4$	160,000	44,000	88,000	200	7550	635		1400	18,200	20	67.	8.55
Runoff														
4[e] Silver Ck. at Panoche Road	55	NaMgCa-SO$_4$	6,200	845	3,700	19	300	475		205	95	—	3.0	7.8
5[e] Panoche Ck. at highway I-5	57	NaCaMg-SO$_4$	3,700	495	2,400	17	170	410		150	125	—	2.0	7.9
6 Silver Ck. at Moreno landslide	18	NaMg-SO$_4$	6,500	1,000	3,950	15	490	275		525	240	12	10.	8.20
Integrated watershed														
7 Silver Ck. headwaters seep I	140	NaMg-SO$_4$	3,600	575	1,900	6	245	160		555	80	26	5.5	7.70
8 Griswold Ck. headwaters	155	Na-SO$_4$	4,350	920	2,350	7	145	185		440	250	6	9.7	7.95
Well														
9 Panoche Valley domestic at 400 ft	56	CaMgNa-SO$_4$	3,000	295	1,650	4	195	360		220	270	38	4.0	7.20

[a] Data are reported in significant figures (2.5 maximum).
[b] Total dissolved solids.
[c] Total acidity as H.
[d] Total alkalinity as HCO$_3$.
[e] California Department of Water Resources data except for Se.

Source: Modified from Ref. 31 with permission.

sodium and magnesium sulfate). Selenate in solid solution with the soluble sulfate phase could then hypothetically determine the Se concentration in the waters.

VII. TRANSPORT MECHANISMS

A. Waters

Transport waters representative of Se mobilized by these processes indicate a significant hydrologic reservoir of Se (Table 3) [9]. Runoff from winter flushing events from the two major drainages, Silver and Panoche Creeks, near the valley floor contains approximately 60 µg Se/L. The $\delta^{18}O/^{16}O$ and $\delta D/H$ analyses of surface water samples collected at higher elevations 12 days into the storm hydrograph remained on the isotopic evaporative trend line for prerainfall samples. Concentrations of Se (up to 19 µg/L) also remained elevated in these waters, providing evidence for the importance of subsurface flow in making Se available for transport. Samples of perennial groundwater associated with structural synclines at the heads of Griswold and Silver Creeks (Figure 1), in the upper reaches of the Panoche Creek drainage, represent integrated waters encompassing the geologic section from Jurassic to Miocene. These waters contain elevated Se at an average concentration of 148 µg Se/L. Water samples from deeper wells (120 m) in Panoche Valley contain up to 56 µg Se/L.

B. Mass Wasting and Suspended Sediment

The steeply dipping marine shales, especially the Moreno Shale, show characteristic instability [33], which leads to extensive mud and debris flows, slumps, piping, and deeply incised gullies. Low effective rainfall (25–35 cm/yr) supports little vegetation on steep, poorly consolidated slopes, which allows severe erosion in the higher hills and produces rapid alluvial fan deposition below [34]. Selenium concentrations of bed sediments in ephemeral drainages in Moreno Shale–derived materials range form 3.6 to 8.5 ppm Se, with a median of 6.8 ppm.

Data from analysis of 48 sediment samples from ephemeral streams emerging from the foothills near the point of discharge onto the alluvial fans [9] (Figure 1) show a median of 0.8 ppm Se downslope from the Kreyenhagen Shale and 2.8 ppm Se downslope from the Moreno Shale. The sediments from the Moreno Shale were observed to include recent, more localized, debris-flow materials. A survey of 21 channel sediment samples in a longitudinal profile of the main drainage of the study area (San Carlos, Silver, lower Panoche Creeks, Figure 1) shows a uniform background of 1–2 ppm Se regardless of adjacent formations. This survey includes an iron hydroxide mat (<0.5 ppm Se) deposited in San Carlos Creek downstream from a tailings pond of the New Idria Mercury Mining District. A second profile after a rainfall event shows a range of 1.7–2.9 ppm Se, with a median of 2.4 ppm. In the final deposition stage of the Panoche Creek system, before issuance onto the alluvial fan, the fine-grained residual deposit of suspended load contains 95% of the low-level Se in the insoluble fraction.

VIII. SUMMARY AND MODEL

The irrigated farmlands giving rise to the most contaminated shallow groundwater collected as subsurface drainage and delivered to Kesterson NWR lie just downslope of the interfan area between Panoche and Cantua Creek alluvial fans and at the southern distal end of the Panoche Creek fan. This is where Se and sulfur in soils and soil salinities were highest in 1984. A reconnaissance of the rock outcroppings in the drainage basins to this area from the surrounding Coast Ranges showed elevated concentrations of Se in the extensive surficial exposures of the marine Upper Cretaceous-Paleocene Moreno and Eocene-Oligocene Kreyenhagen Shales. These shales have decreasing exposures to the north and to the south. Alternative source materials investigated in the California Coast Ranges, including Cretaceous and Tertiary sandstones, Pliocene-Pleistocene continental rocks, and waters from the Franciscan assemblage, serpentinites, and the New Idria Mercury Mining District, are comparatively barren of Se.

The occurrence of sodium-magnesium-sulfate waters and minerals associated with the seleniferous rocks and the significant correlation of Se, sulfur, and salinity in alluvial fan soils and of sodium, sulfate, and Se in the inflow waters to Kesterson NWR, led to the proposal that reduced Se in elevated concentrations in these marine pyritic shales is weathered (oxidized) with sulfur, concentrated by evaporation in soluble sulfate (SO_4^{-2}) salts on farmland soils, and mobilized as selenate (SeO_4^{-2}) by irrigation into subsurface agricultural drains.

Within these source formations, two mechanisms are inferred for overall Se transport from the Coast Ranges to the San Joaquin Valley. In the first mechanism, Se moves in solution load, delivering selenate in pulses or episodes of surface runoff. This movement occurs as a result of both rainfall-induced groundwater throughflow and stripping of soluble salts from surface exposures or recently deposited materials during hydrologic events. The arid climate promotes accumulation through many evaporative concentration–dissolution cycles on semiarid, mass-wasted slopes. This system, consequently, acts as almost a closed basin, open only intermittently during large storms, in developing the elevated Se levels delivered to the fans and, especially, the interfans. Because samples of deeply weathered shales contain 95% of the Se as soluble selenate, mass wasting of accumulated salts may also result in the transport of soluble Se.

Integral to the second mechanism is extensive mass wasting in the form of landslides, slumps, or mudflows, which causes mechanical disintegration of the parent geologic materials that then can be carried and deposited during hydrologic events. Both this material and that eroded from channel banks and beds can transport insoluble Se compounds downstream. These particulate Se compounds may be primary selenides or secondary elemental Se and selenite complexes generated during intermediate weathering. Insoluble forms also may be produced if selenate comes into contact with anoxic environments and is re-reduced. Selenium is not concentrated in stream sediments but rather is lost as it is weathered into the soluble fraction and is washed out. Hence, the Se concentrations associated with stream sediment movement are consistent with those measured in soils on the valley floor [2], and these insoluble species provide a comparatively low-level, solid source concentration of Se, although of large mass, to the alluvial fans. Once dispersed on the fans, this relict form of Se is then oxidized or leached out over time, providing a continuing background of soluble Se contamination to the valley.

IX. CONCLUDING PERSPECTIVES

As suggested by lithological information, the Se-enriched Upper Cretaceous-Paleocene Moreno and Eocene-Oligocene Kreyenhagen Shales are similar in depositional environment, that is, in areas of the continental shelf edge and adjacent continental slopes. These areas are often associated with marine upwellings of nutrient-rich water that lead to periodic plankton blooms. The Miocene Monterey Formation, of similar biogenic nature, is not present in the study area but is a suspected source of Se in the Tulare Lake Bed Area evaporation ponds, further south in the San Joaquin Valley, where deformed birds are now found [35]. Recent studies show values of Se concentrations in the Monterey Formation comparable to those in the Moreno and Kreyenhagen Shales [36].

If this nutrient-rich water is subjected to occasions when there is limited access to the ocean, conditions of anaerobic stagnation may evolve. "Black shales" (i.e., reduced shales rich in organic matter and sulfides, especially pyrite) that contain an enrichment of trace elements, including

Se, greater than that which is normally found in seawater, could form [37,38]. Accumulations similar to those of other bioreactive elements (such as chromium and vanadium) may occur as organic material settles to the sea floor bottom and concentrates these elements in the oxygen minimum zone [39]. The refuge ponds may function as modern analogs of ancient marine seas in which elevated concentrations of Se were known to be deposited in reduced bottom sediments. From such a present-day environment, however, high concentrations of Se are hypothesized to be mobilized and bioaccumulated in the food chain, thus causing aquatic bird deformities.

The correlation of Se and sulfur in the waters derived from the postulated source Cretaceous marine shales led to certain relations being observed. These relations represent a penecontemporaneous occurrence; however, a cogenetic relation has not been proven. It has been generalized that deposits of sulfur-containing minerals are often geochemically secondary, and Se in them could have been redeposited from other sources. Biogeochemically enriched shales, rather than just volcanically enriched Cretaceous shales, are apparently potential source areas [39]. Though not all primary associations of Se have been investigated in this reconnaissance study, the direct interconnection between geologic and ecologic cycles in the San Joaquin Valley through the extensive mobilization of Se from a disseminated source is evident.

ACKNOWLEDGMENTS

Members of the U.S. Geological Survey who deserve thanks are W. C. Swain, R. R. Tidball, and R. C. Severson for assistance with selection and collection of water and geologic samples; R. H. Mariner for assistance with X-ray diffraction; Alan Bartow for his insights into his work on the evolution of the San Joaquin Valley; David Piper for his thoughts on the Monterey Formation; and David Jones and F. J. Heimes for help with geologic map illustration. Coast Range study area work was prepared in cooperation with the San Joaquin Valley Drainage Program.

REFERENCES

1. T. S. Presser and H. M. Ohlendorf, *Environ. Manage.* 11: 805–821 (1987).
2. R. R. Tidball, R. C. Severson, C. A. Gent, and G. O. Riddle, U.S. Geol. Surv. Open-File Rep. 86-583, 1986, pp. 1–15.
3. S. J. Deverel and S. K. Gallanthine, *J. Hydrol.* 109: 125–149 (1989).
4. K. Belitz and F. J. Heimes, U.S. Geol. Surv. Water-Supply Paper, 2348, 1988, pp. 1–28.

5. R. J. Gilliom, K. Belitz, F. S. Heimes, N. M. Dubrovsky, S. J. Deverel, J. L. Fio, R. Fujü, and D. G. Clifton, U.S. Geol. Surv. Water-Resour. Invest. Rep. 88-4186, 1988, pp. 13-31.
6. G. H. Davis and J. F. Poland, U.S. Geol. Surv. Water-Supply Paper, 1360-G, 1957, pp. 409-587.
7. T. S. Presser and I. Barnes, U.S. Geol. Surv. Water-Resour. Invest. Rep. 85-4220, 1985, pp. 1-73.
8. R. A. Berner, *Geochim. Cosmochim. Acta 48*: 605-615 (1984).
9. T. S. Presser, W. C. Swain, R. R. Tidball, and R. C. Severson, U.S. Geol. Surv. Water-Resour. Inv. Rep. 90-4070, 1990, pp. 1-66.
10. I. Barnes, Sources of selenium, *Proc. Second Selenium Symp.*, Mar. 23, 1985, Berkeley, Calif., 1986, pp. 41-51.
11. D. F. Davidson and H. A. Powers, U.S. Geol. Surv. Bull. 1084-C, 1959, pp. 1-81.
12. J. D. Hem, U.S. Geol. Surv. Water-Supply Paper 2254, 1985, pp. 1-263.
13. M. S. Anderson, H. W. Lakin, K. C. Beeson, F. F. Smith, E. Thacker, *U.S. Dept. Agric. Handbook 200*: 12-24 (1961).
14. T. S. Presser, M. A. Sylvester, and W. H. Low, *Environ. Manage. 18*(3) (1994).
15. T. S. Presser and I. Barnes, U.S. Geol. Surv. Water Resour. Invest. Rep. 84-4122, 1984, pp. 1-26.
16. T. W. Dibblee, U.S. Geol. Surv. Geol. Map, 1971.
17. T. W. Dibblee, U.S. Geol. Surv. Open-File Rep. 75-0394, 1975.
18. J. A. Bartow, U.S. Geol. Surv. Open-File Rep. 88-49, 1 sheet, scale 1:24,000, 1988.
19. J. A. Bartow, U.S. Geol. Surv. Open-File Rep. 88-528, 1 sheet, scale 1:24,000, 1988.
20. R. M. Norris and R. W. Webb, *Geology of California*, Wiley, New York, 1976.
21. W. P. Irwin and I. Barnes, *Geology 3*: 713-716 (1975).
22. R. K. Linn, New Idria Mining District, in *Ore Deposits of the United States, 1933-1967, The Graton-Sales Volume* (J. D. Ridge, Ed.) American Institute of Mining, Metallurgical and Petroleum Engineers, New York, 1968, pp. 1624-1649.
23. M. B. Payne, Type Moreno Formation and Overlying Eocene Strata on the West Side of the San Joaquin Valley, Fresno and Merced, California, Div. Mines Geol. Spec. Rep. 9, State of California Division of Mines, San Francisco, Calif., 1951, pp. 3-29.
24. D. J. McGuire, Stratigraphy, depositional history and hydrocarbon source-rock potential of the Upper Cretaceous-Lower Tertiary Moreno Formation, Central San Joaquin basin, California, Ph.D. Thesis, Stanford Univ., Stanford, Calif., 1988.
25. J. E. Schoellhamer and D. M. Kinney, U.S. Geol. Surv. Oil Gas Invest. Map OM 128, scale 1:24,000, 1953.
26. R. W. Milam, Biostratigraphy and sedimentation of the Eocene and Oligocene Kreyenhagen Formation, Central California, Ph.D. Thesis, Stanford Univ., Stanford, Calif., 1985.
27. J. Zimmerman Jr., *Am. Assoc. Petrol. Geol. Bull. 28*: 953-976 (1944).
28. M. Fleischer, *Econ. Geol., 50th Ann. Vol.*: 970-1024 (1955).
29. R. G. Coleman and M. Delevaux, *Econ. Geol. 52*: 499-527 (1957).

30. K. B. Krauskopf, *Introduction to Geochemistry*, McGraw-Hill, New York, 1967.
31. T. S. Presser and W. C. Swain, *Appl. Geochem.* 5: 703–713 (1990).
32. Y. K. Kharaka, W. D. Gunter, P. K. Aggarwal, E. H. Perkins, and J. D. DeBraal, U.S. Geol. Water-Resour. Invest. Rep. 88-4227, 1988, pp. 1–420.
33. R. J. Chorley, S. A. Schumm, and D. E. Sugden, *Geomorphology*, Methuen, New York, 1984, pp. 605.
34. W. B. Bull, U.S. Geol. Prof. Paper 437-A, 1964, pp. 1–71.
35. R. A. Schroeder, D. U. Palawski, J. P. Skorupa, U.S. Geol. Surv. Water-Resour. Invest. Rep. 88-4001, 1988, pp. 1–86.
36. D. Z. Piper and C. M. Isaacs, Minor element sources in the Monterey Formation, California: seawater chemistry of deposition (in press).
37. J. D. Vine and E. B. Tourtelot, *Econ. Geol.* 65: 253–272 (1970).
38. H. D. Holland, Econ. Geol. 74: 1676–1680 (1979).
39. T. S. Presser, *Environ. Manage.* 18(3) (1994).

7
Distribution and Mobility of Selenium in Groundwater in the Western San Joaquin Valley of California

Steven J. Deverel, John L. Fio, and Neil M. Dubrovsky

U.S. Geological Survey
Sacramento, California

I. INTRODUCTION AND BACKGROUND

Selenium concentrations in groundwater underlying agricultural areas of the western San Joaquin Valley, California, result from the complex interactions among irrigated agricultural practices and physical and chemical processes. Although irrigated agriculture in the western valley began in the late 1800s, most of the area remained unirrigated until the 1930s and 1940s. The total acreage of land irrigated with pumped groundwater increased more than threefold in 1924 and continued to increase through 1955 [1]. Surface water imported from northern California replaced groundwater for irrigation from the early 1950s to the late 1960s and further increased the amount of irrigated acreage.

Irrigation using surface water and groundwater in the western San Joaquin Valley has substantially altered the physical and chemical nature of the groundwater flow system. Irrigation has changed the recharge and discharge mechanisms and the distribution of hydraulic head by redistributing chemical solutes (including selenium) associated with valley deposits. The amount of irrigation water presently applied is more than five times the annual precipitation of about 200 mm. Figure 1 shows the location and geographic features of the western San Joaquin Valley, as described by Gilliom et al. [2].

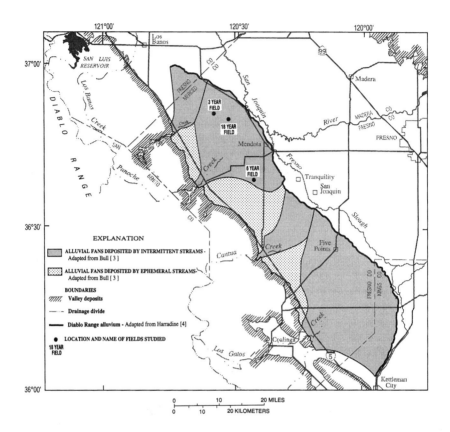

Figure 1 Location and geographic features of the western San Joaquin Valley. (From Ref. 2.)

Studies of the geohydrology of the San Joaquin Valley throughout this century have increased understanding of the physical and chemical processes affecting groundwater flow and the distribution and mobility of groundwater constituents. The purpose of this chapter is to present an integrated understanding of the processes affecting the distribution and mobility of selenium in groundwater at three scales of observation. At the regional scale, the mobility and distribution of selenium are related to alluvial fan geomorphology, the location of shallow groundwater, history of irrigation, water use, pumping, and geologic source material. At the local (farm-field) scale, the distribution and mobility of selenium are affected primarily by groundwater flow to drain laterals, historic water level changes, and evapoconcentration of shallow groundwater. At the subregional scale, which includes dozens of drainage systems, selenium mobility and distribution are affected by the interactions of regional groundwater flow with drainage systems.

II. GEOHYDROLOGIC SETTING

The deposits of the San Joaquin Valley are derived from the Sierra Nevada on the east and the Coast Ranges on the West. The Sierra Nevada, a fault block that dips southwestward, is composed of igneous and metamorphic rock of mostly pre-Tertiary age. The Diablo Range of the California Coast Range, which borders the study area to the west, consists of an exposed core assemblage of Cretaceous and Upper Jurassic age of marine origin overlain by and juxtaposed with marine deposits of Cretaceous age and marine and continental deposits of Tertiary age. The alluvial fan deposits of the western valley are derived from the Diablo Range. The fans deposited by ephemeral streams are smaller than and encircled by the four major alluvial fans deposited by intermittent streams (Little Panoche, Panoche, Cantua, and Los Gatos Creeks) as shown in Figure 1. Shallow deposits of the major alluvial fans are typically coarse-grained in the upper and middle fan areas and fine-grained in the lower fan areas, including the margins of the fans [3]. The fine-grained soils at the margins of the fans generally are poorly drained and often saline.

Under natural conditions, groundwater recharge was primarily from infiltration of water from intermittent streams in the upper parts of the alluvial fans [5]. Groundwater flowed from southwest to northeast, reflecting the general topographic trend of the area. Discharge from the system was by evapotranspiration and streamflow along the valley trough [6,7]. Flowing wells mapped by Mendenhall et al. [6] indicate that artesian conditions prevailed along a broad stretch of the valley trough. The predevelopment groundwater flow system was characterized by horizontal flow from the recharge area to the valley trough. The pumping of groundwater for irrigation, which began in the late 1800s and reached a maximum in the 1960s, substantially altered the groundwater flow system.

Figure 2 depicts the changes in the groundwater flow system that occurred from predevelopment through the mid-1980s in a hypothetical cross section of the Panoche Creek alluvial fan. Prior to the onset of irrigation, hydraulic gradients were primarily horizontal and flow paths extended from the recharge areas at the head of the alluvial fans to discharge into the San Joaquin River and adjacent wetlands (Figure 2A). The

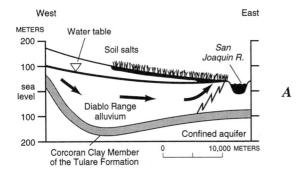
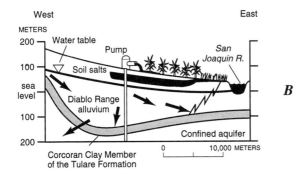
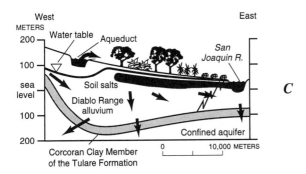

Figure 2 Geohydrologic sections through Panoche Creek alluvial fan illustrating the evolution of the groundwater flow system in the western San Joaquin Valley. Arrows indicate direction of flow. (A) Shallow distribution of soil salts and primarily horizontal direction of groundwater flow between recharge areas in the upper part of the fan and discharge areas along the San Joaquin River during predevelopment. (B) Changes in groundwater flow direction and distribution of soil salts from the 1930s through the 1960s. (C) Discontinuation of pumping in the late 1960s caused a rise in the water table. Irrigation of low-lying areas and continued irrigation of middle- and upper-fan areas caused further downward displacement of soil salts.

dark areas on Figure 2A delineate the approximate distribution of soil salts close to land surface prior to development. After irrigation began, groundwater pumpage and crop evapotranspiration replaced natural discharge and irrigation water became the primary source of recharge (Figure 2B). Groundwater flow directions, mostly horizontal across the valley during predevelopment, changed to primarily downward during the years of groundwater pumping (Figures 2A and 2B).

Pumping of groundwater increased the depth to the water table, and irrigation with pumped groundwater caused downward displacement of soil salts (Figure 2B). Replacement of groundwater with surface water imported from northern California in the 1960s caused hydraulic pressures to increase and the water table to rise, creating a need for drainage in low-lying areas of the valley (Figure 2C). Soil salts were further displaced in the valley trough where irrigation water generally had not been applied previously. Distribution of selenium in the groundwater is related to the hydraulic changes depicted in Figure 2. Application of irrigation water caused leaching of soil salts containing selenium and increased selenium concentrations in the groundwater. Subsequent evaporation of shallow groundwater increased salt and selenium concentrations in the valley trough and lower alluvial fans. Drainage systems were installed to prevent evaporation and to leach accumulated salts. These processes are discussed in relation to the measured distributions of selenium and other constituents at the regional, local, and subregional scales.

III. DISTRIBUTION AND MOBILITY OF SELENIUM IN GROUNDWATER

A. Regional Scale

An understanding of the distribution of selenium in groundwater on a regional scale led to the development of hypotheses about hydrologic and geochemical processes and prompted the need for testing these hypotheses at the local and subregional scales. In general, the spatial variability of selenium concentrations in the regional aquifer in the western valley has been affected by hydrologic and geomorphic processes that occurred over geologic and recent time: evapoconcentration of groundwater and soil water, salt accumulation, salt leaching, oxidation–reduction processes, and groundwater and solute movement to agricultural drainage systems. This information was developed from evaluation of the analyses of groundwater samples collected throughout the western valley in relation to information about the distribution of soil salinity and soil selenium, geomorphology, geologic source material distribution, and land use.

An overview of the spatial distribution of selenium in shallow groundwater (defined herein as groundwater within 5–10 m of land surface) and the depth distribution, which extends several hundred meters below land surface, is presented in this section.

Reconnaissance studies in 1984 and 1986 of the regional distribution of selenium in shallow groundwater showed that high selenium concentrations occur only in Coast Ranges alluvial sediments and that samples of shallow groundwater in Sierra Nevada sediments contained less than 1 μg/L of selenium per liter [8–10].

1. Selenium in Groundwater Underlying Diablo Range Alluvium

Concentrations of selenium in shallow groundwater generally were less than 20 μg/L in the middle fan areas of the alluvial fans deposited by Cantua, Little Panoche, Los Gatos, and Panoche Creeks [3,10] (Figure 3). In the lower fan areas, particularly at the northern and southern margins of the Panoche Creek fan, selenium concentrations were as much as several hundred micrograms per liter. Patterns of selenium distribution are more difficult to generalize for smaller fans of ephemeral streams, because only a small area of these fans is underlain by shallow groundwater. However, some of the highest selenium concentrations were determined in samples collected at the lowest altitudes of the ephemeral stream fans between the Panoche Creek and Cantua Creek fans.

The distribution of dissolved solids in shallow groundwater is similar to that of selenium. Dissolved solids concentrations in groundwater generally are less than 3000 mg/L in the middle fan area of the Panoche Creek alluvial fan. Dissolved solids concentrations commonly are greater than 5000 mg/L; concentrations are greater than 10,000 mg/L in some areas along the fan margins. Dissolved solids are similarly distributed in the Cantua Creek and Los Gatos Creek alluvial fans. Like selenium, concentrations of dissolved solids in areas of ephemeral stream fans are higher than in corresponding topographic locations on adjacent fans of the major streams.

Before much of the study area was irrigated, the distribution of soil salinity resembled the distributions of dissolved solids and selenium in present-day shallow groundwater. Figure 4 shows the distribution of subsurface soil salinity determined in the mid-1940s [4]. The area of artesian wells defined by Mendenhall et al. [6] approximates the natural groundwater discharge area in the valley trough. Soil salinity was highest along the margins of the alluvial fans, where groundwater discharge by

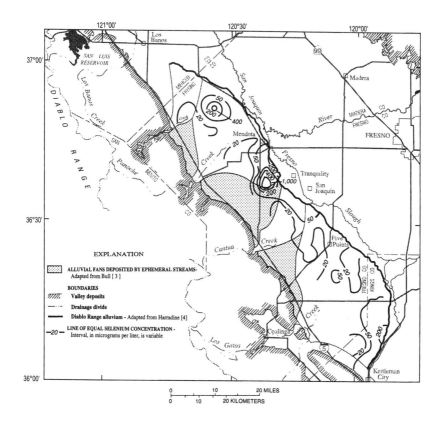

Figure 3 Distribution of selenium concentrations in shallow groundwater. (From Ref. 10.)

evapotranspiration brought solutes to the soil surface, probably for several thousand years.

Similar historical distributions of soil salinity and shallow groundwater salinity indicate that dissolved solids in shallow groundwater were leached from saline soils by irrigation water. However, distributions are not similar in areas where the soils were irrigated before the soil salinity study of the 1940s [4]. The first few decades of irrigation probably leached most of the readily soluble forms of soil selenium and other salts into the shallow groundwater. Since irrigation began, there has been relatively little horizontal movement of shallow groundwater on a regional scale. Thus, the present-day areal distribution of selenium in groundwater is

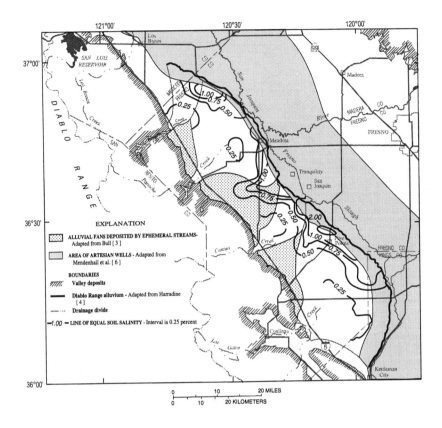

Figure 4 Distribution of percentage of soil salinity as determined by Harradine [4] in the early 1940s. (From Ref. 10.)

similar to the distribution of natural soil salinity in the Diablo Range alluvial sediments. The highest selenium concentrations are at varying depths below the water table, depending on historical irrigation and drainage practices and vertical hydraulic gradients.

In areas where the water table is less than 1.5 m below land surface, the evaporative concentration of dissolved solids in groundwater can increase salinity and selenium concentrations far above the levels resulting from leaching of soil salts by irrigation. Under natural conditions, when little or no recharge of groundwater occurred through the lower fan soils, groundwater discharge by evapotranspiration resulted in the accumulation of salts in the soil. During irrigated conditions, loss of water by evapotranspiration tends to concentrate salts in the groundwater rather than in the soil because soils regularly are flushed by downward percolating irrigation water, and net groundwater movement generally is downward.

Figure 5 shows the relation of δD and $\delta^{18}O$ in 46 samples of shallow groundwater collected from the western valley. Samples collected at the lowest land surface altitudes, where the water table is shallowest, are most enriched in the heavy isotopes and follow a trend line indicating evaporative concentrations. Isotropic data for shallow groundwater of three agricultural fields that were studied in detail fit a similar evaporative trend line. Selenium and dissolved solids concentrations were significantly correlated ($\alpha = 0.05$) with $\delta^{18}O$ because of evaporative concentration [12]. At higher land surface altitudes, where selenium concentrations are lower and the water table is deeper, the shallow groundwater is less enriched in ^{18}O, indicating little or no evaporation.

The distribution of selenium along geohydrologic sections in relation to historical hydrology and geochemical conditions adds a vertical dimension to understanding the spatial distribution of selenium in groundwater and the historical movement of groundwater with high selenium

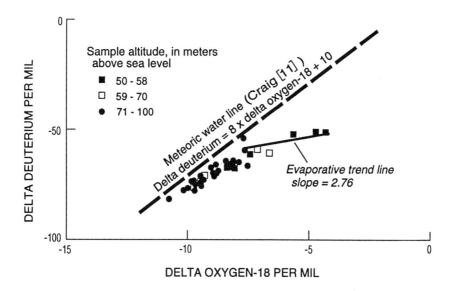

Figure 5 Relation of δD and $\delta^{18}O$ for 46 shallow groundwater samples collected throughout the western San Joaquin Valley. (From Ref. 10.)

concentrations. Observation wells were installed at various depths at several cluster sites along three geohydrologic sections through the groundwater flow system of the Panoche Creek alluvial fan. The cluster sites are aligned approximately with the direction of natural groundwater flow.

The distribution of tritium in groundwater samples provides information about the depth to which recent (since 1952) irrigation water has moved. The minimum depth to which post-1952 recharge has penetrated ranges from 3 m below the water table in the basal area of the alluvial fan to 23 m at the upper end of the fan. The depth of tritiated groundwater indicates that post-1952 irrigation recharge occupied a zone at the top of the aquifer about 15 m thick. The variation of the recharge depth since 1952 reflects variation in lithology, position in the flow system, and irrigation history. In large parts of the study area there is a volume of tritium-free recharge from early irrigation with groundwater that underlies the tritiated water. The depth of soil salts and selenium leached by initial irrigation is substantially greater than the depth at which tritium is detectable.

Concentrations of selenium in groundwater at the cluster sites along the cross sections range from less than the detection limit of 1 µg/L to a maximum of 2000 µg/L. The highest selenium concentrations are associated with high dissolved solids concentrations in the upper part of the aquifer. Selenium concentrations greater than 350 µg/L generally are restricted to the upper 26 m of the saturated zone; however, concentrations of 100 µg/L extend to depths of 24–65 m below the water table at most sites. All samples from wells screened in Sierra Nevada sediments had selenium concentrations less than 1 µg/L. Samples from within 30.4 m of the water table had a median selenium concentration of 350 µg/L, compared to 11.5 µg/L in deeper samples from Coast Range deposits.

In alluvium derived from the Diablo Range, selenium is associated with dissolved solids. Soil salts with high selenium concentrations that accumulated over geologic time and were leached by irrigation water resulted in high selenium concentrations in groundwater as deep as 65 m below the water table. Subsequent evapoconcentration of shallow groundwater resulted in the highest measured selenium concentrations. In contrast, groundwater in alluvium derived from Sierra Nevada rocks with high dissolved solids concentrations had no detectable selenium.

2. *Selenium in Groundwater Underlying Sierra Nevada Alluvium*

The apparent removal of selenium from saline groundwater (10.7–15.2 m below land surface) in the Sierra Nevada alluvium is probably the result of the reduction of selenate to immobile forms of selenium. Oxidation–reduction processes and the apparent removal of selenium from groundwater were investigated at a site near Mendota on the eastern edge of the Panoche Creek alluvial fan (Figure 1). The redox potential of the groundwater at the study site was assessed using platinum electrode measurements and the relative abundance of several dissolved constituents that are sensitive to redox conditions. Results are discussed in detail by Dubrovsky et al. [13]. High concentrations of nitrate in porewater above the water table indicate that oxidizing conditions dominate in the unsaturated zone.

Below the water table, both nitrate and dissolved oxygen disappear from solution and groundwater becomes moderately reducing, as indicated by the high concentrations of iron and manganese. This pattern is consistent with the decrease in platinum electrode redox potential with increased depth below the water table. Selenium concentrations decrease rapidly at the same depth at which manganese concentrations increase, indicating that the decrease in selenium is due to a process that occurs under reducing conditions.

Selenium removal under reducing conditions may be caused by the respiratory reduction of selenite to elemental selenium by anaerobic bacteria [14,15]. The presence of this type of bacterial activity in selected subsurface sediments from the study site was determined by measuring loss of radioactively labeled [^{75}Se]selenite from solution in samples of slurried core materials [14]. Although all sediments tested had bacteria that will reduce selenite, results showed that after 3 days of incubation there was no loss of selenite from solution in the core materials from 6.9 to 8.5 m [13]. In contrast, 78% of the labeled selenite in solutions was lost in the core from 8.8 m and 20% was lost from the core at 9.1 m. After 8 days, loss of selenite from solution was significant in all nine samples analyzed; however, the samples from 8.5 m and deeper had microbial populations capable of reducing selenite at a much higher rate than did microbial populations from sediments at shallower depths. The depth interval of high reduction capability corresponds to the zone at which the porewaters become reducing.

B. Local Sale

Additional detailed investigations at the local scale (within three artificially drained farm fields) have increased quantitative understanding of factors affecting selenium distribution and mobility. Investigations discussed in the following sections were based on hypotheses developed during the regional scale investigations. Drainage systems had been installed in the three fields for 6, 3, and 18 years at the time of the investigations.

Geochemical and hydrologic data collected for shallow groundwater underlying Diablo Range alluvium (within 20 m of land surface) in the three fields in the western San Joaquin Valley indicate that a common sequence of events caused the measured distributions of selenium. First, irrigation of low-lying saline soils leached selenium into the groundwater. Second, with continuing irrigation, the water table rose to within 2 m of land surface and the shallow groundwater was evapoconcentrated, increasing selenium concentrations. Because groundwater hydraulic gradients were predominantly downward throughout the leaching and water table rise, selenium-enriched groundwater was displaced downward in the aquifer. Third, the installation of drainage systems to prevent further salinization of the root zone caused high-selenium groundwater to move toward drain laterals.

1. Six-Year Field

The field discussed in detail by Leighton et al. [16] was drained for 6 years when sampling began in 1986. The field is at the southeastern part of the Panoche Creek alluvial fan (denoted as the M2 site in Figure 1) in an area that was mapped as a high-salinity area in the 1940s [4]. Data presented in reports by Deverel and Gallanthine [10] and by Leighton et al. [16] showed that groundwater at 6 and 9 m below the land surface during sampling in 1986 and 1987 was subject to evapoconcentration near the land surface in the early 1970s and was subsequently displaced downward.

Figure 6 shows the depth distribution of selenium concentrations in groundwater samples collected in 1986 and 1987. Selenium concentrations were significantly correlated with dissolved solids concentrations and with isotopic enrichment ($\alpha = 0.01$). The data showing the relation of δD and $\delta^{18}O$ for groundwater samples collected in 1986 and 1987 are plotted in Figure 7. The points representing samples collected at 6 and 9 m below land surface are the farthest from the intersection of the evaporative trend line and the meteoric water line and represent the most evapoconcentrated samples. Tritium, an indicator of sample age, generally was not detected in samples collected at depths greater than 6 m, indicating that this groundwater infiltrated before 1970. In 1970, surface water applied to this field contained tritium derived from atmospheric nuclear testing conducted from 1952 until 1968. Isotopic enrichment as well as selenium and salinity data indicate that evapoconcentration and

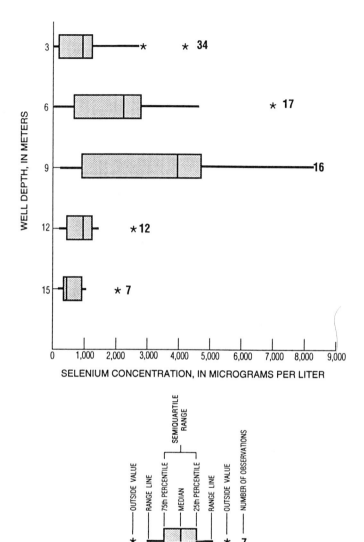

Figure 6 Depth distribution of selenium concentrations in samples collected in observation wells in 1986 and 1987 in the field studied by Leighton, et al. [16]. Outside values are between 1.5 and 3.0 times the semiquartile range from the end of the rectangle. Range lines extend a distance equal to 1.5 times the semiquartile range away from the end of the rectangle or to the limit of the data, whichever is the lesser. Numbers to far right give number of observations.

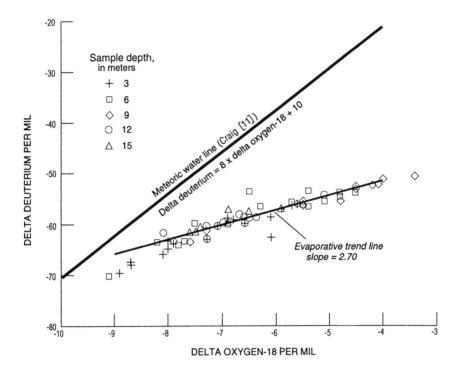

Figure 7 Relation of δD and $\delta^{18}O$ for samples collected in observation wells in the field described in Leighton, et al. [16].

downward displacement are the predominant processes that have resulted in the distribution of selenium.

Samples collected from wells at 12 m are similar in isotopic composition to the samples collected in wells at 3 m. The water at 12 m was applied at the land surface 25–40 years before the study and was subject to minimal isotopic enrichment as it traveled through the unsaturated zone. Normative salt analysis indicated that the high selenium concentrations in this groundwater are the results of leaching of soluble selenium, associated with sodium, and magnesium evaporite salts from the unsaturated zone when this area was irrigated initially [16,17].

A rise in the water table caused the water at the 6- and 9-m depths to become partially evaporated, thus increasing its isotopic enrichment and selenium concentration. This groundwater was within 2 m of land surface in the early 1970s. The highest selenium concentrations and isotopic enrichment in the 6- and 9-m samples (Figures 6 and 7) resulted from evaporation from the shallow water table when it was within 2 m of land surface [16]. After the installation of drains in 1980, the water table lowered so that samples collected at the 3-m depth (near the water table surface) are less concentrated than the 6- and 9-m samples.

2. Three-Year Field

The isotopic composition of groundwater samples collected from wells installed and sampled in 1987 in the 1-year field studied originally by Deverel and Fujii in 1985 (referred to here as the 3-year field) indicated a pattern consistent with the discussion of the 6-year field [12]. The shallowest (3 m) groundwater samples were isotopically enriched relative to the samples collected at the 6-, 12-, and 15-m depths. Because this field had been drained for 3 years at the time of sampling, the shallowest groundwater, which was subject to partial evaporation prior to the drainage system installation, had not been displaced substantially.

3. 18-Year Field

Study results of the field drained for 18 years in 1987 (Figures 1 and 8) were first discussed by Deverel and Fujii [12]. Figure 1 shows the location of the 18-year field, and Figure 8 shows a schematic of the drainage system, locations of observation wells, and geohydrologic cross section A-A'. Deverel and Fio [18,19] reported the results of a quantitative geochemical and hydrologic assessment that led to an understanding of the hydrologic history, present-day flow system, and movement of groundwater and selenium to drain laterals in the 18-year field.

The water table in this field rose to within 1.5 m of land surface in the mid-1960s, and evapotranspiration of groundwater near the water table caused an increase in soil and groundwater salinity. Installation of a drainage system in 1969 reduced evapoconcentration of solutes in shallow groundwater by lowering the water table to below 1.5 m. Prior to installation of the drainage system, the vertical component of groundwater flow was downward with the regional groundwater flow system; therefore saline, evapoconcentrated groundwater near the land surface was slowly displaced downward in the aquifer.

Installation of the drainage system altered hydraulic gradients in the field, and groundwater began to flow upward toward the drain laterals. Willardson et al. [20] reported that the specific conductance of the drain sump water was about 14,000 μS/cm in 1969 when the field was first drained. This is substantially higher than the value of about 9000 μS/cm measured in 1985. Simulated groundwater flow paths and evaluation of

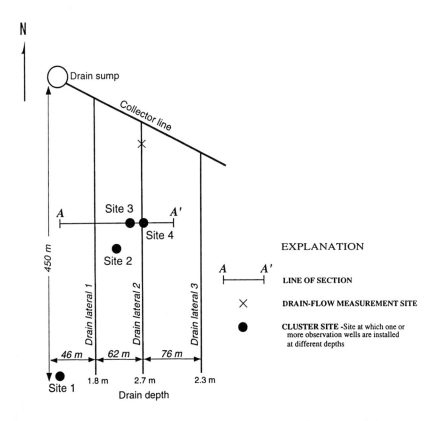

Figure 8 Drainage system in the 18-year field and locations of cluster sites, drain flow-measurement sites, and cross section A-A'.

the chemical and isotopic composition of groundwater samples indicate that a large part of the saline groundwater within about 6 m of land surface was displaced toward the drain laterals by nonsaline irrigation recharge. This displacement occurred within the first decade of the drainage system installation. Displacement of salts from the unsaturated zone by irrigation water also caused a decrease in the specific conductance of water in the drain sump. A mixture of saline, isotopically enriched groundwater from deeper depths and relatively less saline, isotopically depleted groundwater from shallower depths flowed into the drain laterals during the study.

Figure 9 shows simulated groundwater flow paths in cross section A-A' (Figure 8) and the concentrations of selenium in groundwater samples. Groundwater at depths greater than 6 m below land surface had the highest selenium concentrations and moved to the two drain laterals in varying proportions. Drain lateral 2 at 2.7 m below land surface collected substantially deeper high-selenium groundwater than drain lateral 1 at 1.8 m.

An evaluation of isotopic enrichment in drain lateral and groundwater samples during an 18-month period demonstrated that 30–50% and 0–30% of the water flowing in drain laterals 2 and 1, respectively, was deep groundwater (below 6 m). The lower percentages were associated with irrigation events. Increased recharge from irrigation led to increasing selenium loads in both drain laterals. Because the shallow drain lateral

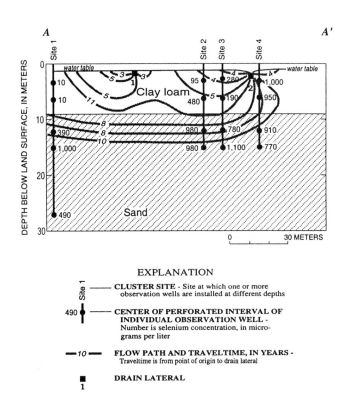

Figure 9 Simulated groundwater flow paths and estimated travel times for irrigated conditions for the 18-year field.

collects less of the deep groundwater, its annual selenium load is about 20% of the load for the deep lateral. The results of the groundwater model simulations generally agreed with the geochemical analysis.

Simulated horizontal flow patterns in the sand layer in Figure 9 demonstrate the interactions of regional and local groundwater flows to drain laterals. The hydrologic analysis of Fio and Deverel [19] indicated that the groundwater in the cross section flowed from upgradient of the A-A' cross section and possibly from within the field. This led to the conclusion that high-selenium water probably would flow into drain laterals for several decades after the study was completed in 1989. The results of this study elucidated a need for additional study of the interaction of groundwater flows with drainage systems at the subregional scale, which consists of dozens of drainage systems.

C. Subregional Scale

Additional research on the distribution and mobility of selenium at a scale intermediate between the regional and local farm-field scales was prompted by questions regarding the movement of selenium associated with the interaction of multiple on-farm drainage systems and the regional groundwater flow system. There was also a need to develop effective management strategies that could be implemented and verified within a single water district. Management strategies to decrease the area requiring drainage and the total volume of drainage water can be implemented most effectively at the water district level of operation [21]. These strategies are dependent on a quantitative understanding of factors affecting the movement of selenium and groundwater to drainage systems. A subregional scale study in the Panoche Water District was designed to extend the regional groundwater flow analysis of Belitz et al. [22] within the subregion and to understand the relationship of groundwater flow with the distribution of selenium in groundwater and concentrations and loads of selenium in drainwater.

To extend the regional flow analyses, hydrologic and geochemical data were collected within the Panoche Water District. Figure 10 shows the location of the Panoche Water District. Three cross sections along general directions of groundwater flow were delineated (Figure 10) for understanding directions of groundwater flow, lithology, and concentrations of selected constituents. Also, time-series data for drain flow and drain water quality were collected and evaluated to understand further the interaction of drainage systems and groundwater.

DISTRIBUTION AND MOBILITY OF SE IN GROUNDWATER 175

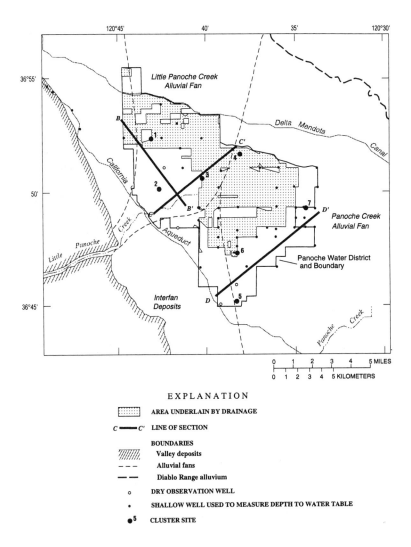

Figure 10 Location of Panoche Water District and geohydrologic cross sections B-B', C-C', and D-D'.

The distribution of selenium concentrations in groundwater samples shown in Figure 11 reflects the interaction of irrigation water, geologic source materials, and directions of groundwater flow. Selenium concentrations are relatively low in samples collected at sites 1 and 2 in the B-B' cross section and site 3 in the C-C' cross section, whereas higher concentrations were found in samples collected at site 4 in the C-C' cross section and sites 5–7 in the D-D' cross section. The higher selenium concentrations were measured at sites located on the Panoche Creek alluvial fan because a substantial area of the exposed rocks in the drainage basin of Panoche Creek consists of the Moreno Formation of Cretaceous age and Kreyenhagen Shale of Tertiary age [23]. These formations contain organic marine shales that generate seleniferous waters [23]. Surface exposures of marine shales of Tertiary age are absent in the drainage basin of Little Panoche Creek, and selenium concentrations in water samples collected from this creek are typically less than those collected from Panoche Creek [24].

The highest selenium concentrations in samples collected at site 4 of cross section C-C' are the result of evaporation from a shallow water table as demonstrated by the high levels of deuterium and ^{18}O in samples. Because of the slow travel times and predominantly upward gradients at site 4 of cross section C-C', high-selenium groundwater was displaced a relatively short distance from the land surface. Similar to the field studied by Deverel and Fio [18], groundwater at these depths (4.1 and 8.3 m below land surface) was close enough to the land surface between 1966 and 1975 for substantial evapoconcentration to take place, probably prior to 1985 when drainage systems were installed at site 6.

In contrast, predominantly downward gradients and faster travel times at site 6 have caused greater displacement of high-selenium, evapoconcentrated groundwater from the land surface. The isotopic compositions of samples collected 13 and 19 m below land surface at site 6 and 27.8 m below land surface at site 3 demonstrate that the groundwater was subject to evapoconcentration [18].

To further understand groundwater and selenium distribution and movement, a three-dimensional steady-state groundwater flow model was developed for the subregion. The results of particle tracking were used to simulate advective transport through the cross sections and are shown in Figure 12 [25]. Travel times and directions of groundwater flow vary substantially within the cross sections. In general, groundwater flows downward at the higher altitudes of the cross sections and upward at the lower altitudes. Moreover, groundwater velocities are slower at the lower

DISTRIBUTION AND MOBILITY OF SE IN GROUNDWATER

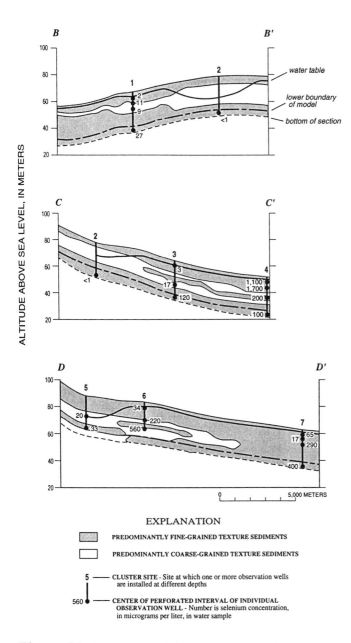

Figure 11 Cross-sectional diagram showing selenium concentrations in observation well samples and texture of deposits.

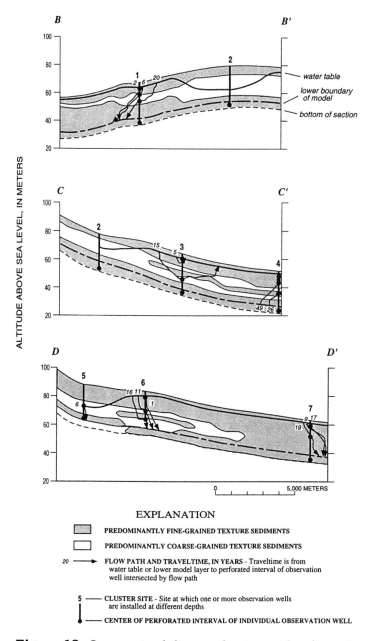

Figure 12 Cross-sectional diagram showing results of particle tracking used to simulate advective transport through the cross sections, flow paths, and travel times.

altitude of the C-C' cross section because of the predominance of fine-grained deposits.

The areal distribution and temporal variability of drainwater selenium concentrations are related to the spatial distribution of selenium concentrations in groundwater. Drainage systems with selenium concentrations between 20 and 100 µg/L are in the Little Panoche Creek alluvial fan where geologic source materials have relatively low concentrations of selenium. Drainage systems with selenium concentrations greater than 100 µg/L are located in the Panoche Creek alluvial fan. The highest selenium concentrations (>1000 µg/L) were in drainwater collected at the most downgradient part of the subregion.

The temporal variability of dissolved solids concentrations are highly correlated with selenium concentrations from drainwater samples throughout the subregion. Figure 13 shows temporal variability of dissolved solids concentrations in drainwater samples collected at cluster sites 3 and 4 and flow in the drainage sumps. For those drainage systems in areas where the concentrations of dissolved solids and selenium are relatively constant with depth (Figure 13A), dissolved solids concentrations vary little even though flow varies substantially over time. Fio and Deverel [19] showed that groundwater flows to drain laterals from different depths during and between irrigation periods. Flow from deeper depths is greater during unirrigated periods and less during irrigated periods. There is little variation in dissolved-solids concentrations in groundwater with depth at site 3 (Figure 10), where the data plotted in Figure 13A were collected. Therefore, variations in drain flow do not result in temporal variations in dissolved solids concentrations. In contrast, because dissolved solids concentrations vary with depth at cluster site 4 (Figure 10), which is adjacent to the drainage system where the data were collected, the dissolved solids and selenium concentrations vary inversely with drain flow (Figure 13B).

Hydrologic and geochemical data provide a better understanding of flow patterns and factors affecting the selenium concentrations in groundwater and drainwater. Modeling of the groundwater flow system effectively simulated the flow system and provided information useful for district water management. Results of groundwater flow modeling and particle tracking delineated probable contributing areas to drainage systems and indicated that 89% of drainwater flow for most drainage systems originated within the drained field, whereas the remainder originated upslope of the drained areas.

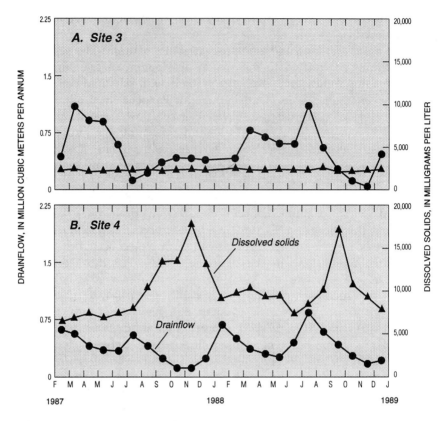

Figure 13 Temporal variability of dissolved solids concentrations and flow in drainage sumps at cluster sites 3 and 4.

IV. SUMMARY AND CONCLUSIONS

The multiscale, interdisciplinary approach to the study of groundwater in the western San Joaquin Valley, California, has increased understanding of the processes affecting the distribution and mobility of selenium. Regional scale studies demonstrated that the distribution of selenium in groundwater underlying alluvium from the Diablo Range is primarily a function of leaching of salts by irrigation and hydraulic changes in the groundwater flow system. The oxidation–reduction reactions in Sierra Nevada sediments affect selenium mobility by transforming selenate to reduced immobile forms. In contrast, selenium is a highly mobile selenate in the oxidized Diablo Range alluvium. The highest selenium concentrations in groundwater are the results of evapoconcentration of shallow groundwater.

Studies of individual farm fields in Diablo Range alluvium provided details about the dissolution of soil salts, evapoconcentration of shallow groundwater, and flow of groundwater to drain laterals. Dissolution of soil evaporite salts containing selenium resulted in transport of selenium to the groundwater during the initial years of irrigation. A subsequent rise in the water table resulted in further evapoconcentration, higher selenium concentrations, and increased soil salinity. Drainage systems were installed to stop the increase of soil salinity. Evapoconcentrated groundwater with high selenium concentrations was displaced downward by higher quality water that infiltrated after the drainage system was installed.

Groundwater flow modeling and geochemical and hydrologic data demonstrate that drainwater is a mixture of older, high-selenium groundwater moving from depths greater than 6 m below land surface and shallow low-selenium groundwater that was more recently recharged. The proportions of these two types of water vary depending on the depth of the drain lateral and recharge. Deeper drain laterals collect deeper groundwater, resulting in larger selenium loads. Higher recharge causes shallower groundwater to flow to the drain laterals.

Subregional observations at a scale between the regional and local scales and in the Panoche Water District provided detailed information about the interaction of the groundwater flow system and drainage systems. Groundwater generally flows downward in the higher altitudes and upwards to drainage systems in the lower altitudes of the subregion. Groundwater flow modeling indicates that over 89% of the drain flow in the drainage systems originates in the drained areas of the District. Variations in dissolved solids and selenium concentrations with depth in groundwater cause temporal variations of dissolved solids and selenium concentrations in drainwater. Varying concentrations are caused by varying depths from which groundwater flows to the drainage systems. These depths are influenced by seasonally variable recharge. Implementation of changes in water and land management strategies occur primarily within individual water districts; therefore, continued development of a quantitative assessment of the groundwater flow system will provide a critical tool for the evaluation and effective development of water and land management alternatives.

REFERENCES

1. W. B. Bull and R. E. Miller, Land subsidence due to ground-water withdrawal in the Los Banos-Kettleman City area, California, Part 1. Changes in the hydrologic environment conducive to subsidence, *U.S. Geol. Surv. Prof. Pap.* 437-E, 1975.

2. R. J. Gilliom and others, Preliminary assessment of sources, distribution, and mobility of selenium in the San Joaquin Valley, California, *U.S. Geol. Surv. Water Resour. Invest. Rep.* 88-4186, 1989.
3. W. B. Bull, Geomorphology of segmented alluvial fans in western Fresno County, California, *U.S. Geol. Surv. Prof. Paper* 352-E, 1964.
4. F. F. Harradine, *Soils of Western Fresno County*, Berkeley, California, Univ. California Press, 1950.
5. K. Belitz and F. J. Heimes, Character and evolution of the ground-water flow system in the central part of the western San Joaquin Valley, California, *U.S. Geol. Surv. Water-Supply Paper* 2348, 1990.
6. W. C. Mendenhall, R. B. Dole, and Herman Stabler, Ground water in San Joaquin Valley, California, *U.S. Geol. Surv. Water-Supply Paper* 398, 1916.
7. G. H. Davis and J. F. Poland, Ground-water conditions in the Mendota-Huron area, Fresno and Kings Counties, California, *U.S. Geol. Surv. Water Supply Paper* 1360-G, 1957.
8. S. J. Deverel, R. J. Gilliom, R. Fujii, J. A. Izbicki, and J. C. Fields, Areal distribution of selenium and other inorganic constituents in shallow ground water of the San Luis Drain service area, San Joaquin Valley, California: A preliminary study, *U.S. Geol. Surv. Water Resour. Invest. Rep.* 84-4319, 1984.
9. S. J. Deverel and S. P. Millard, *Environ. Sci. Technol.* 22(6): 697–702 (1988).
10. S. J. Deverel and S. K. Gallanthine, *J. Hydrol.* 109: 125–149 (1989).
11. H. Craig, *Science* 133: 1702–1703 (1961).
12. S. J. Deverel and R. Fujii, *Water Resour. Res.* 24(4): 516–524 (1988).
13. N. M. Dubrovsky, J. M. Neil, R. Fujii, R. S. Oremland, and J. T. Hollibaugh, Influence of redox potential on selenium distribution in ground water, Mendota, western San Joaquin Valley, California, *U.S. Geol. Surv. Open-File Rep.* 90-138, 1990.
14. R. S. Oremland, J. T. Hollibaugh, A. S. Maest, T. S. Presser, L. G. Miller, and C. W. Culbertson, *Appl. Environ. Microbiol.* 55(9): 2333–2343 (1989).
15. J. M. Macy, T. A. Michael, and D. G. Kirsch, *Fed. Eur. Microbiol. Soc. Microbiol. Lett.* 61: 195–198 (1989).
16. D. A. Leighton, S. J. Deverel, and J. K. Mcdonald, Spatial distribution of selenium and other inorganic constituents in ground water underlying a drained agricultural field, western San Joaquin Valley, California, *U.S. Geol. Surv. Water Resour. Invest. Rep.* 91-4119, 1992.
17. M. W. Bodine, Jr., and B. F. Jones, The salt norm; a quantitative chemical-mineralogical characterization of natural waters, *U.S. Geol. Surv. Water Resour. Invest. Rep.* 86-4086, 1986.
18. S. J. Deverel and J. L. Fio, *Water Resour. Res.*, 27(9): 2233–2246 (1991).
19. J. L. Fio and S. J. Deverel, *Water Resour. Res.* 27(9): 2247–2257 (1991).
20. L. S. Willardson, B. D. Meek, L. B. Grass, G. L. Arkey, and J. W. Baily, *Groundwater* 8: 11–13 (1970).
21. San Joaquin Valley Drainage Program, A management plan for agricultural subsurface drainage and related problems on the westside of San Joaquin Valley, *San Joaquin Valley Drainage Program*, 1990.

22. K. Belitz, S. P. Phillips, and J. M. Gronberg, Numerical simulation of groundwater flow in the central part of the western San Joaquin Valley, California, U.S. Geol. Surv. Water Supply Paper 2396, 1993.
23. T. W. Dibblee, Jr., Geologic maps of the Pacheco Pass, Hollister, Quien Sabe, Ortigalita Peak, San Benito, Panoche Valley, and Tumey Hills quadrangles; San Benito, Santa Clara, Merced and Fresno Counties, California, U.S. Geol. Surv. Open-File Rep. 75-0394, 1975.
24. T. S. Presser, W. C. Swain, R. R. Tidball, and R. C. Severson, Geologic sources, mobilization, and transport of selenium from the California Coast Ranges to the western San Joaquin Valley: A reconnaissance study, U.S. Geol. Surv. Water Resour. Invest. Rep. 90-4070, 1990.
25. D. W. Pollock, Documentation of computer programs to compute and display pathlines using results from the U.S. Geological Survey modular three-dimensional finite-difference ground-water flow model, U.S. Geol. Surv. Open-File Rep. 89-381, 1989.

8
Chemical Oxidation–Reduction Controls on Selenium Mobility in Groundwater Systems

Arthur F. White

U.S. Geological Survey
Menlo Park, California

Neil M. Dubrovsky

U.S. Geological Survey
Sacramento, California

I. INTRODUCTION

Selenium (Se) toxicity has recently been associated with anthropogenic activities including disposal of agricultural drainage water, mine tailings, and coal-generated fly ash [1–3]. The complex chemistry of Se has complicated efforts to predict its behavior and to remediate its detrimental effects on water quality. In the environment, Se occurs in multiple oxidation states that are subject to nonequilibrium inorganic oxidation–reduction (redox) processes and complex biogeochemical cycling. Intensive studies on the behavior of Se have centered on issues of surface water and groundwater contamination in the San Joaquin Valley of California. The geologic setting and climate of this region have created soil salinization problems accentuated by irrigation. The leaching and transport of indigenous Se have caused serious biological effects in wildlife habitats, curtailment in agricultural productivity, and contamination of aquifers.

This chapter focuses on the geochemistry of Se in groundwater systems and processes affecting transport and retardation. The geochemistry of such systems is subject to wide variations in redox states and therefore aqueous speciation and microbial processes. The chapter evaluates the effects of these variables on Se transport in different geochemical environments. Two case studies, one in the agricultural source area in the San Joaquin Valley and the other in the engineered discharge area at Kesterson Reservoir, illustrate the interdependency of many of these redox processes and their controls on Se movement in groundwater.

II. SELENIUM SPECIATION AND REDOX CHEMISTRY

Mobility of Se in groundwater systems is directly related to speciation in aqueous solution, sorption properties of the aquifer substrate, and solubility with respect to solid phases. The chemistry of Se, which can occur in four oxidation states under natural conditions, is analogous to the chemistry of sulfur. Like sulfur, Se can exist in the Se(VI) oxidation state as selenate (SeO_4^{2-}), in the Se(IV) oxidation state as selenite (SeO_3^{2-}), in the zero oxidation state as elemental Se(O), and in the Se(−II) oxidation state as selenide (Se^{2-}). In its two higher oxidation states (VI and IV), selenium occurs in the natural environment as aqueous species or solid compounds. Selenium also exists as crystalline and amorphous polymorphs in the elemental state. Finally, reduced selenide can be present as aqueous species, as solids bound with transition metals, and as gaseous forms of hydrogen selenide. As discussed by Presser and Swain [4], Se can be viewed as cycling geochemically between species in these different oxidation states, depending on external biogeochemical constraints. At ambient conditions, Se or metallic selenide (FeSe) in parent rocks weather to soluble selenite under acidic oxidizing conditions and to selenate under alkaline oxidizing conditions. These species can then be reconverted to reduced forms of Se in anoxic groundwaters from which dissolved oxygen (DO) and other oxidizing compounds have been excluded.

A. Aqueous Speciation

The stabilities of the different states of Se are related to the electrochemical potential of the solution. This potential is expressed as millivolts (mV) relative to the standard hydrogen electrode [5]. In addition, the various aqueous oxidation states of Se exist as base species (i.e., SeO_4^{2-}) or protonated forms of these species (i.e., $H_2SeO_4^{2-}$) and as a variety of inorganic complexes, (i.e., $CaSeO_4^0$) and organic complexes. The thermodynamic properties of the inorganic aqueous species and their solubilities relative to end-member solid phases have been tabulated by Elrashidi et al. [6]. Using these data, the major characteristics of aqueous Se speciation for typical drainage water in the San Joaquin Valley of California (Table 1) can be calculated [7]. Figure 1a shows the log concentration of aqueous species versus the electrochemical potential of the solution expressed in millivolts. The calculation corresponds to a total Se concentration of 300 µg/L at a solution pH of 7. Under the most oxidizing conditions, indicative of oxygen-saturated groundwater (>400 mV), dissociated selenate (SeO_4^{2-}) and $CaSeO_4^0$ ion pairs are the dominant species in solution. In moderately oxidizing groundwater (400–0 mV), selenite stability is dominated by $CaSeO_3^0$ and $HSeO_3^-$. Under strongly reducing and neutral pH conditions (<0 mV), selenide stability is dominated by HSe^-. Specific ion pairs and the extent of protonation of Se oxyanions vary with different water chemistry. However, the general oxidation state distributions, as functions of the electrochemical potential, remain comparable.

Besides inorganic forms of Se, nonvolatile organic selenides such as selenoamino acids and volatile methylated forms of selenides, principally dimethyldiselenide, can occur in surface water and groundwater [8]. Documented occurrences of nonvolatile organic species in groundwater are generally limited to dimethylselenonium, an organic decomposition product. Dimethylselenonium has been reported to make up as much as 20% of the total Se in several measurements of groundwater at Kesterson [9]. Volatile dimethylselenide was reported to account for up to 20% of total Se in the Kesterson ponds and 35% in one groundwater well [9]. Sampling of additional wells revealed lower proportions of dimethyldiselenide, generally <10% in surface waters and <1% in monitoring wells [10,11]. Recent experimental work shows that dimethyldiselenide is unstable in the presence of sulfate-respiring bacteria that occur in anoxic muds and wetlands [12,13].

B. Solubility Controls

Mineral phases that equilibrate with aqueous Se species include elemental Se and alkaline earth and transition metal selenates, selenites, and selenides. Figure 1b shows the saturation index of specific Se-containing minerals plotted as a function of electrochemical potential. The saturation index is the ratio of the ionic activity product (IAP) of the aqueous species divided by the solubility product of the Se-containing mineral (K_s). Chemical conditions in the calculations are the same as for the aqueous speciation computations in Figure 1a. If the log ratio of IAP to K_s is less than zero, the groundwater is unsaturated with respect to the specific solid, and thermodynamics predicts that the mineral will dissolve. If the ratio is greater than zero, the water is supersaturated and Se would be expected to

Table 1 Representative Chemical Compositions of Waters Discussed in Text[a]

Site	Drainage water Pond 2	Mendota groundwater P1-24.5	Mendota groundwater P1-45	Mendota groundwater P1-60	Mendota groundwater D9W2	Mendota groundwater LBL 31	Kesterson groundwater D2W2	Kesterson groundwater KR31	Kesterson groundwater LBL 38	Kesterson groundwater LBL 14
Date	4/24/85	4/23/87	4/22/87	4/23/87	4/2/85	6/9/86	3/10/86	6/25/86	10/3/85	10/4/85
Depth	—	8.1	10.8	20.1	16.2	10.0	4.5	8.2	7.1	7.1
Temp	17	21.0	21.5	22.0	16	14	14	14	18	21
Na	2470	2200	4500	560	2860	2430	2910	3500	1020	2130
K	7.4	4.6	13	2.7	11.1	5.9	6.8	6.9	4.5	7.3
Ca	574	130	580	34	418	408	763	264	97	250
Mg	316	130	790	3	39	336	415	243	174	576
Cl	1600	2600	8500	670	2800	1580	2020	1980	920	2750
HCO_3	198	316	180	134	750	202	380	311	530	376
SO_4	5900	1600	5100	360	3100	4330	5930	5160	950	1690
B	15.7	4.6	7.0	0.9	14.9	14.7	18.3	16.1	0.8	0.6
SeO_3	0.052	0.019	0.000	0.000	0.001	0.001	0.003	0.011	0.001	0.001
Se_{tot}	0.314	0.480	0.000	0.000	0.001	0.002	0.080	0.094	0.001	0.001
pH	8.75	7.85	7.23	8.38	7.65	6.86	6.80	7.60	7.35	6.80
Eh	+613	+391	+130	+97	+209	+414	+442	+442	+299	+134
Fe^{2+}	0.000	0.020[b]	14.000[b]	0.250[b]	0.410	0.000	0.000	0.000	1.02	1.50
Fe^{3+}	0.000	—	—	—	0.000	0.000	0.000	0.000	0.000	0.000
Mn	0.001	0.45	11.0	0.21	8.20	10.80	6.78	4.24	0.88	4.70
DO	9.30	0.2[c]	0.2[c]	0.2[c]	0.00	0.10	0.20	0.09	0.00	0.02
H_2S	0.000	—	—	—	0.000	<0.4	0.000	0.000	0.010	0.009
NO_3	43.7	1.5	<0.1	<0.1	<0.4	<0.4	43.0	23.5	<0.4	<0.4

[a]Drainwater analyses correspond to agricultural drainage input to the Kesterson Reservoir from the San Luis Canal. Mendota and Kesterson analyses are for pumped water samples from nested wells. Concentrations in milligrams per liter, depths in meters, temperature in degrees Celsius, and Eh in millivolts relative to the hydrogen electrode. Selenate concentrations discussed in text are calculated as the difference between analyzed concentrations of selenite (SeO_3) and total selenium (Se_{total}).
[b]Samples analyzed for total Fe only.
[c]Samples analyzed by oxygen electrode; other DO analyses by colorimetric methods.

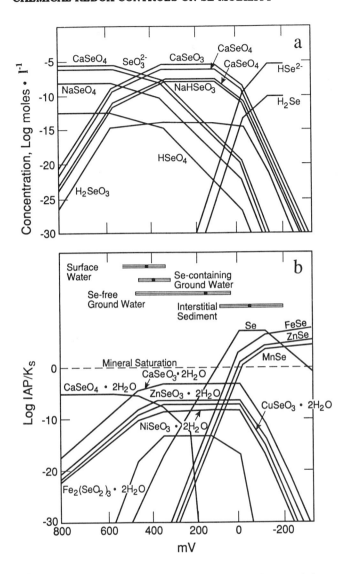

Figure 1 (a) Aqueous Se speciation and (b) degree of chemical saturation with respect to selenium minerals plotted as a function of electrochemical potential. The horizontal dashed line represents chemical equilibrium between the aqueous and solid phases defined as the ratio where the ionic activity product in solution (IAP) equals the solubility product of the specific mineral phase (K_s). Calculations are for agricultural drainwater at Kesterson Reservoir (Table 1) assuming a concentration of 300 μg total Se per liter at pH 7. Horizontal bars correspond to measured Eh values for different geochemical environments. (Modified from Ref. 7.)

precipitate out of solution. Figure 1b shows that the selenate phase, $CaSeO_4 \cdot 2H_2O$, is undersaturated by approximately five orders of magnitude under strongly oxidizing conditions. Comparable undersaturation exists for $CaSeO_3 \cdot 2H_2O$, the most insoluble selenite phase. These calculations indicate that selenate and selenite concentrations in these oxic groundwaters are not controlled by solubility constraints. Under differing oxidizing chemical conditions, other selenate or selenite phases may be more insoluble. For example, the insoluble basic ferric selenite $[Fe_2(OH)_4SeO_3]$ is reported in selenium-rich alkaline soils of Hawaii [14].

Aqueous Se concentrations are strongly limited by solubility controls under reducing conditions. At concentrations comparable to those of the oxidized species, reduced aqueous Se is significantly supersaturated with respect to elemental Se (Figure 1b) and metal selenide mineral phases. Selenium would therefore be expected to precipitate from solution. The reduction of selenate to crystalline elemental Se can be written as the electrochemical half-cell

$$SeO_4^{2-} + 6\,e^- + 8\,H^+ \rightarrow Se_{(c)} + 4\,H_2O \quad (+270\text{ mV, pH 7}) \tag{1}$$

with a standard potential of +874 mV [6]. The potential shown in parentheses in reaction (1) is calculated for water containing a total of 300 μgSe/L at pH 7. Reaction (1) predicts that solid-state elemental Se predominates at potentials less than +270 mV and that aqueous selenate and selenite would be stable at potentials greater than +270 mV. The ranges in platinum electrode redox potentials for waters at the Kesterson Reservoir are plotted in Figure 1b. The oxidizing surface waters in the ponds plot at potentials higher than elemental Se saturation. Groundwaters, which are generally anoxic, approach Se saturation at their lowest measured potentials, and interstitial waters in the strongly anoxic muds in the pond bottoms are clearly oversaturated with respect to Se.

The above calculations are based solely on pure Se mineral phases. Selenium can also substitute as a trace element in sulfur-containing phases such as thenardite (Na_2SO_4), mirabilite ($Na_2SO_4 \cdot 10\,H_2O$), and bloedite $[Na_2Mg(SO_4)_2 \cdot 4\,H_2O]$ [4] and in metal sulfides. In addition, solid-state Se can exist as Se(0) and Se(–II) in organic compounds as documented by Weres et al. [15] for organic-rich sediments.

C. Sorption Controls

In addition to solubility constraints, Se can be removed from groundwater by sorption processes that are also dependent on speciation. Experiments show that selenate adsorbs only weakly on clays, iron oxides, and other soil components at neutral pH [16]. In addition, this limited sorption capability is found to be strongly suppressed by competitive effects with SO_4^{2-}. Sorption is therefore a poor candidate for retarding SeO_4^{2-} in sulfate-rich groundwaters commonly associated with elevated Se concentrations. Other experimental studies have shown that selenite adsorbs more strongly onto soil substrates that selenate and is essentially independent of SO_4^{2-} concentrations [17–19]. This difference in sorption affinities explains why the selenate/selenite ratios are often higher in groundwaters than in surface waters. An example of these phenomena was shown by White et al. [7] when they compared surface drainage water and infiltrated drainage waters in the shallow alluvial aquifer beneath Kesterson Reservoir (Figure 2). The low selenite concentrations in the groundwater resulted from a strong sorption affinity in the aquifer.

D. Relationship to Groundwater Redox Couples

The actual redox state of a groundwater system, as represented by the potentials shown in Figure 1, is often difficult to assess quantitatively.

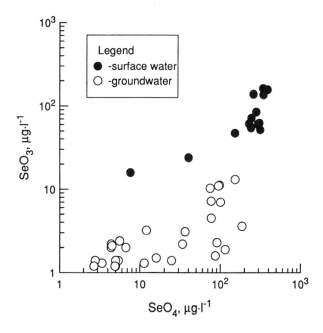

Figure 2 Comparison of log concentrations of selenate and selenite in surface water and groundwaters from Kesterson Reservoir. (Modified from Ref. 7.)

Apparent redox disequilibrium occurs in groundwater due to (a) chemical and/or biological kinetics, (b) mixed potentials caused by different redox couples that are not in thermodynamic equilibrium, and (c) precipitation or oxidation reactions at the surface of the platinum (Pt) electrode that is commonly employed as the redox measurement technique [20,21]. Measured abundances of redox-sensitive species, therefore, often do not correspond quantitatively to speciation predicted from thermodynamic calculations. For example, White et al. [7] showed a general lack of correlation between the observed selenate–selenite speciation in oxic groundwater and measured redox potentials using a Pt electrode. The half-cell reaction describing this relationship is

$$SeO_4^{2-} + 2\,e^- + 3\,H^+ \rightarrow HSeO_3^- + H_2O \quad (+450\,mV,\,pH\,7) \quad (2)$$

with a standard potential of +895 mV [22]. Potentials based on measured SeO_4^{2-}/SeO_3^{2-} concentration ratios and corrected for complexation and activities using the EQ3NR speciation model [23] are compared to measured Eh potentials in Figure 3. The potentials for the SeO_4^{2-}–SeO_3^{2-} couple [reaction (2)] generally fall below the diagonal equipotential line (Figure 3). This relationship implies that the ratio of SeO_4^{2-} to SeO_3^{2-} measured in groundwater is higher than predicted on the basis of the Pt electrode measurements. Redox disequilibrium is the result of slow rates of reaction between aqueous species [24].

According to thermodynamics, the speciation of Se in groundwater should be related quantitatively to the speciation of other redox-sensitive elements in solution. The presence of dissolved oxygen (DO) and NO_3^- are indicative of oxidizing conditions in groundwater. The reduction potential for DO can be described by the half-cell reaction [25]

$$1/2\,O_{2(aq)} + 2\,H^+ + 2\,e^- \rightarrow H_2O \quad (+800\,mV,\,pH\,7) \quad (3)$$

The expected cell potential of +800 mV, based on air-saturated groundwater (8 mg/L) at neutral pH and 25°C, is generally much more oxidizing than Eh values measured using a Pt electrode [7,26].

The NO_3^- ion can be reduced to several species including N_2O, N_2, and NH_4^+. The half-cell reaction for reduction to ammonium ion can be written as

$$NO_3^- + 10\,H^+ + 8\,e^- \rightarrow NH_4^+ + 3\,H_2O \quad (+350\,mV,\,pH\,7) \quad (4)$$

Both this potential and the half-cell potentials for DO [reaction (3)] are more positive than the potential describing the reduction of SeO_4^{2-} [reaction (1)]. Therefore, thermodynamic considerations, excluding kinetic factors, would predict that the presence of either species in appreciable

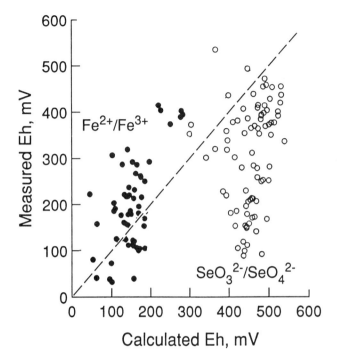

Figure 3 Comparison of electrochemical potentials measured by the platinum electrode and calculated for (○) the selenate–selenite couple and (●) the ferrous iron–ferric hydroxide couple for groundwater at Kesterson Reservoir. A direct correlation between the two methods is represented by the dashed diagonal line. (Modified from Ref. 7.)

quantities would prevent the reduction of SeO_4^{2-} and the precipitation of elemental Se.

Other chemical species indicative of reducing conditions in groundwater are Fe^{2+}, Mn^{2+}, and H_2S. The corresponding reducing reactions involving these species are

$$Fe(OH)_3 + 3\,H^+ + 3\,e^- \rightarrow Fe^{2+} + 3\,H_2O \quad (+120\text{ mV, pH 7}) \quad (5)$$

$$MnOOH + 3\,H^+ + e^- \rightarrow Mn^{2+} + 2\,H_2O \quad (-196\text{ mV, pH 7}) \quad (6)$$

and

$$SO_4^{2-} + 9\,H^+ + 8\,e^- \rightarrow HS^- + 4\,H_2O \quad (-200\text{ mV, pH 7}) \quad (7)$$

Potentials for the reaction (5) half-cell assume equilibrium between aqueous Fe^{2+} and Fe^{3+} in equilibrium with iron hydroxide. At higher measured Fe^{2+} concentrations, this relationship often approximates the redox potentials measured by the Pt electrode (Figure 3). This correlation implies that Fe^{2+} is the electrochemically active species controlling Eh measurements under reducing groundwater conditions. Owing to the insolubility of iron sulfide, dissolved sulfide is not expected to be detected when measurable concentrations of Fe^{2+} are present. The potentials generated by the half-cell reactions for Fe^{2+}, Mn^{2+}, and H_2S [reactions (5)–(7)] are more negative than the half-cell potentials describing the reduction of SeO_4^{2-} to elemental Se [reaction (1)]. Therefore, these species are capable thermodynamically of reducing SeO_4^{2-} to elemental Se. However, their ultimate impact on Se mobility in groundwater depends on the rates at which these redox reactions occur.

III. CASE STUDIES OF SELENIUM MOBILITY IN THE SAN JOAQUIN VALLEY

The geochemical processes described in the preceding section have important ramifications in understanding Se mobility in groundwater systems. Specific issues related to redox chemistry will be illustrated by considering two case studies of Se mobility in aquifers in the San Joaquin Valley of California (Figure 4).

A. Selenium Distributions in the Agricultural Source Areas

Total selenium concentrations, as high as 3000 µg/L, occur in shallow saline groundwaters in the Panoche Creek alluvial fan that extends from the eastern flanks of the Coast Range eastward into the San Joaquin Valley (Figure 4). These waters have evolved chemically by leaching of natural soil salts and subsequent concentration by evapotranspiration [27]. Application of large quantities of irrigation water has caused a rapid expansion of areas affected by shallow water tables and salt accumulation [28]. An important concern for water management in the area is the potential for downward movement of these shallow Se-rich waters into deeper aquifers that are pumped extensively for agricultural and domestic purposes.

1. Hydrogeology at Mendota

Dubrovsky et al. [29] characterized the vertical distribution of Se in shallow groundwater at a site at the eastern edge of the Panoche Creek alluvial fan at Mendota less than 1 km southwest of the San Joaquin River

CHEMICAL REDOX CONTROLS ON SE MOBILITY 195

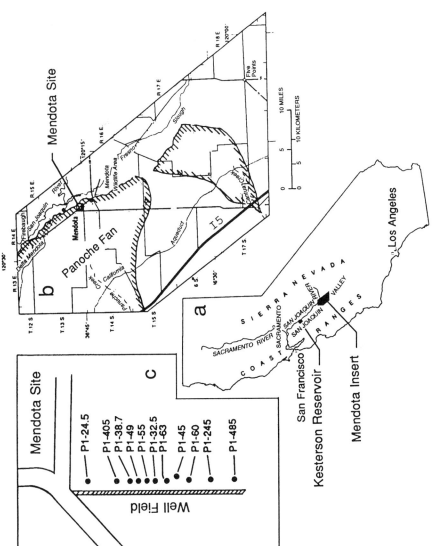

Figure 4 (a) Location map for the Mendota and Kesterson study areas; (b) regional map of the western part of the San Joaquin Valley; and (c) well field map of the Mendota site (not to scale).

(Figure 4). Historically, this area was a discharge zone with high salinities resulting from near-surface evapotranspiration. Subsequent groundwater exploitation depressed water levels more than 20 m below land surface. Since 1967, water levels have recovered partially owing to decreased pumping resulting from the importation of surface water for irrigation. The site is currently a recharge zone, with seasonal water use resulting in water table fluctuations of greater than 2 m [29].

Surficial geology of the site consists of fine-grained Coast Range sediments deposited at the toe of the Panoche Creek alluvial fan [30]. The upper 4 m of sediments at the site are fine-grained layers of clay and silt alternating with thin layers of sand, some of which show strong iron(III) hydroxide staining. With increasing depth, mottling occurs in reduced, gray, medium- to coarse-grained sands predominating at depths greater than 8 m. On the basis of high salinity, geomorphic location, and mineralogy, Dubrovsky et al. [29] concluded that most of the shallow fine-grained sediments were derived from Coast Range material. The deeper sands were a mixture of river channel and alluvial fan sediments derived principally from the Sierra Nevada. Coast Range sediments are both higher in Se concentrations and more oxidizing than the Sierra Nevada sediments. The difference in redox conditions is attributed partly to the higher proportion of reduced Fe-containing minerals, including biotite, hornblende, and magnetite in the deeper, dominantly granitic sands.

2. Selenium Distributions

Selenium concentrations, salinity, and other chemical parameters were characterized at the Mendota site from chemical sampling of groundwater. The samples were pumped from a series of wells screened between 8 and 80 m and from hydraulic extraction of porewater from three cores at the site using the method of Manheim [31]. Selected groundwater chemical data from the monitoring wells are listed in Table 1. Both groundwater and extracted porewaters (Figure 5) show a zone of variable salinity with depth at the site. This feature is caused by the infiltration of shallow surface water that has been concentrated chemically by evapotranspiration and redissolution of soil salts produced by earlier cycles of desiccation and fluctuations in water level [29]. The highest salinity occurs in the upper zone of the Sierran sediments at a depth of approximately 15 m. The decrease in dissolved solids at greater depths is the result of mixing of the shallow, saline groundwater with deep groundwater that has lower dissolved solids concentrations [29].

Elevated Se concentrations were detected only in the shallowest monitoring well with a midpoint screen depth of 8.1 m (Table 1). Selenium

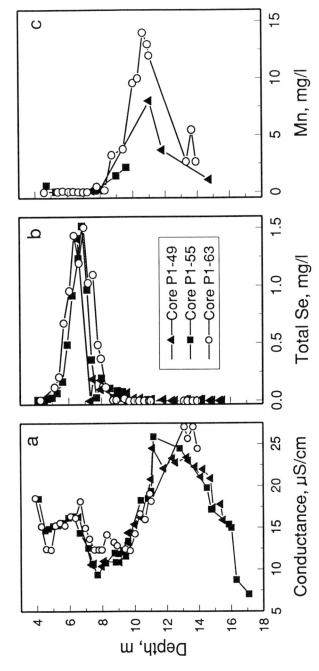

Figure 5 Vertical distributions of (a) specific conductance, (b) total dissolved Se, and (c) Mn in porewaters at the Mendota site. (Modified from Ref. 29.)

concentrations from monitoring wells at depths of greater than 8 m were less than the detection limit of 1 μg/L. More detailed analyses for hydraulically extracted porewater showed that concentrations of total Se exceeded 1500 μg/L in a narrow zone at a depth of 7 m (Figure 5). Selenium concentrations decreased rapidly to less than 100 μg/L at depths greater than 8 m. Analysis of Se species showed that essentially all of the Se in solution was SeO_4^{2-}. The concentration of SeO_3^{2-} was less than 1 μg/L in all of the groundwater samples except for 19 μg/L found in the shallowest well (8.1 m). A concentration of 9 μg/L was determined for a pore sample from a comparable depth (8.3 m). These data, like those for Kesterson (Figure 2), show that SeO_4^{2-} is the most mobile Se species in groundwaters in the San Joaquin Valley. This is due to a high solubility and a low sorption affinity.

In other areas of the Coast Range alluvial fan, Se concentrations in the surface water and groundwater correlate directly with specific conductance measurements and concentrations of the dominant anions, Cl^- and SO_4^{2-} [27]. This correlation is due to concentration effects caused by evaporation and redissolution of these soluble species. High salinities at the Mendota site (Figure 5) also indicate that evaporation has significantly increased chemical concentrations in waters infiltrated to depths less than 15 m. This is also documented by evaporative trends in deuterium-^{18}O data [29]. However, the lack of corresponding Se below approximately 8 m indicates that SeO_4^{2-} has been removed from the groundwater below this level. The mobility of Se is therefore strongly dependent on specific aquifer chemistry.

3. Correlation with Redox Chemistry

The redox potential of the groundwater at the Mendota site was determined by Dubrovsky et al. [29] from Pt electrode measurements and from the abundance of dissolved constituents sensitive to redox conditions. Platinum electrode potential measurements were made by pumping groundwater samples from wells through a flow-through cell that excluded atmospheric contamination. Measurements showed a maximum of 391 mV at 8 m. This voltage decreased to a minimum of 66 mV at 17 m, indicating that the groundwater became more reducing with depth (Figure 6).

Dissolved oxygen and NO_3^- are electron acceptors and are therefore indicative of oxidizing conditions [reaction (3)]. DO concentrations in groundwater at the Mendota site, measured using a flow-through cell, were <1 mg/L in all wells except the shallowest well at 8 m (Table 1). DO in the shallowest well ranged from 0.2 to 7.9 mg/L between 1985 and 1988.

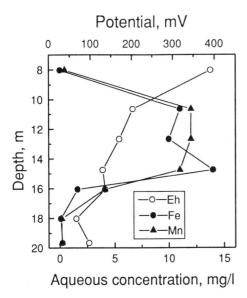

Figure 6 Platinum electrode redox potentials and concentrations of total Fe and Mn in groundwater at the Mendota site. (Modified from Ref. 29.)

This variation was due to water table fluctuations at or near this depth. These fluctuations affected the penetration of atmospheric O_2 and therefore the depth of oxidizing conditions in the sediments. This effect is also evident in the mottling of the sediments by iron oxyhydroxide staining in this zone. Nitrate concentrations were below the detection limit (<0.1 mg/L) in all wells except the shallowest 8-m well, in which NO_3^- varied between 1.0 and 1.5 mg/L. Nitrate in porewater samples taken above the water table showed a maximum concentration of 19.5 mg/L at 3.5 m. Nitrate decreased to less than 0.1 mg/L below 15 m. DO and NO_3^- data showed that the soil water was oxidizing above the water table. However, these electron acceptors became depleted at depths greater than 1 m below the water table. The data also indicated that elevated Se concentrations in groundwater and porewater were associated with measurable DO and NO_3^- concentrations near the water table.

The concentrations of Fe and Mn in oxidizing, near-neutral-pH waters were limited by the low solubilities of Fe(III) and Mn(III) oxyhydroxides [reactions (5) and (6)]. Therefore, the presence of Fe^{2+} and Mn^{2+}, measured as total Fe and Mn, was a good indicator of reducing redox potentials. Concentrations of Fe in groundwater samples showed nearly two orders of magnitude variation in the monitoring wells, ranging from 0.20 mg/L at a depth of 7 m to a maximum of 14 mg/L at 14 m (Figure 6). The vertical distribution of dissolved Mn was similar. A minimum concentration of 0.45 mg/L occurred at the 7-m depth, and a maximum concentration of 12 mg/L at 10 m (Figure 6). The location of the transition zone from oxidizing to reducing conditions was defined most clearly by the profile of Mn concentration in porewater samples (Figure 5). The porewater data showed that Mn concentrations began to increase at depths below 7.5 m, then rose rapidly to 14 mg/L below 11 m. The data showed that the Se concentrations decreased rapidly over the same depth interval as Mn concentrations increased. Decreasing concentrations of Fe and Mn at depths greater than 15 m were due to mixing with more dilute, deeper groundwater.

High salinities persisted to depths of 5–6 m below the interface at which the groundwater became reducing (Figures 5 and 6). However, the elevated Se concentrations initially associated with evaporative concentration of these waters were subsequently removed by reduction [reaction (1)]. Removal of Se from solution implies that Se accumulates in the solid phase in the aquifer due to precipitation and/or sorption. Loss of Se by degassing of methylated SeO_2 is unlikely below the water table at the site owing to its low redox potential and competing anaerobic demethylation reactions [32; see also chapter 15, this volume]. Sediment samples were collected across the oxic–anoxic groundwater interface to determine if higher concentrations of solid-state Se existed in the reducing zone of the aquifer below the peak Se concentration in the groundwater [29]. However, the results indicated that Se in the sediments exhibited only minor variations with depth (Table 2). These Se concentrations were comparable to maximum and minimum concentrations estimated for Sierra Nevada sediments (<0.15 mg/kg) and Coast Range sediments (>0.30 mg/kg), respectively, in the San Joaquin Valley [33]. The lack of an apparent accumulation of solid Se below the redox interface may be due to dispersion in a broad zone in the sediments. In addition, if the redox zone reflects only the transitory water table based on recent water use history, the mass of precipitated Se may be small.

B. Selenium Distributions Beneath the Kesterson Reservoir

The Kesterson Reservoir is located on the western side of the San Joaquin Valley of California (Figure 4). The description of the reservoir and the associated environmental impact of Se are discussed in detail elsewhere in

Table 2 Comparisons of Selenium Concentrations in Solids and Porewaters from Mendota Alluvium

Depth (m)	Selenium total in solid (µg/g)	Selenium total in porewater (µg/L)
6.58	0.4	1410
6.86	0.6	1520
7.35	0.4	412
7.71	0.2	33
8.11	0.3	104
8.99	0.1	93
9.63	0.2	9
11.13	0.1	13
13.38	0.4	5
15.51	0.2	2

Source: After Ref. 29.

this volume [Chapters 4 and 5]. Briefly, the Kesterson Reservoir was constructed by the U.S. Bureau of Reclamation in 1972 as a regulating reservoir for the San Luis Drain, the master drain for transporting agricultural drainage water from the west side of the San Joaquin Valley. Completion of the drain to a final discharge point into the Sacramento Delta was stopped because of financial constraints and environmental concerns. A series of shallow unlined ponds, the Kesterson Reservoir became the final storage facility for the drainage water. Agricultural discharge to the reservoir was stopped in 1986 because of the high waterfowl mortality rates associated with dissolved Se and the potential for surface water and groundwater contamination. The ponds were subsequently flooded with local low-selenium groundwater pumped from wells along the eastern periphery of the reservoir. In 1989, pumping was stopped, the ponds were dried out, and the containment structures were leveled. A detailed well-monitoring program characterized Se concentrations in the groundwater system and the redox controls on Se mobility during this period [7,10,11,34].

1. Hydrogeology at Kesterson

Due to a more easterly location relative to the Mendota site, Coastal Range sediments are generally absent at Kesterson. Sierra-derived deposits in the shallow aquifer beneath the ponds consist predominantly of well-sorted micaceous sands [34]. These sandy units represent channel and overbank deposits from the San Joaquin River system and are similar to the deeper sands beneath the Mendota site. Mineralogic compositions of the Sierra sands were dominated by quartz and feldspars with lesser amounts of hornblende, pyroxene, sphene, and allanite. Thin layers of darker interbedded fine sand contained higher concentrations of Fe-containing biotite and magnetite. The sandy units are separated by discontinuous beds of silt and clays, classified as sandy loams or clay loams with clay components consisting of smectite and kaolinites [34].

Extensive pumping and infiltration tests were performed on the shallow aquifer [34]. Results suggested that the permeability of the clay units was about three orders of magnitude less than that of the sand units. Therefore, drainage water infiltration rates from the overlying ponds were expected to be more rapid where clay layers were thin or absent. One such area was located on the western side of Pond 2 (Figure 7). Here, maximum penetration depths of drainage water exceeded 50 m based on distributions of tritium and conservative chemical components including chlorine (Cl) and boron (B) [7]. During operation of the reservoir, a groundwater mound formed under the flooded ponds. This recharge caused groundwater to migrate away from the reservoir in all directions and to overcome the regional hydrologic gradient. The lateral extent of this drainage water infiltration was assessed by Benson et al. [34]. Based on the 10 mg/L boron isograd shown in Figure 7, most of the ponds were underlain by drainage water. Areas northeast of the reservoir had boron concentrations indicative of uncontaminated native groundwater (Figure 7). Inaccessibility to private land to the east of the reservoir precluded groundwater monitoring. However, geophysical resistivity measurements did indicate a plume of high-salinity water extending eastward approximately 300 m.

2. Selenium Distributions in Groundwater

The chemistry of the shallow groundwater system at Kesterson was classified into two types based on origin [7]: (a) native groundwater indigenous to the local area and (b) drainage groundwater that infiltrated from the overlying ponds. The chemical concentrations of the drainage waters were consistently higher than those of native groundwater due to evaporation in the agricultural source areas and subsequently in the Kesterson ponds. Linear correlations were apparent between major chemical species in drainage water, both in the surface ponds and in that which infiltrated into the alluvial aquifer. An example of the linear relationship between SO_4^{2-} and B concentrations is shown in Figure 8a. However, no

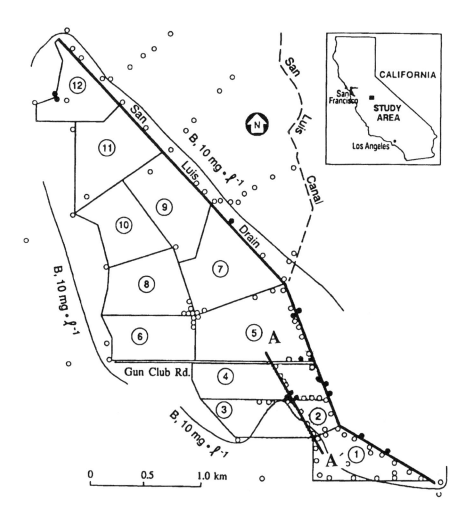

Figure 7 Location of Kesterson Reservoir. Numbers refer to pond designations. Solid circles were monitoring wells that contained in excess of 10 μg total Se per liter. Open circles represent wells in which Se concentrations were consistently below 10 μg/L. The 10 mg/L boron isograd defines the lateral extent of drainage water infiltration. Line A-A' corresponds to the alignment of the cross sections shown in Figure 9 (not to scale).

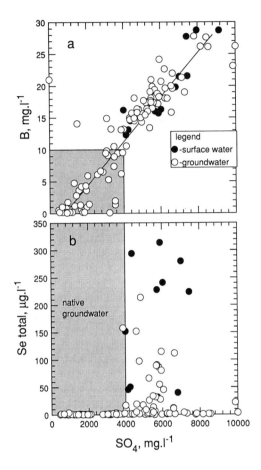

Figure 8 Concentrations of (a) boron and (b) total selenium plotted against SO_4 in Kesterson surface water and groundwater. Shaded boxes define concentration ranges that are exclusive of surface water compositions and denote either native groundwater or mixtures of native groundwater and drainage water. (Modified from Ref. 7.)

linear correlations existed between the concentration of total dissolved Se species and major chemical species such as SO_4^{2-} (Figure 8b). As at the Mendota site, this lack of correlation with evaporative increases in major chemical species indicates the preferential loss of Se from groundwater. The loss explains the limited and discontinuous distributions of Se [7] in drainage water beneath Kesterson (Figure 7). Less than 10% of the Kesterson monitoring wells ever contained Se concentrations greater than

10 µg/L, the EPA drinking water standard. These wells are differentiated with respect to location in Figure 7.

The most consistent occurrence of Se was along the western periphery of Pond 2 (Figure 7). Vertical distributions of Se and B along a NW–SE-tending cross section on the west side of Pond 2 are shown in Figure 9. Deep penetration of drainage water in this area, as evidenced from the B distributions, may have been due to local discontinuity in the clay layers. Selenium concentrations decreased in the groundwater as a function of depth and always reached background Se levels of <1 µg/L at depths of >20 m (Figure 9a). The limited vertical and horizontal distribution of Se compared with extensive drainage water infiltration at this site is additional evidence for the immobilization of Se in shallow groundwater.

3. Temporal Variations in Selenium Concentrations

Selenium and boron data for drainage water flowing in and out of Pond 2 from mid-1984 to 1989 are plotted in Figures 10a and 11a, respectively. Surface waters generally contained 300–450 µg Se/L and 10–20 mg B/L from 1984 to the initiation of groundwater pumping in July 1986. Fluctuations in composition were due principally to seasonal effects of irrigation and evaporation of the drainage water. The replacement of drainage water by pumped groundwater in the ponds is indicated by the precipitous drop in surface water Se concentrations after July 1986 (Figure 10a). This decrease indicated that minimal Se was being remobilized from the underlying Se-contaminated sediments. The B concentrations exhibited a corresponding rapid drop, approaching pumped source water concentrations of approximately 5 mg/L (Figure 11a).

Selenium and boron distributions in groundwater responded over time to changes in the surface water chemistry (Figures 10b and 11b). Nested wells located on the western margin of Pond 2 exhibited B concentrations comparable to those for surface drainage in Pond 2 from March 1984 to July 1986 (Figures 11a and b). In contrast, the earliest detection of significant Se concentrations at a groundwater site did not occur until November 1984 (Figure 10b), approximately 6 months after installation of the monitoring wells and more than $1^1/_2$ years after high surface water Se concentrations had been first reported [35]. After the initial breakthrough, Se concentrations increased most rapidly in the shallowest well, more slowly in the deeper wells, and did not reach the deepest well until June 1985. Se concentrations in deeper wells also remained consistently lower than the corresponding shallower wells, indicating that Se was being immobilized with increasing depth of groundwater.

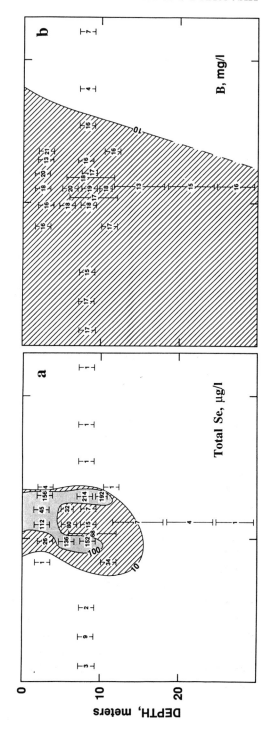

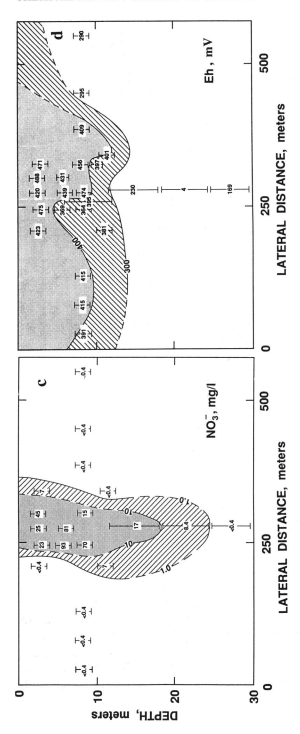

Figure 9 Vertical distributions for (a) total selenium, (b) boron, (c) NO$_3$, and (d) Eh for groundwaters sampled along the western edge of Pond 2 in May 1986. The horizontal axis is centered on the intersection of Ponds 2, 3, and 4 and oriented along the line segment A-A' shown in Figure 7 (not to scale). Vertical bracketed lines correspond to well screen intervals. (Modified from Ref. 7.)

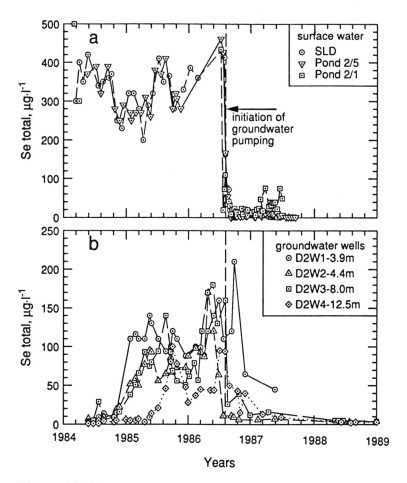

Figure 10 Temporal variations in selenium concentrations (a) in surface water discharged from the San Luis Drain and Pond 2 and (b) in groundwater along the western edge of Pond 2 at Kesterson Reservoir. (Modified from Ref. 7.)

The same wells responded very rapidly to the replacement of drainage water in the ponds after July 1986. Except for the shallowest well, Se concentrations had dropped by at least 75% during the first two months and all wells had concentrations below 10 µg/L within 1 year (Figure 10b). These response times to decreasing surface water Se inputs are much shorter than the lag times of 1–2 years between the introduction of drainage water to the ponds and detection of Se in the wells. This

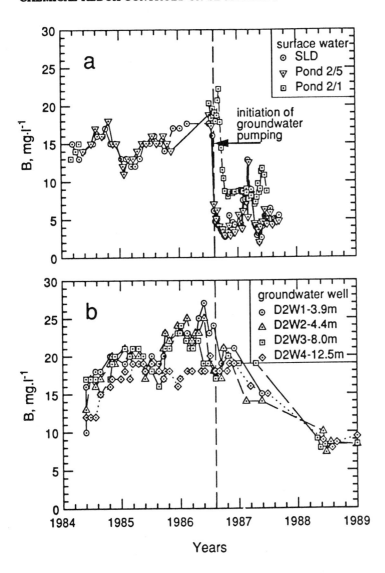

Figure 11 Temporal variations in boron concentrations in (a) surface water discharged from the San Luis Drain and Pond 2 and (b) in groundwater along the western edge of Pond 2. (Modified from Ref. 7.)

difference in response times is indicative of a strong retardation of Se in the aquifer. Boron concentrations in the monitoring wells exhibited a more delayed response to drainage water replacement, taking more than a year to reach B concentration levels in comparable surface water (Figure 11b). This delayed response suggested a residual source of B in the shallow aquifer that did not exist for Se, most probably desorption of B from clays.

4. Correlation with Redox Chemistry

As discussed by White et al. [7], two redox systems have operated in the Kesterson system: a shallow redox zone in the muds beneath the pond bottoms and a second deeper redox zone in the underlying alluvial aquifer. During operation of the Kesterson ponds and under preceding seasonal wetlands conditions, a layer of organic-rich sediments accumulated over much of the pond bottoms. Most of these interstitial waters, less than 1 m below the pond bottoms, were anoxic. This condition was indicated by the complete removal of DO and NO_3^-. The presence of up to 100 mg of sulfide per liter was indicative of anaerobic reduction of sulfate. Platinum electrode measurements in these muds were often less than –200 mV. Selenium concentrations in most of these waters were often near background levels, indicating that Se was effectively immobilized in the pond sediment. Solid-state Se distributions in these shallow sediments and mechanisms of Se uptake by microbial processes are discussed in detail by Weres et al. [15].

Much of the Se initially in the drainage water was removed in these pond bottom sediments before infiltrating into the underlying aquifer. Selenium penetrated into the shallow alluvial aquifer only in areas where either the bottom sediments were thin or the infiltration rates were sufficiently high [7]. Such a plume existed beneath the west side of Pond 2 (Figure 9a). Once in this environment, correlations were evident between Se and other chemical parameters that were indicative of oxidizing conditions in the groundwater.

In the alluvium, Pt electrode potentials ranged from 0 to approximately +500 mV relative to the hydrogen electrode standard (Figure 12a). This range in potentials was indicative of mildly reducing to highly oxidizing conditions and was comparable to potentials measured at Mendota (Figure 6). Negligible Se was observed in groundwater with Eh < +200 mV (Figure 12a). Above this Eh, the maximum observed Se concentrations generally increased with more oxidizing potentials. The vertical front of the Se plume along the west side of Pond 2 in 1984 was constrained at depth by Eh potentials more reducing than +300 mV (Figure 9). Groundwaters with more negative potentials at greater depths

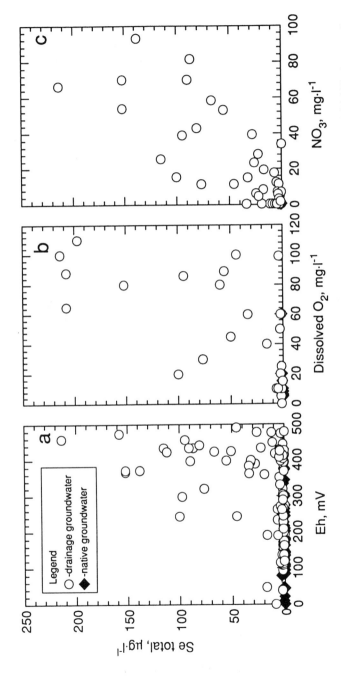

Figure 12 Correlations between total dissolved selenium and (a) measured Eh, (b) dissolved oxygen, and (c) NO_3 in groundwater and surface waters. (Modified from Ref. 7.)

beneath this front were devoid of Se, even though they contained high B concentrations indicative of drainage water.

The presence of DO and NO_3^- is an indicator of oxidizing conditions [reactions (3) and (4)]. Maximum groundwater DO concentrations of 118 µg/L were much lower than for air-saturated conditions in the surface water. However, groundwater samples containing elevated Se concentrations always contained measurable DO (Figure 12b). White et al. [36] also documented a close correlation between penetration depths of DO and Se in groundwaters along the margin of Pond 2 (Figure 13). Maximum groundwater NO_3^- concentrations derived from agricultural drainage were approximately 100 mg/L. Where NO_3^- was detected in groundwater, elevated concentrations of Se were also found (Figure 12c). A close correlation also existed between the spatial distribution of Se and NO_3^- plumes along the west side of Pond 2 in 1986 (Figure 9). Thermodynamic considerations [reactions (1)–(3)] would predict that the presence of DO and NO_3^- would prevent the reduction of SeO_4^{2-} and the precipitation of elemental Se.

The chemical species Fe^{2+}, Mn^{2+}, and H_2S are indicative of reducing groundwater conditions. Significant Fe^{2+} concentrations were detected in both reducing native and drainage groundwaters at Kesterson. Due to the insolubility of iron sulfide, dissolved sulfide was not detected when measurable concentrations of Fe^{2+} were present. Dissolved sulfide was

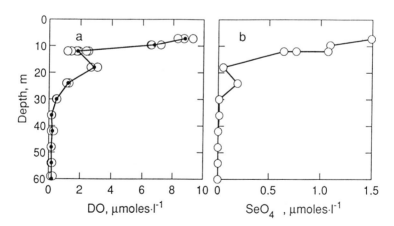

Figure 13 Distribution of (a) dissolved oxygen and (b) selenate as a function of depth near Pond 2 at Kesterson Reservoir. Solid and open circles correspond to individual and average data, respectively. (Modified from Ref. 36.)

generally confined to more reducing conditions in the anoxic muds in the pond bottoms. Exclusionary relationships generally existed between measurable concentrations of Se and the presence of dissolved Fe^{2+}, Mn^{2+}, and H_2S (Figure 14a–c). This relationship is expected because the electrochemical potentials for reduction of the respective oxidized species to Fe^{2+}, Mn^{2+}, and H_2S [reactions (5)–(7)] are more stable thermodynamically than the half-cell potentials describing the reduction of SeO_4^{2-} to elemental Se [reaction (1)].

The vertical distribution of measurable Fe^{2+} (>0.05 mg/L) at the Pond 2 site corresponded closely to the groundwater region with a measured redox potential below +300 mV (Figure 9d). As previously discussed, this correlation is expected because dissolved Fe^{2+} is the active redox species determining Pt electrode responses (Figure 3). Fe^{2+} was not present in the initial oxidizing surface water. Therefore, Fe contained in the drainage water at depth must have been contributed from the aquifer. Potential sources of Fe include dissolution of abundant ferrous oxides and silicates in the granitic sands [37] and microbial reduction of ferric oxyhydroxide [38] produced from weathering of these minerals during transport and deposition. Fe^{2+} is not stable in the presence of DO at neutral pH. Therefore, the redox front, established within the zone of drainage water infiltration, occurs when the rate of Fe release from the aquifer exceeds the rate of DO transport from the surface water. The removal of DO from the system permits the reduction of NO_3^- and Se at the redox front.

IV. MECHANISMS OF SELENIUM REDUCTION

The major mechanism for Se uptake from groundwater systems is related to its immobilization under reducing conditions. The solubility calculations indicated that SeO_4^{2-} and SeO_3^{2-} were several orders of magnitude undersaturated with respect to solid mineral phases. Therefore, elemental Se and metal selenides were the only solid phases expected to precipitate. Redox-sensitive species, including DO, NO_3^-, Fe^{2+}, Mn^{2+}, and H_2S, could directly control the solubility of Se via electrochemical potentials generated by reactions (1)–(7). However, several studies have documented that the rates of such inorganic redox reactions, notably for NO_3^- and SO_4^{2-}, are very slow [39]. Similar behavior of Se is supported by the apparent lack of agreement between analyzed Se speciation and the predicted thermodynamic distributions [7,24,40]. Therefore, Se speciation in groundwater is related not only to the thermodynamics of the redox reactions but also to the rates at which these reactions occur.

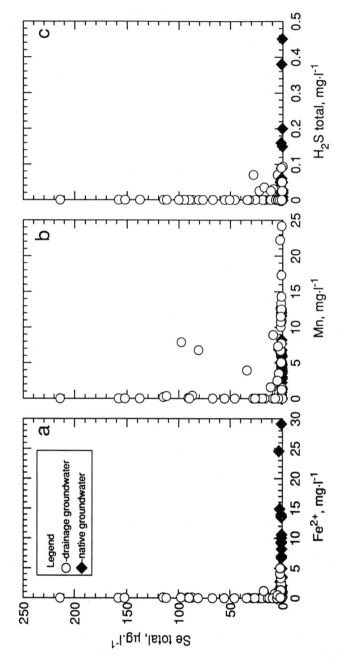

Figure 14 Correlations between total dissolved Se and (a) Fe^{2+}, (b) Mn^{2+}, and (c) total sulfide in groundwater and surface waters. (Modified from Ref. 7.)

A. Rates of Reaction

The rates of SeO_4^{2-} reduction in tracer injection tests in Kesterson groundwater are discussed in detail in Chapter 5 of this volume. One such study, directly applicable to the geochemical issues addressed in the present chapter, was performed in the reducing groundwater regime at Pond 2 (Figure 7). This groundwater contained no measurable DO, low B, and significant concentrations of Fe^{2+} and Mn^{2+}. The closed-system pumping test involved the removal of groundwater through a packer from a perforated zone of a 10-m-deep supply well. Before reinjection into a corresponding zone of the aquifer in an adjacent well, 1000 µg/L of SeO_4^{2-}, SeO_3^{2-}, and fluorescein, a nonreactive dye, was added to the fluid. Fluid samples were then removed periodically from the injection well over a 30-day period and analyzed to determine the rate of Se loss in the groundwater. The effects of aquifer dispersion were eliminated by comparing the ratio of the resulting Se concentrations to the fluorescein content.

Additional experiments assessed the relative importance of the aqueous versus substrate environment on rates of Se reduction in the aquifer. Samples were collected from the supply well under closed-system conditions in 125-mL glass flow-through flasks. These flasks were then injected into the field with SeO_4^{2-} and SeO_3^{2-} in amounts equivalent to those used in the pump test. The flasks were sampled periodically in the laboratory over a 30-day period [7].

The results of the downhole injection test indicated that the initial 1000-µg/L amounts of SeO_3^{2-} and SeO_4^{2-} were removed from the groundwater in less than 10 and 25 days, respectively (Figure 15). In contrast, the reduction rates of SeO_4^{2-} and SeO_3^{2-} in the closed-flask experiments were very slow over the same period. The aqueous chemical conditions downhole and in the sealed flasks were identical. Therefore, the more rapid rates of Se reduction in the aquifer must be controlled by the substrate environment and not directly by the aqueous solution. Although aqueous species such as Fe^{2+} and Mn^{2+} are indicators of the aquifer redox environment at both Kesterson and Mendota, these species do not directly reduce Se.

B. Microbial Processes

The rapid rates of downhole Se reduction in the tracer experiment imply reactions mediated by microbial activity. Such activity would be expected to occur in shallow groundwater systems with abundant nutrients and carbon from agricultural drainage and wetlands such as at Mendota and Kesterson. The importance of microbial activity has long been recognized

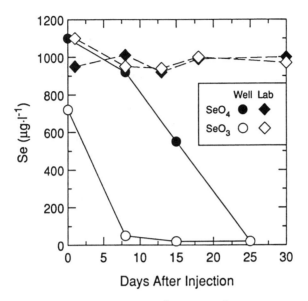

Figure 15 Uptake of SeO_4^{2-} and SeO_3^{2-} as functions of time during an injection test in reducing native groundwater at Site LBL 38 (Table 1). Solid lines correspond to downhole data and dashed lines to closed-system flask experiments. (Modified from Ref. 7.)

in influencing Se chemistry [see Chapters 16, 17]. Column experiments have recently shown the effective removal of SeO_4^{2-} by microbial activity in anoxic wetland sediments [15,42]. Sedentary types of microbes, which attach to mineral surfaces and organic substrates, generally predominate over free-moving forms in groundwater environments [43]. Such surfaces provide both a means of physical support and, in the case of organics, metabolizable compounds for microbial growth. Such a microbial distribution effectively explains the relative importance of the aquifer substrate in controlling the rates of Se reduction observed in the Kesterson tracer injection tests (Figure 15).

The role of aquifer redox conditions in microbial respiratory reduction of SeO_4^{2-} was demonstrated by Dubrovsky et al. [29] in experimental studies using the Mendota core samples. This was achieved by measuring the loss of $[^{75}Se]SeO_4^{2-}$ from slurried core materials using the approach of Oremland et al. [42]. Results in Table 3 show the percent loss of Se from solution after 3 and 8 days. The samples from 8.5 m and deeper in the aquifer showed a higher bacterial activity than the shallower samples.

Table 3 Experimental Rates of Selenium Reduction in Mendota Alluvium Expressed as Percentage of Total $^{75}SeO_4$ Removed from Solution

Depth	Amendment	Time	
		3 days	8 days
6.95	None	0	20
7.25	None	0	21
7.56	None	0	83
7.86	None	0	—
7.86	Acetate	0	12
8.17	None	0	18
8.47	None	0	89
8.78	None	78	90
28.8	Acetate	87	95
9.08	None	20	92
9.38	None	0	92
9.38	Acetate	92	92

Source: After Ref. 29.

This depth interval corresponds to the zone at which the porewaters became reducing (Figure 6), supporting the observation that microbial reduction of SeO_4^{2-} is an anaerobic process. Analyses of the Se solid phase showed that the SeO_4^{2-} was reduced to either amorphous red Se or, less frequently, black crystalline Se [42].

Because of the chemical similarity between the oxidized SeO_4^{2-} and SO_4^{2-} species, the initial supposition was that common aerobic SO_4-reducing bacteria such as *Desulfovibrio desulfuricans* were also using SeO_4^{2-} as an electron acceptor, thereby reducing and removing Se from the drainage waters. However, the experimental work of Zehr and Oremland [13] demonstrated that SO_4^{2-}-respiring bacteria do not preferentially use SeO_4^{2-}. Due to the very high dissolved SO_4^{2-} concentrations present in agricultural drainage water, Se reduction by such microbes would be highly inefficient. Macy et al. [44] isolated at least two active microbe types in anaerobic cultures, one capable of reducing SeO_4^{2-} to SeO_3^{2-} and the other of reducing SeO_4^{2-} to elemental Se.

Additional insight into the nature of microbial controls is inferred from the close correlation between dissolved NO_3^- and SeO_4^{2-} in Kesterson groundwater (Figures 5 and 12). Column experiments [15] and groundwater tracer experiments (Chapter 5) have also demonstrated that

Se uptake is effectively suppressed by the presence of NO_3^-. Because NO_3^- is inactive electrochemically in most organic reactions [39], the correlation with Se must be mediated by microbial processes. In recent detailed culture experiments using pond bottom muds from the agricultural source area, Oremland et al. [42] confirmed that besides NO_3^-, anaerobic microbial reduction was suppressed by O_2 but not by SO_4^{2-} or precipitated FeOOH. The final reduced Se phase produced in the experiments was again elemental Se.

Microbes can use DO, NO_3^-, Fe^{3+}, and SO_4^{2-} as electron acceptors during the oxidation of organic material as an energy source. These reactions are controlled by the net energy generated. Therefore, the nature and extent of microbial activity should be related directly to the thermodynamics describing the inorganic redox potentials. Anaerobic microbes will utilize DO in the groundwater first because it is the most readily reduced species [reaction (2)]. After DO is consumed, NO_3^- becomes the next most readily reduced species [reaction (3)]. Most denitrifying bacteria types can reduce NO_3^- in the absence of oxygen [43]. With the loss of NO_3^-, the microbes will then reduce SeO_4^{2-} [reaction (1)]. Recent work indicates that bacteria under anoxic conditions will then reduce Fe^{3+} in iron oxyhydroxides to Fe^{2+} [reaction (4)] or magnetite [38]. Under strongly anoxic conditions, organotrophic bacteria can use SO_4^{2-} as an electron acceptor to produce H_2S [reaction (6)]. This sequence of redox environments is described by the potentials of the above reactions and is comparable to that proposed by Berner [45] in classifying the vertical reduction sequence associated with oxidation of organic carbon in marine sediments. Therefore, microbial activity represents a mechanism that controls the rates for the redox reactions described in the preceding sections of this chapter.

V. CONCLUSIONS

The behavior of Se in groundwater systems is controlled by several geochemical processes, including aqueous speciation, sorption, and precipitation. The relative importance of a specific process is in turn strongly influenced by the oxidation state of Se and therefore the electrochemical potential of the groundwater. Selenate, Se(VI), the most oxidized Se species, is the most mobile form of Se in groundwater and is not significantly retarded by either sorption or precipitation. This species occurs under oxic conditions in the presence of DO and NO_3^-. Although also highly soluble in groundwater, the more reduced SeO_3^{2-} species, Se(IV), occurs at lower concentrations than SeO_4^{2-} as it has a much stronger sorptive affinity in aquifers. In contrast, the reduced forms Se(0) and Se(II) are highly insoluble, forming elemental selenium and metal selenides. Anoxic groundwaters, denoted by high concentrations of H_2S, Fe^{2+}, and Mn^{2+}, therefore contain low concentrations of dissolved Se.

The relationship between Se oxidation and reduction and the presence of redox-sensitive couples agree with predicted electrochemical potentials. However, measured Se distributions between SeO_4^{2-} and SeO_3^{2-} species appear out of equilibrium due to relatively slow reactions kinetics. The results of field tracer tests indicating rapid Se immobilization in aquifers under reducing conditions imply acceleration by microbial activity. Such microbes utilize the energies of the redox reactions in the same reduction sequence predicted from thermodynamics, sequentially consuming dissolved O_2, NO_3^-, SeO_4^{2-} and finally SO_4^{2-}. The observed Se distribution is controlled strongly by the rates at which this microbial reduction occurs.

Based on the field studies described, the ultimate fate of dissolved Se in groundwater can be deduced from the concept of groundwater redox barriers. These barriers control the location of the oxic–anoxic interface in the aquifer. In the Mendota example, this interface and corresponding Se distributions were controlled by water table elevation and a lithology that was more oxidizing at shallow depth due to a lower abundance of Fe^{2+}-containing minerals. Below this interface, Se was reduced and immobilized. The native groundwater beneath the Kesterson Reservoir, in contrast, was naturally reducing. Selenium containing oxic groundwater plumes resulted from accelerated infiltration rates caused by construction of the overlying ponds. A redox barrier was established in a zone in the aquifer where the rates of DO and NO_3- reduction by microbial activity and reaction with the aquifer minerals exceeded this infiltration rates. This zone corresponded to the maximum penetration depth of Se into the groundwater. Results of the field studies showed that the mobility of Se can be predicted in groundwater systems on the basis of careful analysis of chemical redox conditions.

REFERENCES

1. T. S. and H. M. Ohlendorf, *Environ. Manage.* 11:805 (1984).
2. D. G. Grisafe, E. E. Angino, and S. S. Smith, *Appl. Geochem.* 3:601 (1988).
3. D. L. Naftz and J. A. Rice, *Appl. Geochem.* 4:565 (1989).
4. T. S. Presser and W. C. Swain, *Appl. Geochem.* 5:703 (1990).
5. R. M. Garrels and C. L. Christ, *Solutions, Minerals and Equilibria*, Cooper, San Francisco, 1976.
6. M. A. Elrashidi, D. C. Adriano, S. M. Workman, and W. L. Lindsay, *Soil Sci.* 144:141 (1987).

7. A. F. White, S. M. Benson, A. W. Yee, H. A. Wollenberg, Jr., and S. Flexser, *Water Resour. Res.* 27:1085 (1991).
8. G. A. Cutter, *Science* 217:829 (1982).
9. T. D. Cook and K. W. Bruland, *Environ. Sci. Technol.* 21:1214 (1987).
10. O. Weres, A. R. Janouni, and L. Tsao, *Appl. Geochem.* 4:543 (1989).
11. R. H. Long, Lawrence Berkeley Rep. 25874, Univ. California, Berkeley, 1988.
12. R. S. Oremland and J. P. Zehr, *Appl. Environ. Microbiol.* 52:1031 (1986).
13. J. P. Zehr and R. S. Oremland, *Appl. Environ. Microbiol.* 53:1365 (1987).
14. H. W. Lakin, in *Selenium in Agriculture* (M. S. Anderson, H. W. Lakin, K. C. Benson, F. F. Smith, and E. Thatchers, Eds.), U.S. Dept. Agriculture, Washington, D.C., 1961, p. 3.
15. O. Weres, H. R. Bowman, A. Goldstein, A. Smith, and L. Tsao, *Air Water Soil Pollut.* 49:251 (1990).
16. R. H. Neal and G. Sposito, *Soil Sci. Soc. Am. J.* 53:70 (1989).
17. L. S. Balistrieri and T. T. Chao, *Soil Sci. Am. J.* 51:1145 (1987).
18. B. Bar-Yosef and D. Meek, *Soil Sci.* 144:11 (1987).
19. K. F. Hays, A. L. Roe, G. E. Brown, K. O. Hogson, J. O. Leckie, and G. A. Parks, *Science* 238:783 (1987).
20. D. Langmuir, Eh-pH determination, in *Procedures in Sedimentary Petrology* (R. E. Carver, Ed.), Wiley, New York, 1971, p. 597.
21. D. Thorstenson, U.S. Geol. Surv. Open File Rep. 84-072, 1984.
22. S. I. Zhdanov, in *Standard Potentials in Aqueous Solution* (A. J. Bard, R. Parsons, and J. Jordan, Eds.), Marcel Dekker, New York, 1985, p. 93.
23. T. J. Wolery, Lawrence Livermore Natl. Lab. Rep. UCRL-53414, 1984.
24. D. D. Runnels and R. D. Lindberg, *Geology* 18:212 (1990).
25. A. J. Bard, R. Parson, and J. Jordan, *Standard Potentials in Aqueous Solution*, Marcel Dekker, New York, 1985, p. 356.
26. M. Sato, *Econ. Geol.* 55:928 (1960).
27. S. J. Deverel and R. Fujii, *Water Resour. Res.* 24:516 (1988).
28. K. Berlitz and F. J. Heimes, U.S. Geol. Surv. Water Supply Paper 2348, 1990.
29. N. M. Dubrovsky, J. M. Neil, R. Fujii, R. S. Oremland, and J. T. Hollibaugh, U.S. Geol. Surv. Open File Rep. 90-138, 1990.
30. W. R. Lettis, U.S. Geol. Surv. Open File Rep. 82-526, 1982.
31. F. T. Manheim, U.S. Geol. Surv. Prof. Paper 550-C, C256, 1966.
32. R. S. Oremland and J. P. Zehr, *Appl. Environ. Microbiol.* 52:1031 (1986).
33. R. R. Tidall, W. D. Grundy, and D. L. Sawatzky, Kriging techniques applied to element distribution in soils of the San Joaquin Valley, *HAZTECH International Conf. Proc.*, Denver, Colo., 1986, pp. 992-1009.
34. S. M. Benson, A. F. White, S. Halfman, S. Flexser, and M. Alavi, *Water Resour. Res.* 27:1071 (1991).
35. T. S. Presser and I. Barnes, U.S. Geol. Surv. Invest. Rep. 84-4122, 1984.
36. A. F. White, M. L. Peterson, and R. D. Solbau, *Ground Water* 28: 584 (1900).
37. A. F. White and A. Yee, *Geochim. Cosmochim. Acta* 49:1263 (1985).
38. D. R. Lovely, *Geomicrobiology* 5:375 (1987).

39. J. I. Drever, *The Geochemistry of Natural Waters*, Prentice-Hall, Englewood Cliffs, N.J., 1982.
40. I. Rosenfeld and O. A. Beath, *Selenium*, Academic, San Diego, Calif., 1964, p. 321.
41. J. W. Doran, *Adv. Microb. Ecol.* 6:1 (1982).
42. R. S. Oremland, N. S. Steinberg, A. S. Maest, L. G. Miller, and J. T. Hollibaugh, *Environ. Sci. Technol.* 24:1157 (1990).
43. E. A. Paul and F. F. Clark, *Soil Microbiology and Biochemistry*, Academic, San Diego, Calif., 1989.
44. J. M. Macy, T. A. Michel, and D. G. Kirsch, *Microbiol. Lett.* 61:195 (1989).
45. R. A. Berner, *J. Sediment Petrol.* 51:359 (1981).

9
Kinetics of Selenium Uptake and Loss and Seasonal Cycling of Selenium by the Aquatic Microbial Community in the Kesterson Wetlands

Alexander J. Horne

*University of California
Berkeley, California*

I. INTRODUCTION

Animals and plants in Kesterson Reservoir, a large marsh in the San Joaquin Valley of California, became contaminated with up to 300 ppm dry weight of selenium (Se) after several years of exposure to agricultural drainwater rich in dissolved Se. The drainwater was also rich in plant nutrients that supported a very productive marsh where plants, invertebrates, and mosquitofish flourished, apparently unaffected by high Se. In contrast, several bird species suffered serious reproductive failures that were characteristic of Se toxicity.

To reduce the toxicity, several cleanup scenarios were proposed. One proposal was to permanently flood the wetland with clean, low-Se water. The chemistry of Se is much like that of sulfur, and, given the productivity of the wetlands, flooding would create permanent anoxia in the sediments and chemically immobilize Se in its reduced or zero oxidation states. Research reported elsewhere has shown that permanent flooding also creates conditions for biological immobilization of Se [1,2]. The main route of Se to the sediments was through rapid sedimentation of dead material [2]. Loss of Se from living organisms was much slower. Thus the rate-limiting step in the cleanup is the rate of depuration or loss of Se by living organisms.

A relatively rapid rate of depuration of Se by animals and plants is crucial for the cleanup of Se-polluted sites using the permanent flooding technique. If the natural rates of Se loss are slow, then other, more costly methods such as physical excavation or burying [3–5], chemical binding, accelerated biological loss [6], or cropping become necessary. In contrast to the rapid rates of uptake, the literature gives little guidance on the rates of Se loss, especially for polluted natural wetland systems.

The purpose of this study was to estimate the kinetics of Se flux, both loss and any uptake, during the cleanup of the Kesterson Reservoir marsh ecosystem. The longest lived animal or plant was the mosquitofish, which survives almost 2 years. The 1200-acre (480-ha) wetland was originally contaminated with Se in 1978, and thus all biotic components were in equilibrium by 1986 when these measurements were made. The equilibrium was disturbed by the cleanup, which substituted low-Se water (<2 ppb) for the former supply of Se-rich water (up to 300 ppb).

Although Se declined throughout the wetland, this report is limited to decreases measured in a 1-acre (0.5-ha) experimental enclosure, the P5A mesocosm, situated in the heart of the wetter section of Kesterson Reservoir. The inflow of groundwater and other variables were well controlled in the mesocosm, in contrast with the rest of the system, where periodic drying up of the wetland complicated the depuration and uptake kinetics. Mesocosms have been shown to be appropriate in both time and space scales for medium-term ecological experiments [7]. Only details of Se kinetics are described here. Full details of the extensive studies on groundwater, sediment chemistry, and the ecology of the Se-contaminated Kesterson wetlands can be found elsewhere [3,7]. Details of the biomass, biology, and Se fluxes in 23 species or parts of the wetlands ecosystem at Kesterson are reported by Horne and Roth [2] and Horne [1].

II. METHODS

The methods of sample collection, processing, and biomass and Se analysis have been presented elsewhere [1,2]. Kinetics based on Michaelis-Menten or Monod equations are fully applicable only under steady-state conditions such as in laboratory cultures and seem not to apply to long-term fluxes in natural wetlands. However, the rate of change in both biotic and soluble Se in the mesocosm was quite slow relative to the depuration rates of most organisms. Thus, a quasi-steady-state model can be used that gives an apparent depuration half-saturation constant (the equivalent of K_s) for the conditions in the mesocosm. In laboratory studies, the loss of a substance over time from an individual organism is often used to calculate depuration kinetics. This was not possible in the mesocosm where the average depuration of a population was measured. However, many of the important organisms at Kesterson Reservoir are very small, and some are microscopic or multicellular algae. Even under laboratory conditions these must be studied as a population. Again, the difference between the depuration rates in individuals and populations is encompassed in the use of an apparent half-saturation constant.

A linear transformation of the Michaelis-Menten equation gives

$$S = V_{max}\,(S/v) - K_s$$

where S is the soluble Se concentration (ppb); V_{max}, the maximum depuration velocity; v, the measured depuration rate (ppb/day); and K_s, the apparent half-depuration constant. This plot is more accurate than the more normally used ($1/v$ versus $1/S$) Lineweaver-Burke plot [10]. Because other processes operate in the real world (e.g., recycling, changes in growth rates), only the initial decline period produced suitable data for this model. In addition, soluble or sediment Se can be plotted against Se in biota in various trophic levels and an approximate depuration coefficient estimated.

The loss of Se for all species where there were sufficient data can be somewhat arbitrarily divided into three phases. Phase 1 is the steep initial depuration that occurred in the first few months of the experiment. This decline followed the initial drop in soluble Se by at least a month. Phase 2 is the longer, slower decline that coincided with the slowly dropping level of soluble Se and a small winter rise. Phase 3 includes some regular seasonal rises and falls in tissue Se for almost all organisms. These permitted both uptake and depuration kinetics to be studied.

For brevity, only the kinetics of five types of organisms are reported here; further details are available elsewhere [2]. These organisms, which represent a wide range of common wetland biota, are *Chara* (a large freshwater macroalga), the attached microbial community (primarily bacteria and microalgae, but also fungi, protozoans, and rotifers), epifaunal chironimid larvae (living among the *Chara* stands), benthic chironomid larvae (living in the sediments), and the only aquatic vertebrate that could tolerate the high Se levels, high temperatures, low dissolved oxygen, and high total dissolved solids in this wetland, the small mosquitofish *Gambusia affinis*.

III. RESULTS

Depuration of Se in the mesocosm began when the supply of Se-rich water (up to 300 ppb) was replaced by low-Se groundwater (<2 ppb). In all except the first phase of rapid decline, both depuration and uptake rates were species-specific. In addition, rates were usually not constant within any phase, and periods of rapid loss would be interrupted by an occasional period of Se uptake. The differences in the kinetics of Se depuration between species (or between trophic levels) is illustrated by the shape of the smoothed depuration and uptake curves shown in Figure 1. The curves range from the steep and smooth trend for soluble Se to progressively flatter and more erratic depuration kinetics as one moves from *Chara* (an autotroph) through mixed trophic levels (aufwuchs) to the carnivorous mosquitofish.

A. Selenium Depuration Kinetics in Phase I: Initial Rapid Selenium Loss and Slower Secondary Decline

The initial Se loss had two distinct periods: (a) an initial rapid decline that lasted for about 4 months and was not species-specific and (b) a much slower decline during which differences between species were evident. During the initial period of steep decline, rates were very similar for all trophic levels (−0.39 to −0.46 ppb/day, see Table 1). The initial steep decline phase followed a hyperbola typical of enzyme–substrate interactions (Figure 2). For *Chara*, a linear transformation gave an apparent half-depuration constant, K_d, of 6.5 ppb soluble Se (the intercept). The regression on this line had a coefficient of determination of 0.77. An estimate of V_{max} of 3.3 ppb/day was calculated from the slope. The K_d values for the other most common biota in this wetland were 4.5 ppb (aufwuchs) and 3.5 ppb (chironomids and mosquitofish).

Depuration kinetics for Phase 1 can be modeled in several ways. We tested four models for decline: linear, negative exponential, log-linear, and a power function. Although all models explain some of the results, the best and most statistically significant fit is the negative exponential model.

The initial steep exponential loss over 4 months was followed by a slower decline that persisted for several months until the first winter period. Unlike the steep initial losses, these later rates were less at the lowest trophic levels. The pure plant stand (the macroalga *Chara*) showed an average rate of −0.03 ppb/day; the mixed microbe–microalgae–protozoan aufwuchs system had rates of −0.08 ppb/day; the mostly herbivorous chironomid larvae showed the highest rates, −0.24 ppb/day; and the mostly carnivorous mosquitofish showed an average rate of −0.18 ppb/day.

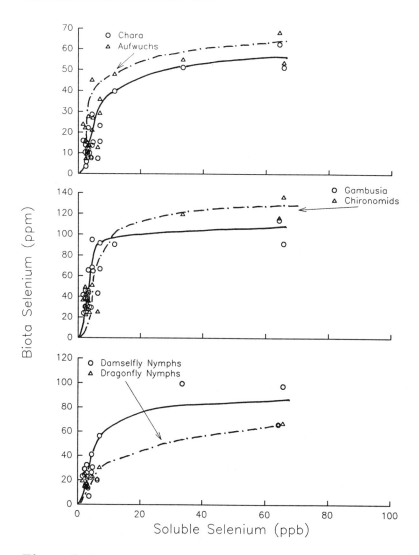

Figure 1 Substrate uptake and depuration (Monod-type) plots for selenium in biota compared with soluble Se in the water during the initial 4-month steep depuration period in the Kesterson mesocosm experiment. The most common organisms are shown. The more gentle uptake curves for the main photosynthetic organisms (the macroalga *Chara* and the microalgae in the aufwuchs community) relative to the animals are reflected in higher depuration coefficients K_d and predict their more rapid progression to low tissue Se over the 2.3 years of the experiment.

Table 1 Selenium Depuration or Average Uptake Rates (ppm/day) for the Three Phases of Se Change in the Experimental P5A Mesocosm at Kesterson Reservoir[a]

Taxon/group	Selenium flux (ppm/day)			
	Phase 1		Phase 2	Phase 3
	Initial	Main		
Water	−1.71	−0.03	0	+0.1
Chara	−0.44	−0.03	+0.04	−0.11
Aufwuchs	−0.39	−0.08	+0.06	−0.29
Chironomids (epifaunal)	nd	−0.24	+0.14	−0.14
Chironomids (benthic)	nd	—	—	—
Gambusia	−0.46	−0.18	+0.07	−0.12

[a]Values for water and the five most common organisms are shown. nd = no data when too few organisms were present for Se analysis. Benthic chironomids were not present during the early period of the experiment. Note the initial high depuration, which was similar for all species, the lower depuration rates (late Phase 1 and Phase 3), and small uptake rates in winter–spring (Phase 2), all of which showed trends toward higher rates at higher trophic levels.

B. Phase 2 Kinetics: Winter Increases

For most organisms, the first set of minimum values for biotic Se were found in June–July 1987, a year after the introduction of low-Se water. This nadir was followed by a small but persistent increase in Se in most animal and plant groups through the autumn, and a definite winter peak was reached in the winter of 1987–1988 (Figure 3). Although mean Se concentrations rose by relatively small amounts, the increases were statistically significant for *Chara* ($t = 6.65$, df $= 19$, $P > 0.01$) and aufwuchs ($t = 5.53$, df $= 16$, $P > 0.01$). Emergent plants that did not follow the same seasonal cycle are described separately below.

Unlike the biota, soluble Se in the water column continued to decline, showing no winter peak. Thus, the second, plateau phase was not an obvious reflection of direct contamination. The kinetics of this winter increase in biotic Se are shown in Figure 1 and Table 1. Selenium fluxes switched from slow depuration in the later part of depuration in Phase 1 to slight uptake in Phase 2 (0.04–0.14 ppb/day). Unlike Phase 1, the timing of the changes in Phase 2 was similar for all common species, although the loss rates were highest for epifaunal chironomids and lowest for *Chara*

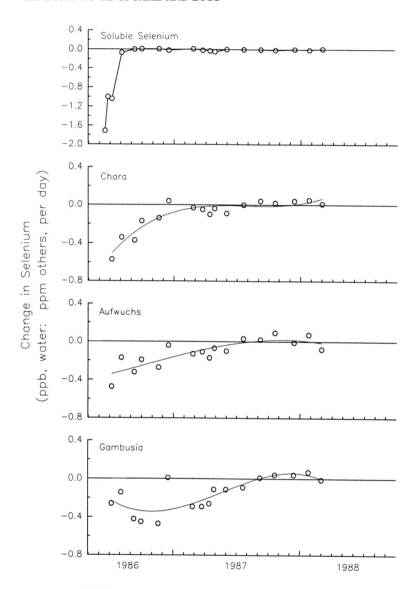

Figure 2 Selenium depuration kinetics for three common species in the Kesterson mesocosm. The fluctuations for soluble Se in the water are also shown. The uptake and depuration curves over the three phases described in the text are composed of data collected over 2.3 years. Values shown are smoothed daily values.

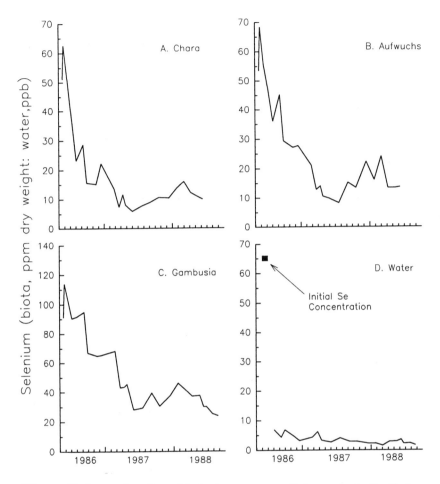

Figure 3 Seasonal cycling of Se in three common biota species in the Kesterson mesocosm over 2.3 years. The winter–spring peak is shown well for 1987–1988 but is partially obscured by large variations during the rapid initial decline period in the winter of 1986–1987. Note that variation in the now low soluble Se values did not correspond to the changes in biota, which were due presumably to decreases in depuration rates in winter rather than to changes in uptake rates. Chironomids (not shown and scarce in the autumn following the annual outmigration to the aerial phase) also showed a similar but less well-defined seasonal cycle.

(Table 1). No corresponding increase in the rate of supply of soluble Se occurred just prior to the increase in biotic Se.

Monod-type uptake kinetics were not calculated for this phase because the changes in soluble Se concentrations were small. Nevertheless, the small winter peak in 1987–1988 was long enough to correspond to the saturation level at that substrate concentration (7-month average, 2.9 ± 0.8 ppb soluble Se). This value can be compared with the concentration in the previous 6-month period (3.9 ± 1.4 ppb Se). Apparently, different enzymatic systems are operating, and Se concentrations in biota at any one time of year are not a simple function of the soluble Se content in the medium. This is even more obvious since all trophic levels are responding even though the uptake mechanism operates via the soluble pool for *Chara* and mostly via the food chain for mosquitofish.

C. Phase 3 Kinetics: Seasonal Cycling

Seasonal cycling of Se, with a winter peak and summer depression, was initially imposed on the overall decline in biotic Se. As can be seen in Figure 3, the soluble Se varied only slightly and had no discernible pattern. The first winter–spring peak, in 1986–1987, was obscured by variations in the steep initial decline of Phase 1. However, the second winter–spring peak, in 1987–1988 was quite clear for three major species (*Chara*, aufwuchs, and mosquitofish). Because the aquatic larvae grew into aerial adults in summer, the tiny young chironomid larvae were not collected in large enough numbers for Se analysis in some seasons. However, the chironomid data do support the concept of a winter peak in Se (see Figure 2 in Ref. 8).

Net Se depuration resumed for most species by early spring of the second year, 1988 (Table 1), and was also seen in the smoothed values of Figure 1. Fifteen species or species components (roots, stems, etc.) for which there were sufficient continuous data followed this pattern. Interestingly, seasonal cycling was found for the water snail *Physa*, which invaded the mesocosm in March 1988, after the main decline in soluble Se. The presence of *Physa* was attributed to the decreased salinity at this time since the low-Se groundwater also was fresher than the original agricultural drainwater supply. Also, both epiphytic and benthic chironomids followed the spring decline in Se even though total Se in the surface sediments continued to increase. Like the rise in winter 1987–1988, this overall decline was not due to variations in soluble Se concentrations, which remained low and even rose slightly in early summer. Three groups, adult dytiscid beetles, *Berosus* larvae, and *Scirpus* roots, did not show an increase in Se until summer. Thirteen groups were not common enough for conclusions to be drawn.

Depuration rates in the spring of 1988 (Phase 3, Table 1) were similar to the long-term decline rates measured in most of Phase 1 and ranged from 0.11 to 0.14 ppb/day for the most common species (*Chara*, aufwuchs, mosquitofish, epifaunal chironomids). Like the earlier slow depuration and uptake rates, there was a trend toward higher depuration rates at higher trophic levels. The trend was not as definite as in previous phases. Photosynthetic tissue (*Chara*) depurated at only –0.11 ppb/day, while mixed microbe–algae–protozoan aufwuchs lost as much as 0.29 ppb/day. However, the chironomids and mosquitofish had rates quite similar to those of *Chara* (–0.12 to –0.14 ppb/day).

IV. DISCUSSION

A. Seasonal Cycling

Seasonal cycling of Se in the Kesterson wetland has been well defined in these studies. Other studies at Kesterson have suggested a seasonal cycle, but this conclusion was necessarily tentative, as it was based on collections made only three times during the year [11]. Global patterns of Se cycling resulting from changes in the output of gaseous dimethylselenide are well documented [12,13], and it is reasonable to suppose that this would be reflected in the concentrations of plants and the animals that feed on them. However, the detection limit for Se in uncontaminated organisms is too high for this seasonal change to be very obvious. The winter–spring maximum described for the biota at Kesterson Reservoir correlates well with the global seasonal cycle, which also shows a maximum in spring.

B. Selenium Kinetics

Previous studies on selenium kinetics have been concerned with the most obvious pollution problem of how Se gets into aquatic biota [14–17]. Such studies have usually focused on laboratory cultures grown in defined media [18]. In these cases, uncontaminated organisms are exposed to a constant concentration of one chemical form of Se. Food is supplied with a fixed concentration of Se if particulate pollution is to be assayed and in a Se-free form if soluble Se uptake is to be measured. Where depuration has been measured, it has been in a system where both the source of Se, and the Se recycled by excretion from the test organism are totally removed at one time [19,20]. The standard flow-through bioassays ensure that no Se excreted by the organism remains to contaminate the test species.

However, in polluted ecosystems, many forms of Se, both inorganic and organic, are present, and the concentrations of Se vary with time and space [2]. Most important, the Se excreted by all organisms is partially available for further uptake. Finally, in real systems, the accumulated Se in the food chain cannot be removed at once, if at all, and must slowly change to biologically unavailable forms. The kinetics of biotic losses of Se under these more likely polluted conditions are likely to be much slower than those found in laboratory cultures but more realistic for the purposes of predicting the "time to cleanup criteria" needed by regulatory agencies.

The three principal routes for loss of Se in aquatic ecosystems are loss to the sediments, to biota, and to the air. In the Kesterson mesocosm, loss to the sediments was very large and relatively rapid [2], but some losses via volatilization of dimethyselenide and dimethydiselenide are possible [21]. Export of Se by insects that have an aerial stage and removal by birds and mammals that feed in the marsh do occur, and contaminated adult insects have been collected near the marsh. However, removal of contaminants by cropping and exporting higher trophic levels indicate that very small percentages are lost each year [22]. Thus, the rate of depuration of Se from living biota to algae [1], which soon die and sink to the anoxic sediments, is the key factor in the short-term (months to years) detoxification in Se-contaminated wetlands. Over longer periods, other routes such as cropping, insect export, and volatilization will eventually spread the contaminant over a much wider area and probably at a concentration that is not toxic to biota.

Laboratory studies using pure cultures or single species indicate that the depuration of Se is apparently much slower than its uptake [20], and the Kesterson marsh studies also indicate that this is true. The apparent depuration constant for Se of 3.5–6.5 ppb for the most common organisms in the Kesterson mesocosm is in the lower range of 2–30 ppb reported for selenomethionine in freshwater phytoplankton [17]. The same authors show that the uptake of selenite and selenate is almost linear. More important, the K_d values found in this study indicate that the biotic depuration systems were always undersaturated relative to soluble Se (and also any food source for those indirectly contaminated). The order of the K_d values predicts the long-term outcome of the experiment in that those species with the higher K_d values (fish and chironomid larvae) took longest to return to low Se concentrations, while those with the highest K_d values (*Chara* and aufwuchs) were most undersaturated and thus lost tissue Se most rapidly.

V. CONCLUSIONS

Biotic depuration of Se by marsh organisms at Kesterson Reservoir was triphasic, an initial steep exponential decline being followed by a plateau (Phase 1), a small winter rise (Phase 2), and then seasonal cycling imposed on a slow overall decline (Phase 3). In the first year, declines ranged between 63% and 92% or original values. Submerged vegetation, *Chara* fell from 57 ppm to 8.2 ppm (10% of original), and common insect larvae fell from 67–170 ppm to 12–28 ppm (13–22% of original). At first, depuration rates for all species were about 0.4 ppm/day, after which most rates slowed by a factor of 4 for the next several months. A reverse Monod-type model gave an apparent half-depuration coefficient, K_d, of 6.3 ppb for *Chara* and 4.5 ppb for aufwuchs, and a V_{max} of 3.3 ppb/day for *Chara*. These two photosynthetic groups made up more than 99% of the submerged biomass. The K_d for mosquitofish and chironomids was 3.5 ppb. Thus, at soluble Se concentrations in Phase 1 (3.2–8.9 ppb), the depuration systems were always undersaturated relative to the soluble Se source, and thus all organisms might be expected to lose body Se. The K_d predicted that mosquitofish and chironomids would lose Se more slowly than the *Chara* and aufwuchs communities. Selenium uptake during the first winter (Phase 2) increased tissue Se to 1.3–3.3 times the Phase 1 minimum. Corresponding Se uptake rates in the fall and winter were mostly slower than the depuration rates measured earlier, ranging from 0.03 ppm/day (*Chara*) to 0.14 ppm/day (epifaunal chironomid larvae). Selenium losses resumed in the first part of Phase 3 in summer, when depuration rates were 0.11–0.24 ppm/day.

The Kesterson marsh data showed for the first time a well-documented seasonal cycle in Se in aquatic biota with a winter–early spring peak. This cycle had not been observed in uncontaminated sites due to the difficulty of measuring Se accurately at values less than 1 ppm.

The population depuration kinetics reported here should be widely applicable because the animals and plants studied have worldwide distribution. Model predictions can now be made of the rate of natural "cleaning" that will occur for each species group if the source of Se pollution in wetlands is terminated. Examples are found in marshes below power station fly ash ponds, agricultural evaporation ponds, and some natural wetlands.

ACKNOWLEDGMENTS

The main funding for this research was supplied by the U.S. Bureau of Reclamation with subsidiary support from the University of California,

Berkeley, Environmental Engineering and Health Sciences Laboratory. I thank Drs. O. Weres and J. C. Roth for assistance with chemical analysis field collections and M. L. Commins for help with graphics.

REFERENCES

1. A. J. Horne, *Water, Air, Soil Pollut.* 57/58: 43 (1991).
2. A. J. Horne and J. C. Roth, Selenium detoxification studies at Kesterson Reservoir wetlands: depuration and biological population dynamics measured using an experimental mesocosm and Pond 5 under permanently flooded conditions, Univ. California, Berkeley Environmental Engineering and Health Sci. Lab., 1989.
3. Lawrence Berkeley Laboratory, Hydrological, geochemical, and ecological characterization of Kesterson Reservoir: Eighth Prog. Rep., April 1–June 30, 1988, LBID-1420.
4. California State Water Resources Control Board, Order WQ 87-3, March 19, 1987. This order requires the adoption of the "on-site disposal" option. This option required removal of soil and vegetation above 4 ppm Se to a depth of 6 in. and storage in a lined, covered mound at Kesterson.
5. California State Water Resources Control Board, Order WQ 88-7, July 5, 1988. This order required filling in of all Se-contaminated seasonal wetland areas and "preferred" the volatilization of any other Se in drier parts of the system.
6. W. T. Frankenburger and U. Karlson, in *Biogeochemistry of Trace Metals* (D. C. Adriano, Ed.), 1992, p. 365.
7. J. Bloesch (Ed.), *Hydrobiologia* 159(3): 221 (1988).
8. Lawrence Berkeley Laboratory, Hydrological, geochemical, and ecological characterization of Kesterson Reservoir: First Ann. Rep., Oct. 1, 1986–Sept. 30, 1987, LBL-24250.
9. Lawrence Berkeley Laboratory, Hydrological, geochemical, and ecological characterization of Kesterson Reservoir: Second Ann. Rep., Oct. 1, 1977–Sept. 30, 1988, LBL-26438.
10. J. E. Dowd and D. S. Riggs, *J. Biol. Chem. 240*: 863 (1956).
11. M. K. Saiki, A field example of selenium contamination on an aquatic food chain, Proc. First Annual Environmental Symp. on Selenium in the Environment, Fresno, June 1985, Cal. Agricultural Tech. Inst., Fresno, 1986, pp. 67–76.
12. National Academy of Sciences, *Selenium*, Natl. Acad. Sci., Washington, D.C., 1976.
13. G. F. Combs and S. B. Combs (Eds.), *The Role of Selenium in Nutrition*, Academic, New York, 1985.
14. M. Sandholm, H. E. Oksanen, and L. Pesonen, *Limnol. Oceanogr. 8*: 496 (1973).
15. A. K. Furr, T. P. Parkinson, J. Ryther, C. A. Bache, W. H. Gutenmann, I. S. Pakkala, and D. L. Lisk, *Bull. Environ. Contam. Toxicol. 26*: 54 (1981).
16. J. W. Hilton, P. V. Hodson, and S. J. Slinger, *Comp. Biochem. Physiol. 71*: 49 (1982).
17. M. B. Gerhardt, F. B. Green, R. D. Newman, T. Lundquist, R. B. Tresan, and W. J. Oswald, *Res. J. Water Pollut. Control Fed. 63*: 799 (1991).

18. P. Kiffney and A. Knight, *Arch. Environ. Contam. Toxicol.* **19**: 488 (1990).
19. R. K. Okazaki and M. H. Panietz, *Marine Biol.* **63**: 113 (1981).
20. G. D. Lemly, *Aquat. Toxicol.* **2**: 235–252 (1982).
21. E. T. Thompson-Eagle, W. T. Frankenburger, and U. Karlson, *J. Environ. Qual.* **19**: 125 (1989).
22. J. E. Burgess, *Proc. Annu. Conf. Southeast Assoc. Game Comm.* **19**: 225 (1966).

10
Agroforestry Farming System for the Management of Selenium and Salt on Irrigated Farmland

Vashek Cervinka

California Department of Food and Agriculture
Sacramento, California

I. INTRODUCTION

Agroforestry farming systems operate on the principle that selenium (Se) and salts in drainage water are resources that have economic value. Agroforestry provides farming practices that facilitate management of Se and salts through alternative cropping systems that rely on salt-tolerant trees and halophyte plants to reduce the volume of drainage water and thereby facilitate the removal of excess Se and salts.

II. MANAGEMENT OF SELENIUM AND SALTS BY AGROFORESTRY TECHNIQUES

The long-term viability of agriculture requires a balance of salts in soils: the removal of salts, including Se, must equal the inflow into the farming system. The theoretical management matrix on which agroforestry techniques are based views the remediation of Se and salts from four perspectives, listed below, and evaluates the technological, environmental, and economic feasibility of each strategy.

1. *Potential Products for New Markets* Advocates of agroforestry explore new markets for Se and salts, such as the use of Se in livestock feed supplements and in deer and rodent repellents for the protection of young trees, and the use of salts as industrial feedstock [1]. Selenium

237

currently has market value in the production of glass, electronic components, and in other industrial applications, but additional research is needed.
2. *Harvestable Resources* Salts and Se are viewed as harvestable resources that can be removed by farm crops, cogeneration facilities, and solar evaporators [1]. The economics of Se management and removal require that the increased concentrations achieved by agroforestry systems operate in combination with solar evaporators. By using harvested plants from Se-rich areas as cattle forage in Se-deficient areas, Se can be transported to locations where it is needed as a resource in the diet.
3. *Disposable Waste Products* Salts and Se are viewed as products that can be disposed of through microbial and plant volatilization (Se), biomass burning (salts and Se), and cattle forage (salts and Se) [1].
4. *Manageable Farming Inputs* In irrigated agroforestry systems, salts and Se are farming inputs whose inflow into the system can be reduced by efficient water management practices, irrigation technology, and cropping systems [1]. Water delivery and application methods that are technologically, environmentally, and economically feasible must be designed for the removal of excess Se and salts.

For example, agroforestry systems use drainage water as a resource, not a waste. Such water is reused on crops that have higher salt tolerance, which increases the water's utility value. In an agroforestry system, salt-sensitive crops use fresh, good quality irrigation water. Crops of progressively higher salt tolerance, such as cotton, certain species of trees, and halophyte plants, reuse drainage water several times before its final disposal or treatment. Salt-tolerant trees and halophyte plants, which partially take up the excess salts and Se, reduce the volume of drainage water to be managed, thereby permitting more economical management in solar evaporators or in other water treatment facilities. Figure 1 illustrates the removal of Se and salts from an agroforestry farming system. Because the method uses irrigation water more efficiently, the concentrations of salts and Se are increased in continuously reduced volumes of drainage water, facilitating their removal. Thus, agroforestry provides a sound ecological and environmental method for sustained productivity on irrigated land.

Figure 2 illustrates a flowchart of an agroforestry system for irrigated land. Salt-tolerant trees and halophyte plants provide the base of the system. Trees receive drainage water from salt-sensitive and salt-tolerant conventional farm crops. The trees also use groundwater from higher ground and intercept the flow of water from upslope areas. Trees and plants both use and evapotranspire the water, which concentrates salts, including Se, into small volumes of drainage water, reducing water treatment costs.

AGROFORESTRY FARMING FOR SE AND SALT MANAGEMENT

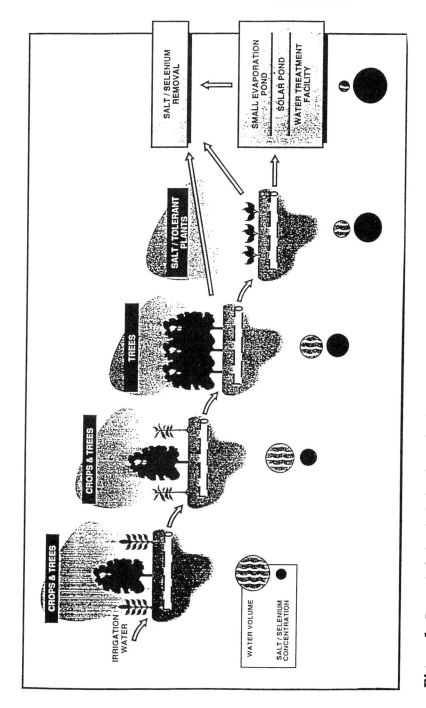

Figure 1 Removal of salt and selenium from a farming system.

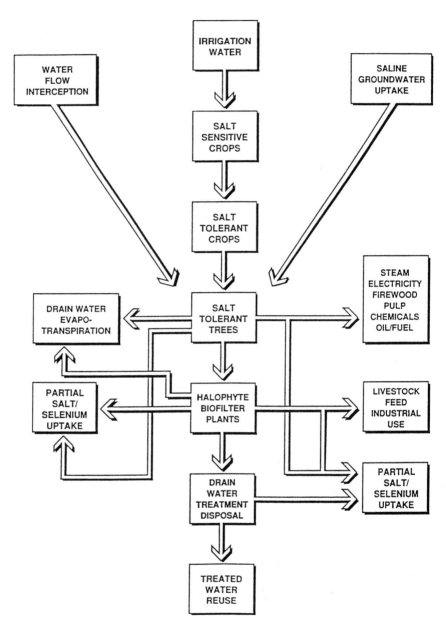

Figure 2 Farming system for the management of irrigated land.

III. AGROFORESTRY RESEARCH PROJECTS AND DEMONSTRATION PROGRAM

The agroforestry program being tested in the San Joaquin Valley requires coordinated research on a number of mutually dependent issues. Several of these ongoing research projects are described briefly in this section.

A. Selection and Propagation of Salt-Tolerant Trees

At the beginning of the agroforestry program, the seeds of salt-tolerant trees were imported mainly from regions of Australia that have salinity problems similar to those in California's San Joaquin Valley. However, because these trees have exhibited uneven growth in California, our research goal is to improve the quality control of planted materials through the propagation of selected trees. The selected trees should (a) have an increased tolerance to salts and boron, (b) evaporate large volumes of water, (c) achieve high yields, (d) have a form suitable for market demand, (e) be frost-tolerant, and (f) provide a healthy environment for wildlife. With respect to these attributes, research has been conducted on eucalyptus and casuarina trees. We are searching for other salt-tolerant tree species to create biological diversity. Trees are selected annually and planted in orchards, mother blocks, and control fields. They are propagated from California feedstock, either from seeds or by cloning or rooting selected cuttings.

B. Selection and Propagation of Halophyte Plants

Halophyte plants use water drained from salt-tolerant trees, further reducing drainage water volume. We have selected highly salt tolerant species of halophyte plants (EC > 30) and are attempting to find a suitable market for these plants. Halophytes typically take up 1.0–3.0 mg Se/kg. Their use as livestock feed supplements in Se-deficient areas has been evaluated [4]. Since halophyte plants can also provide a potential habitat for beneficial insects used in integrated pest management and biological control, the interactions between halophytes and insect pests and crop diseases are being studied. A partial list of halophyte plants tested is presented in Table 1.

C. Biomass Properties and Selenium Concentrations

Selenium uptake and fuel properties have been studied for eucalyptus, casuarina, and tamarisk (athel). Eucalyptus and casuarina samples are taken from agroforestry sites. They have similar Se concentrations,

Table 1 Halophyte Plants[a]

Scientific name	Common name
Agropyron elongatum	'Jose' tall wheatgrass
Agrostis stolonifera	Seaside bent grass
Allenrolfea occidentalis	Iodine bush
Andropogon barbinodis	Cane bluestem
Andropogon ischaemum	Yellow bluestem
Argania spinosa	Argan
Artemisia pycnocephala	Sandhill sage
Atriplex canascens	'Marana' fourwing saltbush
Atriplex lentiformis	'Casa' quailbush
Atriplex patula var. *hastata*	Fat-hen
Bassia hyssopifolia (Poll.)	Fivehook bassia
Brassica sp.	Mustard
Buchloe dactyloides	Buffalo grass
Distichlis spicata	Salt grass
Elymus triticoides	Creeping wild rye
Eriogonum fastciculatum	Wild buckwheat
Euphorbia antisyphilitica	Candelilla
Festuca arundinacea	'Goar' tall fescue
Frankenia grandiola	Alkali heath
Hyparrhenia hirta	Thatch grass
Indigofera miniata v. *leptosepala*	Western indigo
Oryzopsis hymenoides	Indian rice grass
Panicum coloratum	'Selection 75' kleingrass
Panicum coloratum	A-12638 kleingrass
Panicum miliaceum	'Dove' proso millet
Panicum virgatum	'Alamo' switchgrass
Paspalum vaginatum	Siltgrass
Pedilanthus macrocarpus	Waxplant
Pennisetum typhoides	Pearl millet or bajra
Phalaris aquatica	'Perla' koleagrass
Polygonum aviculare	Prostrate knotweed
Populus fremontii	Fremont cottonwood
Salicornia virginica	Pickleweed
Salsola spp.	Russian thistle
Spartina	Priare cordgrass
Spartina alterniflora	Marshgrass
Spartina foliosa	California cordgrass
Spartina gracilis	Cordgrass
Spartina patens	'Flageo' marshhay cordgrass
Sporobolus airoides	'Salado' alkali sacaton

[a]Partial selection for the Agroforestry Program.

averaging about 150–180 µg/kg in the wood and up to 910 µg/kg in the leaves (Figure 3). Ash derived from eucalyptus samples retains about 40% Se. For comparison purposes, tamarisk samples have been selected from natural stands near the agroforestry plantations. Tamarisk samples have higher Se concentrations, up to 3500 µg/kg in leaf tissues [6].

D. Salt and Selenium Balance

The goals of this research project are to (a) investigate the balance of Se and salts in a tile-drained agroforestry system, (b) study the long-term

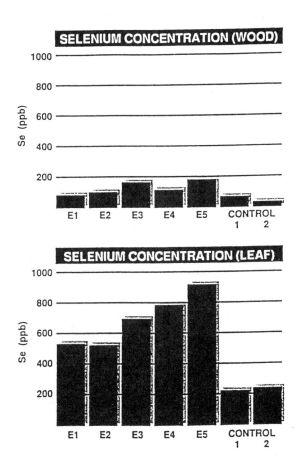

Figure 3 Selenium concentration in eucalyptus wood and leaves.

feasibility of reused saline drainage water by trees and halophytes, and (c) develop a management model for salt and Se balance in agroforestry systems. The measured concentrations of Se and salt in the agroforestry system are presented in Table 2. The achieved Se concentration is about half the salt concentration.

Based on research data obtained from the agroforestry research that started in 1985, an example of calculated salt balance in an agroforestry system is given in Table 3 for a hypothetical farm with a tree area of 10 ha. Using a typical water application rate of 15 ML/(ha·yr) (ML: megaliter = 10 cm of water over an area of 1 ha), the total volume of water applied is 150 ML/hr. At a salt concentration of 7000 mg/L, a salt load of 1050 t (metric tons) is added to the area. The tree plantation has subsurface drains, and the outflow of drainage water is reduced to 43 ML/yr, with a salt concentration of 24,500 mg/L. Because this water is reused by halophytes, the drainage water has a final volume reduced to 23 ML/yr and a final, more concentrated salt load of 46,000 mg/L. The salt load in the final drainage water is discharged into a solar evaporator from which the crystallized salt can be removed. The areas planted to salt-tolerant trees (10 ha) and halophytes (4 ha) receive drainage water from about 650 ha of conventional farm crops.

At present, additional research is needed on Se balance. The drainage water applied to trees in this agroforestry system has a typical Se concentration of 0.50 mg/L. Therefore, the 150 ML of irrigation water (at the rate of 15 ML/ha) used to water the 10 ha of trees brings with it 75 kg of Se. Some of the Se volatilizes in the agroforestry system, some is taken up by trees and halophytes, and the remainder is discharged with salts into a solar evaporator. The ratio of Se in the drainage water applied to trees to the Se in the water drained from the tree area has been measured at 1:1.775

Table 2 Concentration of Salt and Selenium in Agroforestry Systems

Irrigation of trees	Electrical conductivity (ds/m)		Selenium (mg/L)	
	A	B	A	B
Water inflow	10	11.6	0.396	0.578
Water outflow	32.4	37.8	0.7	1.03
Ratio <1 : x>	3.24	3.26	1.77	1.78

A, B: two independent tests.

Table 3 Salt Balance in an Agroforestry System for a Hypothetical Farm with a Tree Area of 10 ha

	Salt-sensitive crops	Salt-tolerant crops	Salt-tolerant trees	Halophyte plants	Solar evaporator
Salt load (tonnes)	1,050	1,050	1,050	1,050	1,050
Area (ha)	560	91	10	4	
IR/W Applic. rate (ML/ha)	7.5	9	15	10	
IR/W volume (ML)	4,200	820	150	43	
IR/W salt conc. (mg/L)	250	1,280	7,000	24,500	
D/W salt conc. (mg/L)	1,280	7,000	24,500	46,000	
D/W volume (ML)	820	150	43	23	23
Leaching fraction	0.2	0.18	0.29	0.53	
Evaporation rate (ML/ha)					13
Area (ha)					1.77
Salt density (kg/m^3)					1,500
Salt volume (m^3)					700
Salt layer height (cm)					4.60

(Table 2). Additional data must be developed on the Se concentration in the drainage water from halophytes and on the volatilization of Se by trees and halophytes.

E. Wildlife Habitat

Agroforestry systems attract wildlife. The interaction of wildlife with agroforestry has been studied for several years to evaluate the effects of Se concentrations on the wildlife food chain and the biological safety of agroforestry farming systems [2].

F. Selenium Transfer

While excess Se is causing ecologists to be concerned about the safety of wildlife in the San Joaquin Valley, problems associated with Se deficiency occur on the range areas of the Sierra Nevada. Deficiency of Se in the diets of cattle is common there and in many other regions [4]. A desirable level of Se in the blood of cattle is 0.08 mg/L. In late summer and early fall, when supplemental feeding of cattle typically begins on the range, average cattle blood Se levels may be as low as 0.01 mg/L. The possibility of a partial transfer of Se from surplus to deficiency regions through supplements in livestock feed has been studied. When the forage harvested from experimental halophyte fields was used as a supplemental feed for cattle, research data indicated that blood Se levels increased compared to when alfalfa hay was used (Figure 4), although cattle weight gain was similar [4].

G. Marketing of Trees and Halophytes

Marketing opportunities for tree biomass harvested in agroforestry systems have also been studied [7]. Potential near-term markets include firewood, paper pulp, and power-generating plants. Production of chemicals and fuels offers a long-term cyclical market sensitive to world oil prices. Currently, eucalyptus oil is used as a food additive, and an established market exists for eucalyptus honey. In the pharmaceutical industry and in other industrial applications, this oil holds promise. To be introduced successfully into agroforestry farming, halophytes must be economically viable. Potential markets include livestock forage (as discussed), oils, and energy biomass.

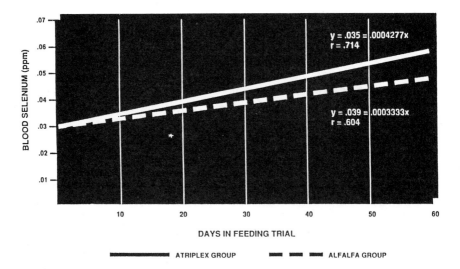

Figure 4 Selenium transfer through feeding halophyte forage. Cattle blood selenium levels versus days in atriplex feeding trial, San Joaquin experimental range.

H. Demonstration Program

The first planting of salt-tolerant trees was conducted on two farms in 1985. During a 7-year period (1985–1991) farmers planted about 700,000 trees in Fresno, Kings, Tulare, and Kern Counties. The primary species planted were *Eucalyptus camaldulensis, Casuarina glauca,* and *Casuarina cunninghamiana*. Mesquite and poplars have also been planted on a limited scale. Farmers have been working with the USDA Soil Conservation Service and the California Department of Food and Agriculture on planting salt-tolerant trees. Halophyte selection is still in the research stage.

IV. SELENIUM FLOW IN AGROFORESTRY FARMING SYSTEMS

Salt-tolerant trees and halophyte plants in agroforestry farming systems assist in the management of Se and salts by concentrating them in reduced volumes of drainage water that can be disposed of in small evaporation basins or removed via technological processes. As shown in Table 3, the agroforestry system reduces drainage water, conventionally disposed of on farms, to about 9% (average value) of its original volume. Table 4 shows the contributions of salt-tolerant trees and halophytes to Se removal. Water applied at the rate of 15 ML/ha with a Se concentration of 1 mg/kg takes up 25 g Se/ha. Because they use multiple cropping that reuses water, agroforestry systems reduce Se and salt loads by reducing the inflow of Se into irrigated farmland. Increases in soil organic matter in agroforestry systems together with proper selection of crops facilitate Se volatilization.

V. ECOLOGY OF AGROFORESTRY SYSTEMS

To build the agroforestry design matrix, an integrated ecological approach is used. Factored into the matrix are Se, salts, soil characteristics, distribution of shallow groundwater tables, water seepage from delivery canals, downslope water flow, characteristics of wildlife habitats, biomass energy content, interactions of the drains with the trees, and wind direction, among other agroecological issues. The integration of salt-tolerant trees and halophyte plants into the agroforestry concept diversifies cropping systems, creates new wildlife habitats, controls soil erosion, improves air quality through absorption of excess carbon dioxide, increases the utility value of water through its multiple reuse, and more effectively manages excess Se and salts.

Shallow groundwater at levels of up to 1.0 m affects large farmland areas in the San Joaquin Valley, inhibiting the production of conventional

Table 4 Selenium Inflow to and Uptake in Agroforestry Sites

Selenium concentration (mg/L)	Inflow of selenium (g/ha) Water application rates (ML/ha)					
	3	6	9	12	15	18
0.001	3	6	9	12	15	18
0.005	15	30	45	60	75	90
0.01	30	60	90	120	150	180
0.05	150	300	450	600	750	900
0.1	300	600	900	1,200	1,500	1,800
0.3	900	1,800	2,700	3,600	4,500	5,400
0.5	1,500	3,000	4,500	6,000	7,500	9,000
0.6	1,800	3,600	5,400	7,200	9,000	10,800
0.8	2,400	4,800	7,200	9,600	12,000	14,400
1	3,000	6,000	9,000	12,000	15,000	18,000

Selenium concentration in biomass (mg/kg)	Selenium uptake in biomass (g/ha)					
	Yield—Halophytes (t/ha)			Yield—Trees (t/ha/yr)		
	6	10	15	25	30	40
0.60	3.60	6.00	9.00	15.00	18.00	24.00
0.80	4.80	8.00	12.00	20.00	24.00	32.00
1.00	6.00	10.00	15.00	25.00	30.00	40.00
1.40	8.40	14.00	21.00	35.00	42.00	56.00
1.80	10.80	18.00	27.00	45.00	54.00	72.00
2.20	13.20	22.00	33.00	55.00	66.00	88.00
2.60	15.60	26.00	39.00	65.00	78.00	104.00
3.00	18.00	30.00	45.00	75.00	90.00	120.00
3.50	21.00	35.00	52.50	87.50	105.00	140.00

farm crops. Trees and halophytes planted in these shallow groundwater areas function as "vertical drains" or "biological pumps." The trees and halophytes take up groundwater, use it for the production of biomass, evaporate it through their foliage, and thereby contribute to the overall management of Se and salts.

Seepage from water delivery canals contributes to the problems of drainage, excess Se, and salts. When salt-tolerant trees and halophytes are planted in these selected areas as part of an agroforestry farming system, they intercept the water seepage and reduce its volume through evapotranspiration.

Water management on upslope land affects farms in low areas in the valley. Water flowing downslope increases management problems of drainage water and salts in the downslope areas. Salt-tolerant trees and deep-rooted halophytes can intercept the subsurface or open-ditch flow of the water and thus reduce the drainage water volume and salt load, including Se.

The ecology of agroforestry systems and wildlife are interrelated because the tree plantations function as "biological magnets" for wildlife, providing both habitat and food sources among surrounding crops. Selected trees and bushes can be planted in blocks or rows of a few hectares dispersed in broader farming areas to control soil erosion.

The San Joaquin Valley experiences air quality problems similar to those in larger metropolitan areas of California. Soil erosion, caused by wind moving freely through a treeless region, also affects some areas of the valley. The large-scale introduction of agroforestry techniques would control the movement of soil particles in the air and would contribute to the enhancement of air quality through the absorption of excess carbon dioxide.

VI. CONCLUSION

The management of salts, including Se, is primarily a farming issue, not a civil engineering problem, although it has been approached as such, with expensive engineering methods being recommended for drainage water management. Instead, a more effective, economical approach is the use of integrated farming techniques like agroforestry that rely on water reuse and diversification of conventional crops, particularly the introduction of salt-tolerant trees and halophyte plants, to manage excess Se and salts by reducing the final volume of drainage water. Agroforestry is economically viable and ecologically sound, providing a long-term solution to farming on irrigated land. Agroforestry techniques should be used in conjunction with irrigation technologies, drainage water treatment, removal of excess Se and salts, and marketing of biomass. Agroforestry's integrated approach will lead to the development of a lasting solution for the management of Se and other salts so that food production can be sustained on irrigated farmland.

REFERENCES

1. V. Cervinka, A Farming System for the Management of Salt and Selenium on Irrigated Land (Agroforestry), California Department of Food and Agriculture, Agric. Resour. Branch Tech. Rep. 1990.

2. D. L. Chesemore, A. R. Dyer, T. D. Kelly, and C. Steggal, Agroforestry Wildlife Study, California State University, Fresno, 1988.
3. J. Cooper and J. Loomis, The Economic Values to Society and Landowners in the San Joaquin Valley Agroforestry Plantations, University of California, Davis, 1988.
4. W. E. Frost, M. W. Thomas, and D. E. Jones, Suitability of Atriplex as Supplemental Feed and Selenium Source for Cattle on Annual Rangeland, California State University, Fresno, 1990.
5. A. F. Heuperman and V. Cervinka, *Integrated Biological and Engineering Systems for Salinity Management in Australia and California, Calif. Dept. Food Agric., Agric. Resour. Branch Techn. Rep., 1992.*
6. B. M. Jenkins, J. F. Happ, and M. Kayhanian, Characterization of Selenium Concentrations and Fuel Properties of Biomass Produced Under Saline Irrigation in the San Joaquin Valley of California, University of California, Davis, 1989.
7. Markets for California Eucalyptus, Calif. Dept. Food Agric., Agric. Resour. Branch Tech. Rep., 1989.

11
The Algal–Bacterial Selenium Removal System: Mechanisms and Field Study

Tryg J. Lundquist, F. Bailey Green, R. Blake Tresan, Robert D. Newman, and William J. Oswald

University of California, Berkeley
Richmond, California

Matthew B. Gerhardt

Brown and Caldwell
Pleasant Hill, California

I. INTRODUCTION

Irrigation drainage water from parts of the San Joaquin Valley in California contains levels of selenium and other trace contaminants that have been implicated in bird deformities at Kesterson Reservoir [1]. Depending on location and season, the drainage water contains 100–1400 µg selenium (Se) per liter, predominantly as selenate [SeO_4^{2-}, Se(VI)], the most soluble form. The California State Water Resources Control Board [2] has recommended an interim maximum mean monthly selenium concentration of 2–5 µg/L in receiving waters and wetlands. Drainage waters also contain 20–150 mg nitrate nitrogen (NO_3^--N) and 2000–4000 mg sulfate (SO_4^{2-}) per liter, depending on season and location [3].

II. THE ALGAL–BACTERIAL SELENIUM REMOVAL SYSTEM

At the University of California, Berkeley, we proposed and studied a biological treatment process called the Algal-Bacterial Selenium Removal System (ABSRS) to remove selenium and nitrate from drainage water. Although many other selenium removal processes have been proposed,

251

including physicochemical treatment [4–9] and biological treatment [10–12], only the ABSRS produces microalgae that provide sources of carbon and energy for the bacteria capable of reducing nitrate, selenate, and selenite [SeO_3^{2-}, Se(IV)] [13]. In addition to this benefit, ponding systems of the ABSRS type are very economical to build and operate compared with conventional wastewater treatment systems [14].

In the proposed flow scheme of the ABSRS, drainage water is introduced to high-rate ponds where microalgae are grown (Figure 1). A high-rate pond is a shallow, continuously mixed raceway designed to maximize algal productivity [16,17]. Because algae are approximately 9.2% nitrogen by dry weight [18], some NO_3^- nitrogen is taken up by algae from the water in this stage. If necessary, the algae are separated from the high-rate pond effluent by sedimentation or dissolved air flotation. The algae are then pretreated and returned to the main flow. The algae-laden water is then discharged to one or more anoxic pits in downstream ponds, where denitrifying bacteria use the algal biomass as a source of carbon and energy to reduce nitrate to nitrogen gas. Selenate is then reduced to selenite and elemental selenium by bacteria that use the oxidized selenium species as electron acceptors for respiration [13,19]. The reduced selenium incorporated in bacterial sludge is then removed from the effluent by dissolved air flotation enhanced with iron salts.

Large systems may produce algae in excess of that required for nitrate and selenium reduction. This excess could be fermented to methane for power generation in a deep pond equipped with a gas capture system. The waste heat produced in the generation of electricity could be used to pretreat algae before they are introduced to anoxic units. Carbon dioxide from the methane fermentation, reduction, and power generation units

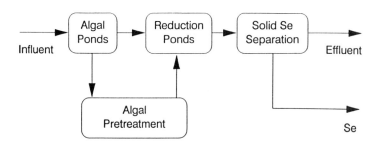

Figure 1 Schematic diagram of proposed process for removing selenium (Se) from drainage water using anoxic reduction units containing bacteria fed with algae. (From Ref. 15.)

could be recycled to the high-rate ponds for pH control. This recycled carbon would also benefit algae production by supplementing the bicarbonate in the influent drainage water.

Based on encouraging laboratory results, a pilot ABSRS was constructed and operated in the field. The field system was used to measure the productivity of algae grown in drainage water, to test removal of nitrate and selenium by bacteria fed with drainage-grown algae, and to study methane fermentation of drainage-grown algae. Further laboratory experiments were conducted to determine the mechanisms of selenium reduction in the system, the effects of two potential inhibitors, nitrate and sulfate, and the quantity of algae required to reduce the nitrate and selenate in the drainage water. Results of these pilot plant and laboratory studies are the subject of this chapter.

III. PILOT PLANT CONFIGURATION AND OPERATION

A. High-Rate Algae Production Ponds

Two high-rate ponds (Figure 2) were constructed adjacent to a subsurface drain at Murietta Farms in Mendota, California. The first pond in the series, denoted HRP 1, was 3.1 m x 15.3 m, and the second pond, denoted HRP 2, was 3.1 m x 9.1 m. When operated at a depth of 22 cm, the lined ponds held approximately 10,000 and 6000 L, respectively. To create a continuous channel in each pond, a dividing wall was installed down the middle to within 1.5 m of each end. A single 0.37-kW motor drove a six-blade paddlewheel in each pond to maintain a linear flow velocity of 15 cm/s.

HRP 1 was operated from July 1987 to June 1988, when farm personnel stopped the influent drainage water supply. HRP 2 was then constructed, and the system was restarted in September 1988 and operated until October 1989. These operation periods were too short for the system to achieve steady state.

To operate the field system, water was pumped from the subsurface drain to a constant-head tank that was metered into HRP 1. HRP 1 effluent was discharged to a 750-L tank for algal sedimentation. From there, water was pumped to HRP 2.

The two ponds were connected in series and were operated differently. Phosphorus and trace nutrients were added to HRP 1 to ensure their presence in the drainage water for optimal algal growth. With the exception of phosphate, though, the nutrients may have already been present in sufficient quantities. No additional nutrients were added to HRP 2. Carbon dioxide was injected into HRP 1 to keep the pH below 8.3, preventing

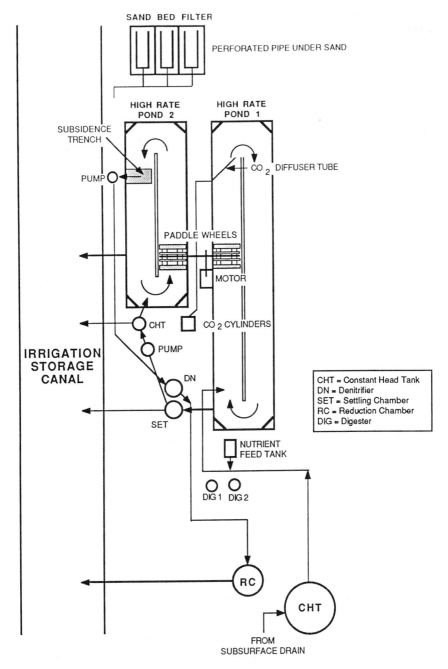

Figure 2 Plan view of ABSRS tested at Mendota, California. CHT, constant-head tank; DN, denitrifier; SET, settling chamber; RC, reduction chamber; DIG, digester. (From Ref. 15.)

precipitation of carbonates, while the pH in HRP 2 was allowed to increase during the day and decrease at night in the normal fashion of active algal culture. Typically, HRP 2 pH varied between 10.5 and 8.75.

B. Nitrate Reduction

A covered 750-L denitrification tank was placed on-line in May 1989. It was loosely filled with 10-cm Intallox saddles to serve as a solid substrate for fixed-film growth and partially mixed with a small submersible pump. Because the nitrate concentration in drainage water at the test site greatly exceeded that of typical drainage water, the pond system could not produce enough algae to satisfy the carbon demand for denitrifying the entire flow. Therefore, 25% of the HRP 2 effluent was sent to the denitrification chamber, while the remainder was discharged out of the system.

Algae harvested from the sedimentation tank, which followed HRP 1, were used as the carbon and energy source for denitrifying bacteria. The naturally settled algae were thickened by centrifugation and, when necessary, rediluted with tapwater to 30 g/L volatile suspended solids (VSS). The concentrated algal sludge was heated to 60°C and then poured through a tube to the bottom of the denitrification tank. This task was performed every 1–2 days.

C. Removal of Soluble Selenium

Effluent from the denitrification chamber was piped to the bottom of a 4000-L reduction chamber, also started in May 1989. The bottom 30% of the chamber contained Intallox saddles. The residence time in the reduction chamber was 3.7–11.1 days. Water left the tank through an overflow pipe at the surface and flowed out of the system.

D. Methane Fermentation of Drainage-Grown Algae

Two digesters were constructed using 3.1-m lengths of 36-cm PVC pipe. The capped pipes were buried to 65% of their length in the ground to minimize temperature variations. Algae were centrifuged and heated as described in the section on nitrate reduction before being fed to the digesters. Outlet tubes allowed displaced effluent to pass into the reduction tank. Sludge residence time in the digesters was approximately 47 days.

IV. STUDIES ON ABSRS SELENIUM REDUCTION MECHANISMS

A. Selenium Analyses

Preliminary studies showed that selenium spike recoveries were highest in the laboratory and field samples analyzed by Method 3500-Se B.4 [20], in which acid permanganate is used to remove organic interferences, selenium is reduced to Se(IV) using HCl, and Se(IV) is determined by hydride atomic adsorption. In particular, when addition of the hydroxylamine hydrochloride was limited to 3 drops per 3-mL sample, quality assurance data showed that the method worked well. All other methods available in our laboratory, including those in which samples were treated only partially so selenium species could be differentiated, yielded poorer quality assurance recoveries. The only speciation method that proved useful involved liquid chromatography plasma mass spectrometry. This method, conducted at the California Department of Health Services, yielded semiquantitative data.

B. Algal Reduction Experiments in the Laboratory

Most of the selenium reduction experiments consisted of combining algae reductant with drainage water in the laboratory. The contact occurred either in batch samples mixed in flasks or in 5 cm x 150 cm acrylic continuous-flow columns. The mixtures were allowed to react for up to 72 h. At appropriate times, samples were drawn and analyzed for soluble selenium. The results were compared with the soluble selenium concentration of the column influent or with that of control samples.

C. Continuous-Flow Stirred-Tank Reactor

A 2-L continuous-flow stirred-tank reactor (CSTR) was constructed for the cultivation of selenium-reducing organisms. Synthetic medium was constituted to simulate the mineral composition of drainage water. It was supplemented with high levels of selenate and acetate [21]. The reaction vessel was inoculated with a culture that had been taken originally from the Mendota digester.

D. Agar Media Experiments

Agar tubes and petri dishes were used to culture selenium-reducing bacteria and to study their microbiology. Small test tubes (1.4 cm x 12.3 cm) were filled with the medium used in the CSTR experiment augmented with 15 g of agar per liter. The tubes were capped with silicone stoppers and autoclaved. Then tubes were inoculated from the CSTR and incubated at room temperature for several days.

V. RESULTS OF STUDIES ON THE MECHANISMS OF SELENIUM REDUCTION

A. Biologically Mediated Reduction

Laboratory and field experiments showed that selenium was removed primarily by a biological process. Both sterilization and spiking with toluene prevented reduction in batch experiments, and inoculation stimulated it.

The time required to reduce soluble selenium was much longer than in known physicochemical selenium reduction processes. Removal required 12 h in some experiments (Figure 3a) and 48 h in others. In contrast, some physicochemical processes require only a few minutes [6,7,22].

Laboratory studies indicated physicochemical reduction with the algal–bacterial process as well. In one batch reduction experiment, soluble selenium declined 27%, from 334 to 243 µg/L, in just 90 min using a reductant concentration of 300 mg/L. It is unlikely that biological reduction could have occurred in such a short period in a noninoculated flask, so some physicochemical reaction may have occurred, possibly sorption onto the reductant biomass. Interestingly, toluene completely inhibited all reduction, including the initial, apparently nonbiological uptake. Soluble selenium was never monitored at such short intervals in subsequent experiments.

B. Inhibition of Selenium Reduction by Atmospheric Oxygen

Several experiments showed that atmospheric oxygen inhibited selenium reduction. Stripping dissolved oxygen from batch reduction flasks decreased the oxidation–reduction potential (Eh) and increased the removal of soluble selenium. These results were expected, since O_2 is the most energetically favorable electron acceptor and its presence would extend the lag period before selenate reduction.

In the agar medium experiments, selenium-reducing bacteria grew well in pour plates, as indicated by the appearance of red reduced selenium, but did not grow on the surface of the same medium in streak plates. Cultures that did form on the streak plates were transferred to agar tubes but showed no ability to reduce selenium. These results showed that the selenium-reducing bacteria are likely to be obligate anaerobes or

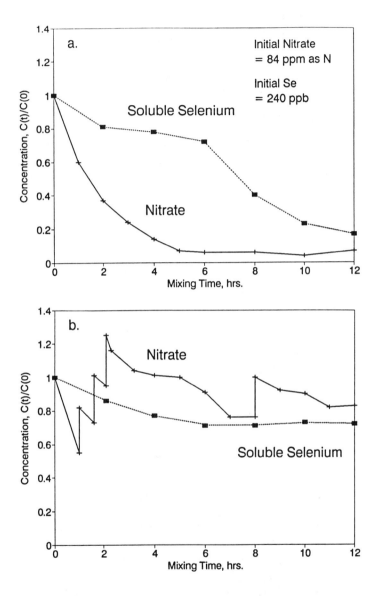

Figure 3 Effect of nitrate on batch reduction of selenium in a mixture of drainage water and digested algae. (a) Nitrate not added during run; (b) nitrate augmented to maintain concentration. (From Ref. 3.)

microaerophilic. This trait distinguishes them from most denitrifying bacteria, which are facultative [23].

C. Possibility of Multiple Organisms

Experiments with agar media in tubes always yielded a pair of red bands near the air–medium interface, regardless of whether the tube was inoculated from the top or the bottom. The red bands were probably selenium reduction products deposited by the cells immobilized in the agar. When tubes with reduced media are exposed to air, an Eh gradient develops as O_2 diffuses into the agar. The two red bands could indicate two organisms with different optimal Eh ranges. An attempt to isolate organisms from each of the two bands was unsuccessful, possibly due to cross-contamination. When material from one or the other of the two bands was inoculated into new tubes, two bands formed again.

In the agar tubes, the sharp red bands always began about 1 mm from the surface and extended down about 10 mm. Below this level, the tubes were less red. Since such a sharp band formed just past the 1-mm depth, it is possible that the bacteria are microaerophilic—that they require some small amount of free molecular oxygen. Other researchers have also reported that selenium-reducing bacteria were microaerophilic [24].

D. Nitrate Inhibition

The laboratory and field data supported the conclusion that nitrate had different effects on the slow and fast mechanisms of selenium reduction. In some laboratory experiments, selenium reduction was not observed until after nitrate had been reduced to about 10 mg/L (Figures 3a and 4). However, in the experiment in which nitrate was periodically added to one flask (Figure 3b), soluble selenium decreased from 240 µg/L to approximately 170 µg/L, a 30% decline, in both the augmented and control flasks during the first 6 h. After 6 h, nitrate in the control flask was very low, and soluble selenium decreased much more rapidly, whereas no further reduction occurred in the flask augmented with nitrate. The change in soluble selenium removal rate at 6 h suggests a change from a slow mechanism unaffected by nitrate to a much faster mechanism inhibited by nitrate. The slow mechanism also appears to be unable to reduce soluble selenium below a certain, rather high level (Figure 3b).

The slower, nitrate-independent mechanism was noted in other experiments. Over 96 h, soluble selenium decreased 20–91% in flasks in which the nitrate concentrations did not fall below 25–80 mg/L as N. In another set of laboratory experiments, selenium reduction continued after 48 h

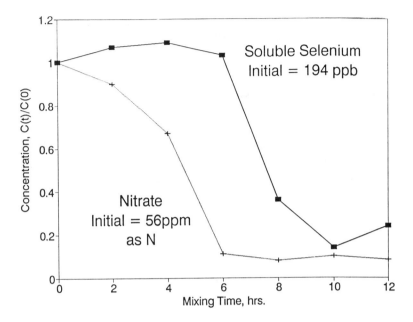

Figure 4 Laboratory batch nitrate and selenium reduction—frequent sampling. (From Ref. 3.)

although nitrate reduction had stopped. Apparently some of the algal biomass not available to denitrifying bacteria was available for selenium reduction.

In the Mendota 750-L reduction chamber, soluble selenium concentration did not decrease in the effluent until nitrate fell below about 10 mg/L, indicating inhibition, but in the 4000-L reduction chamber at Mendota, some selenium reduction did occur in the presence of nitrate. In July 1989, while the nitrate concentration in the reduction chamber effluent was still high (50–75 mg/L as N), speciation analyses showed that Se(VI) was being reduced to Se(IV) in the reduction chamber. Thus, there appears to be a selenium reduction mechanism that is not inhibited by nitrate and a faster mechanism that is inhibited by it.

E. Reduction of Selenium

Experiments conducted with the CSTR that was fed artificial drainage water and with the agar media tubes showed that the main selenium reduction reaction was dissimilatory. The media contained only acetate, selenate, salts, and trace nutrients. All of the usual electron acceptors—oxygen, nitrate, and carbon dioxide—were excluded. The only compound available as an electron acceptor was selenate, so only cells that underwent dissimilatory selenate reduction could grow in the media. It was found that when the CSTR was inoculated with selenium-reducing cultures from the Mendota system, definite bacterial growth and selenium reduction occurred. Many bacterial cells were found in microscopic observations, and, as discussed below, a substance appearing to be red reduced selenium was produced. Therefore, bacteria capable of dissimilatory reduction were definitely present in the Mendota sludges. This result is in agreement with work done by Oremland et al. [19], who reported dissimilatory reduction occurring at Kesterson and in San Francisco Bay sediments.

The inhibition of selenium reduction by nitrate can be explained by the fact that dissimilatory selenate reduction provides less energy than denitrification. The amount of energy an organism can obtain from its carbon/energy source depends on the reduction potential of its electron acceptor (Table 1). An acceptor with a high potential will allow more energy to be transferred from the carbon/energy source than one with a

Table 1 Oxidation–Reduction Potential[a] in Drainage Water in Equilibrium with Various Redox Couples

Redox couple	Equilibrium oxidation–reduction potential[b] (mV)
O_2/H_2O[c]	+750
NO_3^-/N_2[c]	+670
SeO_4^{2-}/SeO_3^{2-}	+380
SeO_3^{2-}/Se^0	+180
SO_4^{2-}/S^0	−280
SO_4^{2-}/HS^-	−280
CO_2/CH_4[c]	−290
CO_2/CH_3COO^-	−340
N_2/NH^{4+}	−350

[a]Calculated from Gibbs free energy of formation data in Ref. 27.
[b]Equilibrium Eh referenced to the standard hydrogen electrode when the molar ratio of the indicated redox couple is unity, unless otherwise indicated. T = 298 K, pH = 8.
[c]Concentrations: O_2 = 0.21 atm, N_2 = 0.79 atm (concentrations in air); NO_3^- = 150 mg/L as N; CH_4 = 0.75 atm, CO_2 = 0.2 atm (approximate concentration in digester gas).
Source: Ref. 3.

lower potential. When O_2 is absent but both nitrate and selenate are available, cultures should tend to use the nitrate before they use the selenate because they can obtain more energy through nitrate reduction. Even if nitrate and selenate are reduced by different bacteria, the nitrate should diminish first owing to the competitive energy edge of the nitrate reducers over the selenate reducers.

An assimilatory selenium reduction mechanism may exist simultaneously with a dissimilatory one. Such a process could conceivably explain the observed nitrate-independent removal of soluble selenium. Sulfate-assimilating bacteria could assimilate selenate into their cells as well, due to its resemblance to sulfur.

Cells will generally produce compounds they need only if the compounds are not already present. To prevent assimilatory sulfate reduction, the products of assimilatory sulfate reduction were added to a flask far in excess of levels that might be required for protein production by the cells. The addition of the sulfur-containing amino acids cysteine and methionine at 4 mM each (490 and 600 mg/L, respectively) stopped selenium reduction but not nitrate reduction, consistent with the conclusion that the initial selenate reduction was assimilatory. However, after nitrate was reduced in the flask, no additional selenium reduction occurred, indicating that the dissimilatory reduction of selenium was also prevented by the sulfur-containing amino acids.

The reason for the inhibition of dissimilatory selenate reduction by the sulfur-containing amino acids is not clear. First, amino acids can chelate metals, so it is possible that minerals essential to the selenium-reducing bacteria were made unavailable. However, another metal chelator, sodium citrate, when added at 4 mM (770 mg/L as citric acid), was found to have no effect on selenium removal. The citrate may have been biodegraded by other organisms present in the flask, though, so this experiment was inconclusive. Second, reduction may have been stopped by sulfide toxicity. At the end of the reduction run, a strong sulfide or mercaptan odor was noticed in the flask. This odor was never found in other reduction flasks and suggests that bacteria were degrading the sulfur-containing amino acids as a carbon source. Since no oxygen was present, the bacteria must have used NO_3- as an electron acceptor in the amino acid decomposition. Third, the Eh was much lower in the presence of the amino acids: the control flask Eh was +175 mV, while the amino acid flask Eh was –155 mV. Cysteine is often used as a reducing agent in anaerobic bacterial growth media [25], and so the cysteine or the sulfide was probably responsible for the reduced Eh. These possibilities leave the existence of an assimilatory pathway uncertain.

F. The Product of Dissimilatory Selenate Reduction

Reduction of selenate in the CSTR produced a red substance insoluble in water. Several analyses of the material confirmed the presence of selenium, and data indicated that it was some form of elemental selenium (Se^0). Of the four forms of amorphous Se^0—powdered red, colloidal red, black, and vitreous—the red material from the CSTR most resembled powdered red amorphous Se^0 in its properties, but it could also have been colloidal selenium. Elemental selenium has been found in other studies of dissimilatory selenate reduction [19], including studies of the Binnie Biosel process [26].

G. Empirical Reactions

Given that the reduction was dissimilatory and that Se(IV) and Se(0) were apparently formed, empirical chemical reactions can be proposed. In the equations below, the reductant is shown as acetate.

$$CH_3COO^- + 4\ SeO_4^{2-} + 5\ H^+ \rightarrow 2\ CO_2 + 4\ HSeO_3^- + 2\ H_2O \quad (1)$$

$$2\ CH_3COO^- + 4\ HSeO_3^- + 6H^+ \rightarrow 4\ CO_2 + 4\ Se^0 + 8\ H_2O \quad (2)$$

$$3\ CH_3COO^- + 4\ SeO_4^{2-} + 11\ H^+ \rightarrow 6\ CO_2 + 4\ Se^0 + 10\ H_2O \quad (3)$$

The overall reaction, equation (3), is similar to dissimilatory sulfate reduction except that the selenium is reduced only to the elemental state [Se(0)] instead of the Se(-II) state, selenide. Note that nitrate is not included in equation (3).

More energy can be obtained in acetate oxidation using Se(VI) rather than Se(IV) as an electron acceptor. The Gibbs free energy change for the reactions shown in equations (1), (2), and (3) is -1024, -884, and -930 kJ/mol acetate, respectively. These values were calculated for pH 8 from thermodynamic data [27], using concentrations of 100 mg/L acetate (3.4×10^{-3} M), 10^{-3} M CO_2, and equimolar concentrations of all selenium species. The presence of Se(IV) in the Mendota reduction chamber and laboratory CSTR is consistent with the calculations that show that more energy per mole of acetate is obtained from reduction of Se(VI) than from reduction of Se(IV). Organisms would tend to utilize Se(VI) first, converting much of it to Se(IV) before further reducing it to Se(0).

More soluble selenium was measured in the CSTR effluent than would be predicted from equations (1) and (2), but the excess may have been due to colloidal selenium passing through the 0.22-μm filters used in the analysis.

H. Quantity of Algae Required

Given the relative concentrations of nitrate and selenium in the drainage water, the amount of reductant needed for denitrification is much greater than that needed for selenium reduction. Unless a nitrate-independent selenium reduction process can be perfected, nitrate concentration determines how much reductant is required.

To calculate the amount of reductant needed, it is necessary to determine the concentration of oxidizing equivalents in the water relative to the desired Eh level. Reduction of selenium without reduction of sulfate requires a final Eh of between +180 and −280 mV (Table 1). Drainage water typical of the San Luis Drain (47 mg nitrate-N and 8 mg O_2 per liter) would require 12.8 meq of reductant per liter to remove the dissolved oxygen, nitrate, and selenate. This reductant can be in the form of any biodegradable compound containing reduced carbon.

Measurement of the reducing equivalents of algae using chemical oxygen demand (COD) analysis yielded a value of 0.180 ± 0.016 meq/mg dry weight. This agrees well with the theoretical value of 0.194 meq/mg dry weight algae calculated using the empirical formula $C_{106}H_{181}O_{45}N_{16}P$ for algae.

Batch experiments showed that 17–31% of the theoretical reducing capacity was available to denitrifying bacteria. Attempts to increase the available fraction using heat and sonic disruption of cells were not successful. Therefore, unless some process is developed to improve algal biodegradability, 190–270 mg algae/L would be required to completely remove dissolved oxygen, nitrate, and soluble selenium from typical drainage water (Figure 5). The algal concentration in the Mendota HRP 1 averaged 178 ± 99 mg VSS/L, slightly lower than the quantity required for complete treatment of typical drainage water. This average concentration might have been higher had dust from the adjacent road not entered the pond. Higher algal production and nitrate uptake should be obtainable, though, by providing carbon dioxide to a second HRP. For example, a series of HRPs could produce 460 mg of algae per liter of typical drainage water while removing 90% of the nitrate. This configuration would leave only 4.7 mg nitrate-N/L to be denitrified.

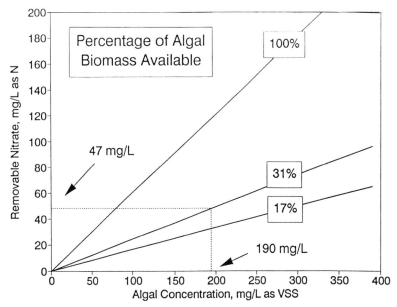

Figure 5 Theoretical nitrate removal as a function of algal concentration and algal availability to denitrifying bacteria. Average drainage water containing 47 mg NO_3^- as N per liter would require 190 mg algae/L at 31% availability. (From Ref. 13.)

I. Continuous-Flow Algae-Fed Columns

Removal of soluble selenium by the continuous laboratory reduction columns was poor when they were fed 320 mg of anaerobically digested algae per liter. Removal of soluble selenium averaged 6% from HRP 1 effluent and 28% from Mendota reduction chamber effluent. The removals were not consistent over time.

When two columns were configured to be operated in series and fed more reductant, selenium removal improved (Figure 6). The first column was fed 960 mg algae/L as VSS, and the second column was fed 900 mg algae/L as VSS. Hydraulic residence time in the system was 48 h at most. Nitrate averaged 100 ± 6 and 6 ± 4 mg/L as N in the system influent and effluent, respectively. Soluble selenium averaged 383 ± 20 and 93 ± 37 µg/L in the system influent and effluent, respectively, with effluent soluble selenium measuring as low as 46 µg/L. Thus, with adequate reductant concentration, the columns achieved high selenium removal.

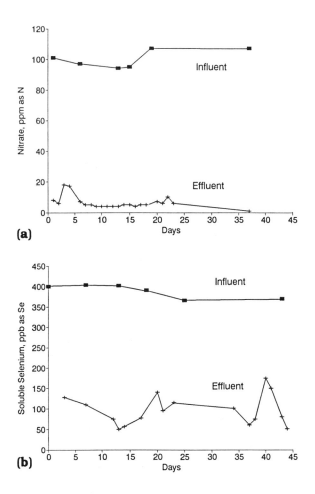

Figure 6 Concentrations of nitrate (a) and soluble selenium (b) in influent and effluent of continuous laboratory reduction columns with high reductant concentration. Reductant (1860 mg/L) added to columns in series. (From Ref. 3.)

VI. MENDOTA FIELD RESULTS

A. Production and Harvesting of Algae

Algal productivity was high in HRP 1, approaching levels measured in municipal wastewater ponds. Algal concentration in HRP 1, measured as VSS, averaged 178 ± 99 mg/L. Seasonally, the averages were 209 ± 103 mg/L in summer, 198 ± 128 mg/L in fall, 121 ± 44 mg/L in winter, and 173 ± 75 mg/L in spring. The annual average productivity, based on the seasonal averages, was 47.5 t/(ha·yr).

The selenium content of the drainage water at the Mendota site, which averaged 330 µg/L, apparently did not inhibit algal growth.

Although the productivity of algae was excellent, the harvesting was poor due to the primitive method used. A simple tank acted as a clarifier to remove algae from the pond effluent. In a larger system, it is envisioned that dissolved air flotation or properly designed sedimentation basins would be employed. Both methods have been shown to be very effective and economical for harvesting algae from HRPs [28–30].

B. Removal of Nitrate

Influent nitrate averaged 121 ± 19 mg/L as N in the drainage water samples. This concentration was 74 mg/L greater than the average in the San Luis Drain prior to its closure. Nitrate was removed in the ABSRS by two mechanisms: direct algal uptake in the ponds and bacterial denitrification in the anoxic units. Overall, removal across the pond system averaged 18 ± 13 mg/L as N. Removal was greater from April to October (21 ± 13 mg/L as N; n = 35) than from November to March (13 ± 10 mg/L as N; n = 16) because algal cultures were more concentrated during the summer.

Uptake of 18 mg/L by a 178-mg/L algal culture represents an algal nitrogen content of 10%, consistent with the reported value of 9.2% nitrogen by dry weight [18]. However, because nitrate and other dissolved solids were greatly concentrated by evaporation, nitrate removal through algal uptake in the algal ponds may have been 20–30 mg/L on a drainage water volume basis.

Removal of nitrate in the anoxic units was successful once a denitrifying culture was established (Figures 7a and 7b). In the denitrification tank, removal was good in late September and mid-October 1989, just before the farm personnel asked for the experiments to be terminated. Some additional nitrate was removed in the reduction chamber, and at the end of the experimental period effluent nitrate was being reduced to less than 10 mg/L as N. A period of several months seems to be required to accumulate sufficient algal biomass and to establish a bacterial culture capable of using algae effectively as a carbon and energy source. The system appears to have been approaching this point when operations were curtailed.

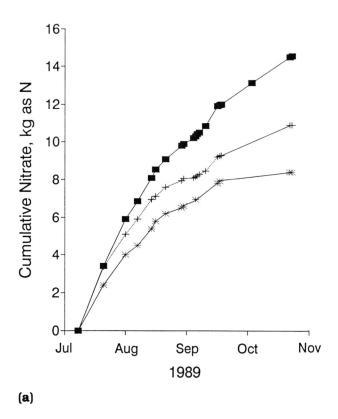

Figure 7 (a) Cumulative mass loading of nitrate to anoxic units. Nitrate was removed in denitrification chamber (difference between denitrification chamber influent and reduction chamber influent) and in reduction chamber. (■) Denitrification chamber influent; (+) reduction chamber influent; (*) reduction chamber effluent. (From Ref. 15.) (b) Denitrification in Mendota ABSRS anoxic units. (From Ref. 13.)

C. Removal of Soluble Selenium

In contrast to nitrate, selenium was removed only in the anoxic units of the ABSRS and not in the ponds. As expected from preliminary laboratory results [31], direct algal uptake of selenium was minimal, and the soluble selenium was at times higher in the pond effluent than in the influent, probably due to evaporative concentration.

Despite nitrate levels of 20–150 mg/L as N in its influent, the reduction tank removed soluble selenium at times (Figures 8a and 8b). In early

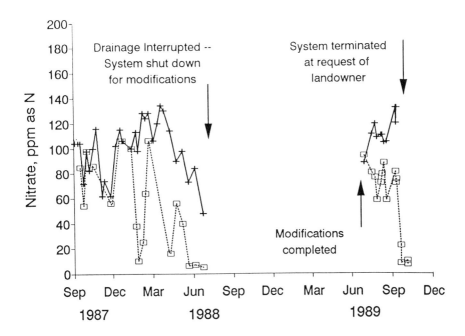

(b)

September 1989, soluble selenium was decreased to 138 µg/L on a daily average and 80 µg/L in a single sample. Unfortunately, the influent data are not available for the period, but in the weeks before and after, the influent soluble selenium was 300–400 µg/L, suggesting that removal did occur in the reduction tank. In October, the pilot plant site had to be vacated, so steady-state nitrate and selenium reduction was never achieved.

Although soluble selenium removal was limited, extensive reduction of selenate was likely to have been occurring. In samples collected in July 1989, when minimal quantities of selenium were being removed in the system, selenate was the predominant species of dissolved selenium found in the drainage water (Figure 9a). A second small peak was found, representing an unidentified species that was possibly organic selenium.

In the analysis of the reduction chamber effluent, a third peak representing selenite was found (Figure 9b), verifying that some selenate

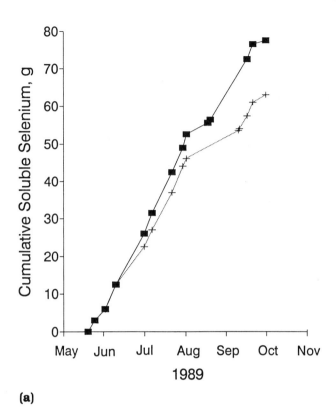

Figure 8 (a) Cumulative mass loading of selenium in reduction chamber. (■) Reduction chamber influent; (+) reduction chamber effluent. (From Ref. 15.) (b) Selenium reduction in Mendota ABSRS reduction chambers. (From Ref. 3.)

reduction occurred. However, most of the effluent soluble selenium was still present as selenate.

In samples taken in September 1989, when the soluble selenium concentration of the reduction tank effluent was 80–100 µg/L, the selenate peak was absent (Figure 10b), indicating that none of the selenate in the drainage water remained. The major peak present was selenite. The total peak area for selenate and selenite was smaller than that of the influent water, indicating that most of the selenate was reduced beyond selenite to elemental selenium, which does not appear in this type of analysis.

Because residual soluble selenium was present as selenite, an initial experiment with ferric chloride ($FeCl_3$) treatment succeeded in reducing

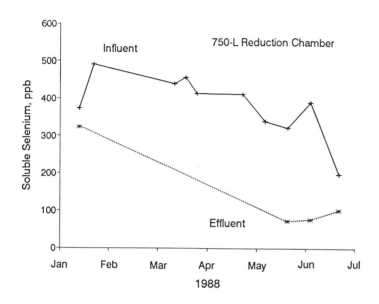

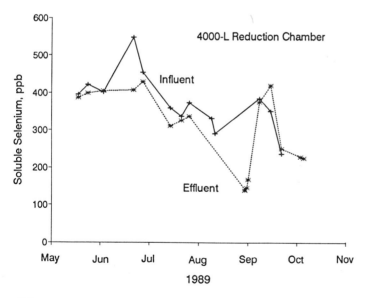

(b)

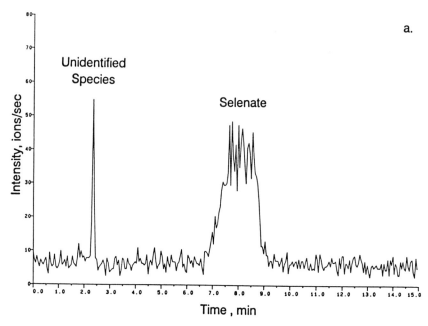

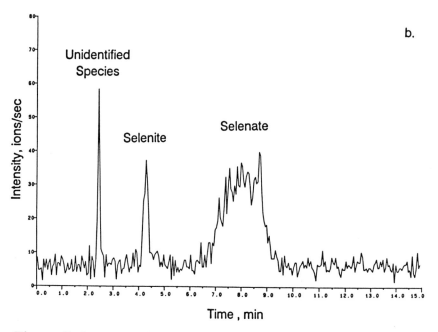

Figure 9 Speciation of soluble selenium in July 1989 samples. (a) Drainage water influent; (b) reduction chamber effluent. (From Ref. 15.)

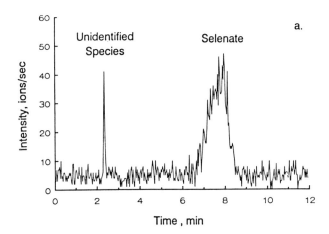

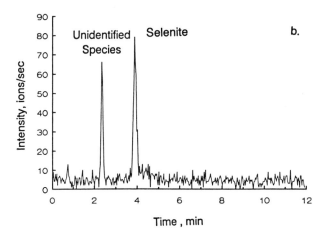

Figure 10 Speciation of soluble selenium in September 1989 samples. (a) Drainage water influent; (b) reduction chamber effluent. (From Ref. 15.)

the soluble selenium concentration to low levels. Soluble selenium was reduced in one sample from 80 to 12 μg/L using 10 mg $FeCl_3$/L and from 72 to 7 μg/L in the other using 20 mg $FeCl_3$/L (Table 2). Even before filtration, clarified supernatants contained only 23 and 9 μg/L, respectively. It is believed that the iron precipitated as ferric hydroxide and selenite sorbed to the precipitate [21]. Thus, the $FeCl_3$ appeared to achieve considerable removal of the soluble selenium that remained after reduction in the Mendota reduction tank. In full-scale systems, $FeCl_3$ could be added in a dissolved air flotation unit treating the reduction tank effluent. The quantities of selenium to be precipitated are so small that minute amounts of iron salts may be effective.

D. Methane Fermentation

Anaerobic digestion of drainage-grown algae was studied as a method for converting excess algae into methane to produce the energy to run the ABSRS. Since high-rate pond systems have very low energy requirements, the methane produced from excess algae may be sufficient to operate the system. Algae grown in drainage water were found to ferment to methane as efficiently as algae grown in sewage ponds. A production rate of 0.16 L of methane per gram of algae as volatile solids fed (L/g) was obtained at ambient temperature in the Mendota digesters. Rates of 0.21–0.25 L/g were obtained at 35°C in the laboratory. Conversion efficiencies of 50–59% of algal energy to methane energy were estimated for drainage-grown algae. Average gas composition (n = 20) for the two digesters over the 3-month period May–July 1989 was 62% CH_4, 22% CO_2, 11% N_2, 3% O_2, and 2% H_2S. The oxygen probably came from air that leaked into samples

Table 2 Removal of Residual Selenium in Mendota Reduction Chamber Water Using Ferric Chloride Treatment

Sample source[a]	Soluble Se before $FeCl_3$ treatment (μg/L)		$FeCl_3$ (mg/L)	Soluble Se after $FeCl_3$ treatment (μg/L)	
	Settled	Filtered[b]		Settled	Filtered[b]
1100 L port	272	80	10	23	12
2200 L port	206	72	20	9	7

[a]Samples drawn from indicated port of reduction chamber.
[b]Filtered through 0.22 μm membrane filter.
Source: Ref. 15.

or the digesters. Therefore, the true methane fraction was probably somewhat higher than 62%.

E. Removal of Other Elements

Many other elements were removed to some extent in the Mendota system. Preliminary studies showed that barium, boron, chromium, cobalt, copper, lead, manganese, nickel, vanadium, and zinc accumulated in the system. Aluminum and silicon also accumulated, but they probably originated in dust blown into the small ponds from the adjacent road.

Sulfate was never reduced in the selenium reduction experiments and was not inhibitory, even though it averaged 3300 ± 600 mg/L as SO_4 in the drainage water.

VII. CONCLUSIONS

Field and laboratory experiments showed that the essential components of the ABSRS—the HRP and the anoxic units for reduction of nitrate and selenium—are feasible and promise high treatment efficiencies. Algae can be grown in drainage water in HRPs to decrease nitrate concentration and provide an inexpensive carbon and energy source for denitrifying and dissimilatory selenium-reducing bacteria. After a sufficient period of time for culture development, bacteria in anoxic zones can reduce nitrate in the drainage water to nitrogen gas and selenate to selenite and elemental selenium. Ferric chloride can be used to remove residual selenite from the water.

Nitrate at concentrations greater than 10 mg/L as N inhibited the reduction of selenate at times, but selenate was reduced in the presence of low levels of nitrate (2–5 mg/L as N). Laboratory experiments suggested that a second, nitrate-independent selenium reduction mechanism also existed.

The reducing power of algae grown in drainage water was 0.180 ± 0.016 meq/mg as measured by chemical oxygen demand, in good agreement with the theoretical value of 0.194 meq/mg.

Of the algal reductant, 17–31% was available for denitrification, so 190–270 mg/L would be required to denitrify typical drainage water. Improved algal productivities should be possible, and any algae not required for reduction can be fermented to methane to provide power for the process.

In field experiments, reduction of nitrate and selenate was nearly complete at times. Although the field studies could not be run for a sufficient length of time, they showed that the ABSRS is a promising process for removing selenium and nitrate from agricultural drainage waters.

It is suggested that future work should include examination of various methods for increasing the availability of algae to nitrate- and selenium-reducing bacteria. Increased nitrate uptake and algal productivity of the HRPs through increased carbon dioxide recycling should be demonstrated. Organic wastes could also contribute reduced carbon. The effects of oxygen on selenium reduction should be clarified to determine whether the organisms are microaerophilic. Exploitation of the possible nitrate-independent selenium reduction mechanism should also be studied. The possibility that the ABSRS could be combined with a ferrous hydroxide reduction process [22] should be examined. Also, another on-site system should be constructed in a location where nitrate and selenium concentrations in the drainage water source more closely resemble those of typical drainage water. The on-site system should be operated for several years without interruption. Liquid and solid effluents should be analyzed for selenium and other toxic elements, and a bioassay study should be conducted on the liquid effluent. Leaching and fermentation studies should be done on sludges. Finally, systems similar to the ABSRS should be explored for reducing other undesirable substances in agricultural drainage water.

REFERENCES

1. K. K. Tanji, A. Lauchli, and J. Meyer, *Environment* 28(6): 6–39 (1986).
2. Water Quality Control Plan (Basin Plan for the San Joaquin River Basin), Amendment Resolution 1989-88, State Water Resources Control Board, Sacramento, Calif., 1989.
3. M. B. Gerhardt and W. J. Oswald, Final Report: Microalgal-Bacterial Treatment for Selenium Removal from San Joaquin Valley Drainage Waters, San Joaquin Valley Drainage Program, U.S. Bureau of Reclamation, Sacramento Calif., Sanit. Eng. Environ. Health Res. Lab. Rep. 90-1, Univ. Calif., Berkeley, 1990.
4. B. J. Marinas, Electrolyte selectivity by thin-film composite reverse osmosis membranes, Ph.D. Thesis, Univ. Calif., Berkeley, 1989.
5. P. Kreft and R. R. Trussell, Removal of inorganic selenium from drinking water using activated alumina, Proc. First Annual Environmental Symposium: Selenium in the Environment, Publ. No. CAT1/860201, Calif. Agric. Technol. Inst., Fresno, 1985, pp. 97–110.
6. W. N. Marchant, U.S. Patent 3,933,635 (1976).
7. R. A. Baldwin, J. C. Stauter, and D. L. Terrell, U.S. Patent 4,405,464 (1983).
8. A. P. Murphy, *Ind. Eng. Chem. Res.* 27: 187–191 (1988).
9. A. P. Murphy, *J. Water Pollut. Control Fed.* 61: 361–362 (1989).

10. E. T. Thompson-Eagle, W. T. Frankenberger, Jr., and U. Karlson, *Appl. Environ.* 55: 1406–1413 (1989).
11. J. W. Kaufman, W. C. Laughlin, and R. A. Baldwin, U.S. Patent 4,519,912 (1985).
12. Epoc Ag, Removal of Selenium from Subsurface Agricultural Drainage by an Anaerobic Bacterial Process, Final Report to Calif. Dept. Water Resources, Sacramento, Calif., 1987.
13. M. B. Gerhardt and W. J. Oswald, Reduction of selenate from agricultural drainage water using anaerobic bacteria grown on algal substrate, presented at Summer Natl. Meeting Am. Inst. Chem. Eng., 1990.
14. D. F. Von Hippel, An analysis of an ethanol-producing solar-bioconversion process using the microalga *Dunaliella* sp. as the biomass crop, Ph.D. Thesis, Univ. California, Berkeley, 1987.
15. M. B. Gerhardt, F. B. Green, R. D. Newman, T. J. Lundquist, R. B. Tresan, and W. J. Oswald, *Res. J. Water Pollut. Control Fed.* 63(5): 799–805 (1991).
16. W. J. Oswald, U.S. Patent 4,005,546 (1978).
17. W. J. Oswald, Micro-algae and wastewater treatment, in *Micro-Algal Biotechnology* (M. A. Borowitzka and L. J. Borowitzka, Eds.), Cambridge Univ. Press, Cambridge, U.K., 1988, pp. 305–328.
18. W. J. Oswald, Large-scale algal culture systems (engineering aspects), in *Micro-Algal Biotechnology* (M. A. Borowitzka and L. J. Borowitzka, Eds.), Cambridge Univ. Press, Cambridge, U.K., 1988, p. 357.
19. R. S. Oremland, J. T. Hollibaugh, A. S. Maest, T. S. Presser, L. G. Miller, and C. W. Culbertson, *Appl. Environ.* 55: 2333–2343 (1989).
20. APHA, *Standard Methods for the Examination of Water and Wastewater*, 17th ed., Am. Public Health Assoc., Washington, D.C., 1989.
21. M. B. Gerhardt, Chemical transformations in an algal-bacterial selenium removal system, Ph.D. Thesis, Univ. Calif., Berkeley, 1990.
22. C. D. Moody and A. P. Murphy, Selenium removal with ferrous hydroxide: identification of competing and interfering solutes, Presented at Second Pan-American Regional Conf., Int. Comm. on Irrigation and Drainage, Ottawa, June 8–9, 1989.
23. R. M. Jeter and J. L. Ingraham, The denitrifying prokaryotes, in *The Prokaryotes*, Vol. 1 (M. P. Starr, H. Stolp, H. G. Truper, A. Balows and H. G. Schlegel, Eds.), Springer-Verlag, Berlin, 1981, pp. 913–925.
24. D. M. Larsen, K. R. Gardner, and P. B. Altringer, A biohydrometallurgical approach to selenium removal, Presented at 23rd Ann. Am. Water Resources Assoc. Conf. Symp., Water Resources Related to Mining and Energy, Salt Lake City, Nov. 1–6, 1987.
25. R. N. Costilow, Biophysical factors in growth, in *Manual of Methods for General Bacteriology*, 2nd ed. (P. Gerhardt et al., Eds.), Am. Soc. Microbiol., Washington, D.C., 1981, pp. 66–78.
26. C. Fraley, N. Elkholy, H. Moehser, E. D. Schroeder, and W. J. C. Pfeiffer, Microbial Removal of Selenium from Agricultural Drainage Water, Report to Dept. of Water Resources, State of California, and the U.S. Bureau of Reclamation, Dept. of Civil Engineering, Univ. California, Davis, 1987.

27. D. D. Wagman, W. H. Evans, V. B. Parker, R. H. Schumm, I. Halow, S. M. Bailey, K. L. Churney, and R. L. Nuttall, *J. Phys. Chem. Ref. Data 11*, Suppl. 2 (1982).
28. Y. Nurdogan, Microalgae separation from high rate ponds, Ph.D. Thesis, Univ. California, Berkeley, 1988.
29. W. J. Oswald, Comprehensive Water Quality Management Program for Laguna de Bay, Sec. 8.1, Vol. 3, Summary Rep. Pt. 8-39 to 8-50, World Health Organization and Laguna Lake Development Authority, Capitol Complex, Pasig, Metro Manila, Philippines, 1978.
30. T. W. Hall, Bioflocculating high rate algal ponds: control and implementation of an innovative wastewater treatment technology, Ph.D. Thesis, Univ. California, Berkeley, 1989.
31. W. J. Oswald, M. B. Gerhardt, F. B. Green, Y. Nurdogan, P. H. Chen, R. D. Newman, L. Shown, C. S. Tam, D. S. Von Hippel, and O. Weres, The role of microalgae in removal of selenate from subsurface tile drainage, in *Biotreatment of Agricultural Wastewater* (M. E. Huntley, Ed.), CRC Press, Boca Raton, Fla., 1989, p. 131.

12
Selenium Accumulation and Colonization of Plants in Soils with Elevated Selenium and Salinity

Lin Wu

*University of California
Davis, California*

I. INTRODUCTION

The distribution of selenium (Se) in rocks and soils of the United States was studied extensively from 1930 to 1950 as part of research on Se toxicity in livestock in the western states [1–3]. Plants have also been studied as indicators of the geographical distribution of Se [4,5]. A generalized summary of Se distribution in soils of the United States [5] showed that soils in areas of the Rocky Mountains and extending into the Great Plains states are high in Se, whereas soils of both the eastern and western coastal states are low in Se. In California, the U.S. Geological Survey pointed out that the rocks of the Coast Range contain iron disulfide (pyrite) [6]. Pyrite-containing rocks of marine origin occur along the entire western margin of the San Joaquin Valley, and some of these contain significant amounts of Se.

High Se concentrations can be expected in many areas with arid climates where the parent soil materials are sedimentary rocks of marine origin. Presser and Barnes [6] reported that the Se concentrations in agricultural drainage in the San Joaquin Valley reached levels as high as 1.4 mg/Se/L. Two fundamental processes are believed to be the cause of the distribution of Se in the soils and shallow groundwaters in the western San Joaquin Valley: (a) High concentrations of soluble salts and Se have resulted from the extent of alluvial deposition of the Diablo Range and the

279

evaporative concentration of shallow groundwaters; and (b) selenium concentrations are highly correlated with soil and groundwater salinity, and the distribution of Se is related to movement of soluble salts to drainage laterals and to the length of the drainage period [7]. Therefore, the Se problem is not just a very localized problem. In the San Joaquin Valley, approximately 160,000 ha of farmland is affected by salinity and high water tables. Salinity and Se are coexisting problems in this region.

Between 1983 and 1985, Ohlendorf et al. [8,9] discovered abnormal rates of embryonic mortality and deformity among several species of water birds nesting at Kesterson Reservoir on the Kesterson National Wildlife Refuge, Merced County, California. These deaths and deformities were attributed to Se that had bioaccumulated and biomagnified in the aquatic food chain [10]. The source of Se was saline irrigation drainwater conveyed by the San Luis Drain to Kesterson Reservoir for disposal by evaporation. Subsequently, Kesterson Reservoir was closed to drainwater in June 1986, and the area was converted into a terrestrial habitat by drying the ponds. Attempts were made to alleviate the deterioration of the habitat at Kesterson Reservoir and to ensure the safety of the food chain in the California Central Valley. The task of restoring natural vegetation requires application of a wide range of knowledge about the chemistry and transformations of Se in soils and its uptake and toxicity in plants and animals.

The progress of research on Se in agriculture and the environment has been reviewed extensively [11–19]. It is not the purpose of this chapter to review the extensive literature on the effects of Se in plants but rather to focus on the recent research conducted in my laboratory regarding the restoration of Se- and salt-contaminated soils and the ecological habitat in the San Joaquin Valley of California. The research activities were supported by the University of California Salinity Drainage Task Force and the U.S. Bureau of Reclamation.

II. SELENIUM AND SALT TOLERANCE IN FORAGE AND TURFGRASS SPECIES

In the 1960s and early 1970s, most of the research work on selenium (Se) in soils and plants emphasized total soil Se concentration and factors affecting Se uptake by plants [15]. More recently, agronomic and general biological interest in Se has focused on its biological functions in animals and humans. Therefore, the chemical form of Se in animal feed and its biological availability have also been subjects of investigation [4,12,20]. The minimum nutritional level for animals is about 0.05–0.10 µg Se/g in dry forage feed; intake below that might cause severe Se deficient diseases

[4,21]. At higher levels, 0.1–1 µg Se/g seems to offer protection against some diseases [22–24], while exposure to even higher levels, 2–5 µg/g, has toxic effects [5,25–27]. Selenium toxicity in animals, resulting from ingesting plants containing from 5 to several thousand milligrams of Se per kilogram dry weight, has also been reported [28].

There are a limited number of reports concerning Se uptake by plants that focus on their Se tolerance. Species known as Se accumulators, such as *Nutunia amplexicaulis* and *Astragalus* sp., are able to accumulate 5000 µg Se/g dry weight. The growth of these plants was found to be stimulated by high concentrations of Se in the culture medium [15]. Other plant species, such as wheat (*Triticum vulgare* L.), may accumulate moderate amounts of Se (up to 30 µg Se/g dry weight) without growth retardation. Selenium- sensitive species such as white clover (*Trifolium repens* L.), red clover (*Trifolium pratense* L.), and perennial ryegrass (*Lolium perenne* L.) may show growth inhibition if over 5 µgSe/g is accumulated in the plant tissue [15,29].

High concentrations of salt and Se are coexisting problems in the soils of the San Joaquin Valley. In order to provide plant materials for soil and water management, the patterns of Se accumulation and Se–salt cotolerance of several forage and turfgrass species were studied in my laboratory.

A. Selenium and Salt Cotolerance in Forage and Turfgrass Species

Crop plant species tolerant to both Se and salinity may be useful for bioextraction of Se from deteriorated agricultural soils. Therefore, the discovery of Se and salinity cotolerant forage and grass species is an important step for water and land management in the San Joaquin Valley. Generally, plants are classified into three groups on the basis of their ability to accumulate Se when grown on seleniferous soils [28]. Primary indicators, or accumulators, can accumulate Se at very high concentrations, up to several thousand milligrams per kilogram; these include *Astragalus, Xylorrhiza, Oonopsis,* and *Stanleya*. These plants also have been found in soils with high salinity. Secondary accumulators may contain a few milligrams of Se per kilogram. Nonaccumulator plants do not normally accumulate Se in excess of 50 mg/kg when grown on seleniferous soils. Typical agricultural crops have a much lower tolerance to Se. Plants on alkaline soils were found to contain 0.01–10 mg Se/kg compared to 0.02–0.2 mg/kg in acid soils. However, some nonaccumulators were found to accumulate up to 15 mg Se/kg [30]. Selenium and salinity cotolerance may exist among the forage and turfgrass species.

1. Symptoms of Selenium Toxicity

Symptoms of Se toxicity in plants have been described only for plants grown in solution culture, not for plants grown in soils with elevated Se concentrations. Both selenium and salinity tolerances and selenium accumulation were investigated for tall fescue (*Festuca arundinacea* Schred.), crested wheatgrass (*Agropyron desertorum* Fich.), buffalograss [*Buchloe dactyloids* (Nutt.)], seaside bentgrass (*Agrostis stolonifera* L.), and bermudagrass (*Cynodon dactylon* L.) [31]. Plants grown in nutrient solution culture supplemented with 1–2 mg Se/L as Na_2SeO_4 displayed two distinct symptoms. In tall fescue, chlorotic spots developed on the older leaves, while young leaves on the same plant remained green with no sign of injury. For crested wheatgrass and the other three species, the leaves showed a bleaching symptom. The entire leaf gradually became pale and chlorotic; this symptom appeared on both young and older leaves and on both young seedlings and plants at a later growth stage. The percentage of plant leaves showing toxic symptoms was positively associated with both the Se concentration in the culture solution and the length of time that the plants were exposed to the Se treatment. Toxicity symptoms found in tall fescue were similar to the yellowing and black spots induced by Se toxicity in cowpea [*Vigna sinensis* (L.) Engl.] [32]. Symptoms of Se toxicity in grasses have also been reported to include snow-white chlorosis of the leaves and pink root tissue [29,33]. Selenium toxicity symptoms of crested wheatgrass and bermudagrass were different from those of tall fescue, suggesting that a different pattern of Se accumulation and distribution in the leaf tissue might exist among the grass species.

2. Selenium Tolerance

Except for seed germination, which does not seem to be inhibited by high Se concentrations, the plant dry weight, root length, and shoot height are highly negatively correlated with an increase in Se concentration in the culture solution, suggesting that for the measurement of Se tolerance, any of these three growth characters can be used [31]. Figures 1a and 1b indicate that 0.5 mg Se/L did not cause noticeable growth inhibition in three of the five species. Only bermudagrass suffered about 50% growth reduction after 5 weeks of exposure to 0.5 mg Se/L. However, the differences in Se tolerance among the five species became more distinct under treatments with 1 and 2 mg Se/L. Tall fescue displayed the greatest Se tolerance among the five species and did not reduce its tolerance ratio (the

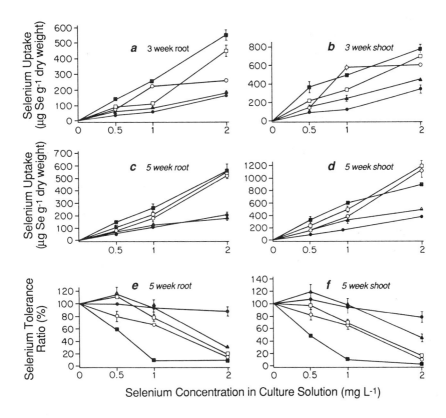

Figure 1 Selenium tolerance ratio and Se uptake of five grass species after 3 or 5 weeks of growth in culture solution supplemented with Se as sodium selenate. (■) Bermudagrass; (▲) buffalograss; (□) crested wheatgrass; (○) seaside bentgrass; (●) tall fescue. Each point represents the mean of three replicates. (From Ref. 31.)

percentage of growth in a Se-containing solution compared to the growth in solution culture without addition of Se) for the three Se concentrations. Bermudagrass showed the least Se tolerance, and its growth was severely inhibited by 1 mg Se/L. The other three species had Se tolerance intermediate between that of the tall fescue and bermudagrass.

The tolerance ratios of tall fescue were tested again under 4 mg Se/L as either selenate or selenite. By 3 weeks of growth, selenate had caused greater growth inhibition than selenite, and this difference increased up to the end of week 5. After 2 weeks of exposure, the tolerance ratio to selenite was reduced from 74% (measured at the end of the third week) to 48% (at the end of the fifth week). The tolerance ratio to selenate was reduced from 86% to 24% during the same period. Selenate is known to be more available to plants than selenite [34], and it is toxic to plants at various concentrations [34,35]. A study on applied Se for agricultural crop plants [25] indicated the selenate is readily removed from the soil by successive harvests of forage. The uptake of elemental Se by plants is extremely small [25].

The increase in growth inhibition corresponded to the increase in Se injury symptoms during the last 2 weeks of the growth period due to increased Se accumulation in the plant tissue. The tolerance ratio is not a constant value throughout the growth stages of the plant; it depends on the stage of growth and length of exposure to Se. However, the relative difference in Se tolerance between species can be detected in a relatively short period of exposure to Se treatment.

Selenium uptake by the five grass species grown in three levels of selenate showed that Se accumulation in both root and shoot tissues corresponded positively with Se concentration in the culture solution (Figures 1c–f). Shoot tissue accumulated greater amounts of Se than did the roots. The amounts of Se accumulated by the plants were inversely related to the Se tolerance of the plant species. For example, Se concentrations in tissues of tall fescue and buffalograss were less than the concentrations in crested wheatgrass, seaside bentgrass, and bermudagrass, where considerable amounts of Se accumulated in the plant tissue, especially in the shoots. An inverse relationship between Se accumulation and Se tolerance has also been found in a species of green algae (*Selenastrum capricornutum*) [36]. This inverse relationship indicates that the uptake of Se by the plants with greater Se tolerance might be restricted by an exclusion mechanism. Tall fescue did accumulate Se up to 200 µg Se/g in the roots and 400 µg Se/g dry weight in the shoots with very little reduction in its plant dry weight production.

The five grass species were grown in solution culture supplemented with 0 (control), 0.5, 1.0, or 2.0 mg selenate per liter. After 5 weeks of growth, for the control treatment, crested wheatgrass and bermudagrass produced nearly twice the amount of dry weight, 6.7 and 7.3 g dry weight (10 plants per pot), respectively, as tall fescue and seaside bermudagrass (4 g), and four times the amount of dry weight as buffalograss. The highest total amount of Se taken up by each of the five species over the three Se concentrations was associated with the Se tolerance of each species and was found at a Se concentration that induced a moderate growth inhibition for that species. For example, bermudagrass exhibited the least Se tolerance among the five species; it had a total Se uptake of 235 µg at 0.5 mg Se treatment per liter. Crested wheatgrass, seaside bentgrass, and buffalograss showed a greater Se tolerance and had their highest amounts of total Se uptake (362, 195, and 124 µg, respectively) under a 1 mg Se/L treatment. Tall fescue had a total Se uptake of 278 µg under a 2 mg Se/L treatment, but this Se level may not represent the highest Se uptake capacity for tall fescue, because the tolerance ratio of tall fescue was not seriously reduced by treatment with 2 mg Se/L.

3. Salt Tolerance

The salt tolerance test with 100 mM salt (NaCl) indicated that crested wheatgrass, tall fescue, and seaside bentgrass had tolerance ratios of 60–70%. With 200 mM salt, tall fescue had a tolerance ratio of about 40%, and crested wheatgrass and seaside bentgrass had tolerance ratios of 30 and 20%, respectively. Bermudagrass and buffalograss had very low tolerance ratios in 100 mM salt and had virtually no growth in 200 mM salt. For tall fescue, crested wheatgrass, and seaside bentgrass, the tolerance ratios measured by root length, shoot length, and dry matter production were comparable, but for bermudagrass and buffalograss (in 100 mM salt), root length tolerance ratios were much greater than shoot length and dry weight tolerance ratios. These results indicate that for these two species, shoot growth is more sensitive than root growth to salt stress [31].

No association between Se tolerance and salt tolerance among the five species was indicated. For example, buffalograss displayed a relatively high Se tolerance but was very sensitive to salt. Crested wheatgrass and seaside bentgrass showed high tolerance to salt but were low in Se tolerance. Bermudagrass was sensitive to both Se and salt. Tall fescue, however, exhibited high tolerance to both Se and salt [31].

The combined Se and salt test indicates that the toxic effect of these two factors seems to be additive (Table 1). For example, in 100 mM salt, tall fescue had a salt tolerance ratio of 70% and in 1 mg Se/L it had a Se tolerance ratio of 90%, but when the Se and salt concentrations were combined, the tolerance ratio was reduced to about 50%. Selenium uptake was increased in both root and shoot tissues by addition of salt to the culture solution, but Na^+ uptake was not significantly affected by Se treatment. Chloride uptake displayed a pattern similar to that of Na^+ uptake, but the Cl^- concentrations in plants were much lower than the Na^+ concentrations.

Table 1 Selenium and Salt Uptake by Tall Fescue from Culture Solution Supplemented with Various Concentrations of Se (Selenate) and Salt[a]

Se and NaCl combination	Tolerance ratio (%)	Se uptake (µg/g dry wt)		Na uptake (µg/g dry wt)		Cl uptake (µg/g dry wt)	
		Root	Shoot	Root	Shoot	Root	Shoot
Control	100	0.1 ± 0.11	0.2 ± 0.2	19 ± 2	30 ± 4	50 ± 14	102 ± 23
50 mM NaCl	88 ± 5	0.4 ± 0.2	0.9 ± 0.3	1026 ± 77	912 ± 45	373 ± 50	509 ± 47
100 mM NaCl	70 ± 5	0.7 ± 0.2	0.7 ± 0.2	1340 ± 95	1413 ± 48	701 ± 30	874 ± 29
1 mg Se/L	82 ± 5	160 ± 5	243 ± 20	18 ± 2	41 ± 7	53 ± 10	81 ± 4
50 mM NaCl + 1 mg Se/L	47 ± 7	180 ± 3	277 ± 17	814 ± 41	815 ± 45	281 ± 44	432 ± 25
100 mM NaCl + 1 mg Se/L	34 ± 3	289 ± 8	325 ± 15	1198 ± 103	1036 ± 103	562 ± 19	1096 ± 40

[a]All values given as mean ± 1 SD.
Source: Ref. 31.

B. Genetic Variation of Selenium and Salt Tolerance Among Tall Fescue Lines

Tall fescue presents a promising potential for use on soils with relatively high levels of salinity and Se. It is a perennial grass and is frequently used for either forage or turf. It has an extensive deep root system and high transpiration rate. Land planted with such a grass may have a reduced water table level and drainage problems. Due to repeated harvesting of the grass, much of the Se may be removed from the soil. The plant materials may be used as a supplement for livestock feeds deficient in Se. This Se- and salt-tolerant grass also may be used for seed production. Seed production for forage or turf should minimize the risk of food chain contamination.

Selenium and salt tolerances and genetic variation of these characteristics were investigated further for tall fescue lines [37]. Six American commercial tall fescue cultivars including 'Alta,' 'Fawn,' 'Kentucky-21,' 'Mustang,' 'Olympic,' and 'Rebel,' and seven world tall fescue lines including Australia (No. 150156), Chile (No. 427127), Japan (No. 422660), Italy (No. 237559), Israel (No. 200339), South Africa (No. 774975), and Russia (No. 283314) were tested for Se and salt tolerance in nutrient solution culture. The Se tolerance data presented by shoot dry weight (Table 2) indicate that among the six American commercial cultivars, 'Kentucky-31' had the lowest tolerance ratio (25%). The rest of the five lines had tolerance ratios ranging from 34 to 45% and were not significantly different from one another. Among the seven world tall fescue lines, the Russian line had the highest tolerance ratio of 45%, and the Israeli line had the lowest tolerance ratio of 28%. The tolerance ratios of the remaining six lines ranged from 31 to 43%.

Shoot tissue Se concentrations of the six American commercial cultivars ranged from 649 to 755 mg/kg dry weight. Among the seven world lines, the tissue Se concentrations were highly variable, ranging from 843 to 1191 µ/g dry weight. Root tissue Se concentrations were between 150 and 550 mg/kg (Table 2), and the root tissue Se concentrations were correlated positively with the shoot tissue Se concentrations. Shoot tissue ($r = 0.63$, $P < 0.01$) dry weight was negatively correlated with the shoot tissue Se concentration ($r = 0.05$, $P < 0.01$). Among the 13 tall fescue lines, the shoot and root dry weight production, tissue Se concentration, and Se tolerance ratios were significantly different ($P < 0.01$).

Broad-sense heritability, calculated for the shoot and root dry weight, and Se tolerance ratios had values of 0.67 and 0.65, respectively. These heritability ratios include both additive and nonadditive genetic effects.

Table 2 Shoot Dry Weight Production, Se Tolerance Ratio, and Se Accumulation of 13 Tall Fescue Lines after 5 Weeks Growth in Solution Culture Supplemented with 2 mg Se/L

Variety	Dry weight[a] of 10 plants (mg)	Tolerance ratio (%)	Tissue selenium concentration (mg/kg)	Total Se accumulation (µg)
American commercial cultivars				
Alta	2982 a	42 abc	796 de	2138 a
Fawn	3212 a	45 ab	649 ef	2092 a
Ky-31	1450 b	25 f	767 cde	1114 cde
Mustang	2007 b	36 bcde	755 de	1518 b
Olympic	2188 b	45 bcd	733 de	1607 b
Rebel	1943 b	34 bcde	714 de	1388 bd
World lines				
Australia	1042 e	33 cdef	929 b	970 def
Chile	1421 cde	43 abc	927 b	1306 bcd
Israel	936 e	28 ef	1191 a	1106 cde
Italy	1106 e	37 bcd	925 a	1018 de
Japan	1482 cde	39 bcd	843 bcd	1258 bcd
South Africa	932 f	31 def	906 bc	845 ef
Russia	1071 e	54 a	521 f	560 f

[a]Means separated by Duncan's multiple range test, $P = 1\%$.
Source: Wu and Huang [47].

No significant correlation occurred between Se tolerance and the shoot dry weight production.

The salt tolerance test (Table 3) with 0.1 M NaCl showed considerable variation of salt tolerance among the world tall fescue lines. Salt tolerance ratios ranged from 19% for the Chilean line to 66% for the American 'Alta' cultivar. Generally, the American cultivars had higher salt tolerance ratios than the world tall fescue lines but were much less variable, with tolerance ranging from 49 to 60%. Shoot Na^+ concentrations were variable among the world lines. A negative correlation was found between the shoot tissue Na^+ concentrations and salt tolerance ratios of the seven world lines ($r = 0.52$, $P < 0.01$). Chloride uptake displayed a pattern similar to Na^+ uptake, but the Cl^- concentrations in plants were lower than the Na^+ concentrations. Selenium tolerance and salt tolerance of the 13 tall fescue lines were independent. Dry weight production under salt treatment was correlated

Table 3 Shoot Dry Weight, Salt Tolerance Ratio, and Tissue Na and Cl Concentrations of 13 Tall Fescue Lines after 5 Weeks Growth in Solution Culture Supplemented with 0.1 M NaCl

Variety	Dry weight[a] of 10 plants (mg)	Tolerance ratio (%)	Tissue selenium concentration (mg/kg)	Total Se accumulation (µg)
American commercial cultivars				
Alta	631 abc	66 ab	1408 ab	587 d
Fawn	585 abcd	60 bc	1093 cd	698 bc
Ky-31	520 bcd	52 bcd	1638 a	499 a
Mustang	626 abc	54 bcd	1228 bc	975 a
Olympic	648 abcd	61 bc	1230 bc	699 bc
Rebel	489 de	49 cd	1462 ab	803 abc
World lines				
Australia	556 cd	28 ef	1233 bc	933 a
Chile	495 de	19 fg	1244 bc	722 a
Israel	757 abc	43 cde	1184 c	710 bc
Italy	798 a	57 bc	1087 cd	705 bc
Japan	599 cd	35 e	1207 c	703 bc
South Africa	789 ab	56 bcd	1092 cd	664 c
Russia	360 ef	45 cde	843 e	802 abc

[a]Means separated by Duncan's multiple range test, $P = 1\%$.
Source: Ref. 47.

positively with salt tolerance ratios. Broad-sense heritability calculated for the salt tolerance of the 13 tall fescue lines had a value of 0.76. This heritability suggested that the tolerance ratio is of value in screening and selecting Se and salt tolerance in tall fescue.

It is evident that genetic variation of Se tolerance and Se accumulation exists among tall fescue lines. Apparently, Se tolerance and Se accumulation varies as much among different lines of tall fescue as it does among different grass species. The positive correlations between the shoot and root tolerance ratios suggest that the shoot and root growth responses to Se stress are similar. The lack of significant correlation between shoot tolerance ratio and shoot dry weight production suggests that an intrinsic growth rate difference exists among the tall fescue lines.

This study indicated that 'Alta' and 'Fawn' are tolerant varieties and had the lowest tissue Se concentration, but they produced a greater amount of shoot biomass and therefore accumulated a greater quantity of total tissue Se. The Israeli line is the least Se-tolerant. It produced the least amount of shoot biomass but had a higher tissue Se concentration; therefore, it also accumulated a high amount of total Se. The Russian line is Se-tolerant, but it had a low tissue Se concentration and low growth rate and accumulated the lowest amount of total Se. Both Se tolerance and intrinsic growth rate apparently are important factors determining the tissue Se concentration and effectiveness of bioaccumulation of Se.

An exclusion mechanism in salt tolerance is indicated by the negative correlation between the shoot tissue Na^+ and Cl^- concentrations and salt tolerance ratios ($r = -0.52$, $P < 0.01$). The positive correlations between the dry weight production under salt stress and salt tolerance ratios ($r = 0.51$, $P < 0.01$) indicate that the growth rate under salt stress reflects the degree of salt tolerance of the tall fescue lines. The salt and/or Se tolerance of a few tall fescue cultivars such as 'Kentucky-31,' 'Alta,' and 'Olympic' was reported [31,38,39].

Horst and Beadle [40] examined the effects of salinity on seed germination and seedling growth of 16 commercial cultivars. In all of these studies, tall fescue showed high potential for use in saline soils. The independent relationship between Se tolerance and salt tolerance of the 13 tall fescue lines indicates that these two characteristics can be selected independently and combined into desirable genotypes. In cases where elevated soil Se is accompanied by high concentrations of salt, such as at Kesterson, California, salt tolerance becomes critical for plants selected for Se bioaccumulation. Eight of the 13 lines had high salt tolerance; among these, Se tolerance varied significantly. 'Alta,' 'Fawn,' Russia (No. 283314), and 'Olympic' have high Se tolerance, while Australia (No. 150156), Israel (No. 200339), 'Kentucky-31,' and 'Mustang' have low Se tolerance. This diversity of salt and Se tolerance in the tall fescue lines is useful for Se bioaccumulation management in saline soils with variably elevated Se levels.

III. EFFECTS OF SULFATE ON SELENIUM TOLERANCE AND SELENIUM ASSIMILATION

Sulfate and chloride are the major inorganic anions in the shallow groundwater of the San Joaquin Valley, California [7], and they may influence Se uptake by plants [26]. Most research on the effects of sulfate on Se uptake by plants have been short-term kinetic uptake assays using excised roots [41–43]. Pratley and McFarlane [44], working with nutrient solution culture, found that sulfate had a greater effect on ryegrass uptake of selenate than it did on uptake of selenite. Gissel-Nielsen [45] observed similar antagonistic effects existing between sulfate and selenate, and to a lesser extent with selenite, in a soil experiment with barley and clover. Mikkelsen et al. [46] studied interactions of selenate with various constituents and found a similar antagonism between Se and sulfate.

Studies of the effects of chloride and sulfate salts on Se tolerance, accumulation, and assimilation in tall fescue and white clover were conducted using nutrient solution culture as well as under field conditions [37,47].

A. Selenium Accumulation

The nutrient solution culture study [48] showed that the Se uptake of tall fescue was markedly inhibited by the presence of sulfate. The Se concentration of shoot tissue was reduced from 1000 μg Se/g dry weight without the addition of sulfate to only 30 μg/g dry weight with the addition of 10 mM sulfate. The tissue Se concentration was reduced further with the increase to 30 mM sulfate, regardless of the presence of Cl^-. There was a slight increase of tissue Se concentration with the addition of a Cl^- salt (Table 4).

Ulrich and Shrift [49], using excised roots of *Astragalus*, found that sulfate completely inhibited Se absorption at a S/Se concentration (meq) ratio of 1:1. Wu and Huang [37] used 3 mg/L (37 μM) Se and 10–30 mM sulfate, but Se uptake was not inhibited completely, even though it was reduced considerably. The cause of this discrepancy may lie in the length of uptake period and the condition of the plants. The study by Ulrich and Shrift [49] used excised roots, and that of Wu and Huang [37] used whole plants and a 5-week growth period; therefore, the competitive effect of sulfate may not completely inhibit Se uptake by the plants during a long period of exposure to Se concentration.

B. Selenium Assimilation

Selenium tolerance ratios and dry weights of tall fescue (Figures 2a and 2b) and white clover (Figures 3a and 3b) measured after 5 weeks of growth in nutrient solution culture indicated that there are substantial differences in Se tolerance between these two species. The addition of 1.5 mM of Na_2SO_4 was able to alleviate substantially the growth inhibition of white clover and alleviate completely the growth inhibition of tall fescue due to the presence of 2 mg Se/L in the culture solution. Both the shoot and root growth of tall fescue were stimulated by the addition of 1 and 1.5 mM Na_2SO_4, regardless of the presence of Se, while the growth of both the

Table 4 Dry Weight Production and Se Accumulation of Tall Fescue Grown in Nutrient Solution Culture Supplemented with 3 mg Se/L and Different Concentrations of NaCl and Na_2SO_4[a]

Combination of salt concentrations in culture solution Na_2SO_4-NaCl (mM)	Dry weight (mg/pot)		Se concentration (μg/g)	
Control	2067 ± 434	462 ± 88	1.0 ± 0.2	0.5 ± 0.0
0–0	513 ± 64	80 ± 17	1064.9 ± 94	499.0 ± 97
0–20	404 ± 75	78 ± 16	1133.1 ± 92	610.7 ± 79
0–40	375 ± 63	62 ± 15	1182.2 ± 120	691.4 ± 79
0–60	354 ± 66	60 ± 10	1212.0 ± 83	670.4 ± 65
10–60	965 ± 274	279 ± 44	25.3 ± 4.1	17.6 ± 4.6
10–0	1176 ± 68	380 ± 63	26.2 ± 4.3	20.3 ± 4.7
20–0	1528 ± 240	365 ± 64	19.6 ± 2.0	10.7 ± 2.3
30–0	1434 ± 395	345 ± 77	13.8 ± 1.8	10.7 ± 2.3
30–20	1233 ± 330	306 ± 33	13.7 ± 4.8	9.8 ± 1.7
SO_4 vs. Cl	—[b]	—[b]	—[b]	—[b]

[a]Mean and 1 SD.
[b]Significant at the 0.001 level of probability, single degree of freedom contrasts.
Source: Ref. 37.

roots and shoots of white clover was increasingly inhibited when the sulfate concentrations were reduced.

Table 5 shows that increases in sulfate concentrations reduced Se accumulation in both root and shoot tissues for both species. Shoot tissue had twice as much Se as root tissue. The root and shoot tissues of the white clover had a Se concentration three times that of tall fescue.

Increased sulfate concentrations in sand culture reduced both tissue Se concentrations and protein Se concentrations (Table 6). The white clover had significantly higher protein Se concentrations than tall fescue, with mean values of 338 and 94 μg Se/g protein under the 2 mg Se/L and 2 mg Se/L plus 1 mmol Na_2SO_4/L treatment, respectively, while the tall fescue had mean values of 243 and 27 μg Se/g protein. Under the 2 mg Se plus 5 mmol Na_2SO_4/L treatment, both the tissue and protein Se concentrations were lower, and the two species had similar values of about 20 μg Se/g dry weight. For all Se–sulfate combination treatments, the tall fescue produced 90% or more of the dry weight of its control treatment, but the white clover had 50% growth reduction under the 2 mg Se/L treatment

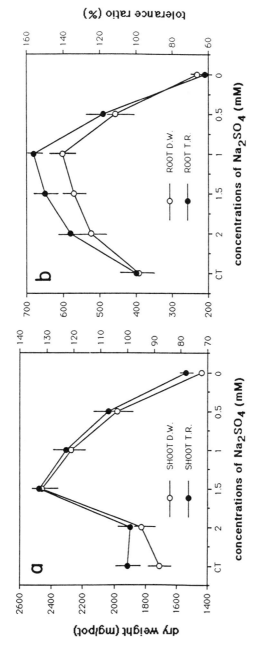

Figure 2 Selenium tolerance ratio (●) and dry weight (○) measured for (a) shoot and (b) root of tall fescue grown in nutrient solution culture supplemented with 2 mg Se/L and different concentrations of Na_2SO_4. (From Ref. 56.)

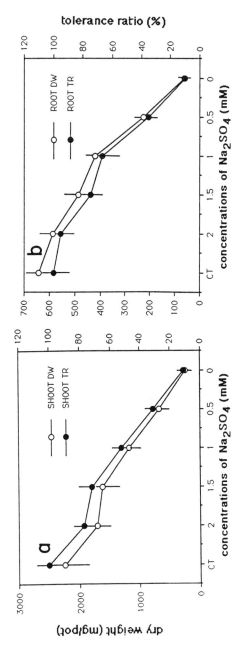

Figure 3 Selenium tolerance ratio (●) and dry weight (○) measured for (a) shoot and (b) root of white clover grown in nutrient solution culture supplemented with 2 mg Se/L and different concentrations of Na$_2$SO$_4$. (From Ref. 56.)

Table 5 Selenium Accumulation in Tall Fescue and White Clover in Nutrient Solution Culture Supplemented with 2 mg Se/L and Different Concentrations of Na_2SO_4

Treatment Na_2SO_4 (mM)	Tall fescue ($\mu g/g$ dry wt)[a]		White clover ($\mu g/g$ dry wt)[a]	
	Shoot	Root	Shoot	Root
0.0	613 + 30	379 + 84	2073 + 215	1394 + 174
0.5	342 + 46	183 + 24	1263 + 109	777 + 87
1.0	210 + 26	81 + 21	597 + 37	491 + 51
1.5	169 + 38	73 + 31	433 + 51	372 + 16
2.0	69 + 6	32 + 6	297 + 83	282 + 79
Control	0.05 + 0.3	0.4 + 0.2	1.8 + 0.4	0.9 + 0.3

[a]Mean +1 SD.
Source: Ref. 56.

Table 6 Selenium Assimilation by Tall Fescue and White Clover in Sand Culture Irrigated with Nutrient Solution Supplemented with 2 mg Se/L and Different Concentrations of Na_2SO_4[a]

Species and Na_2SO_4 treatment (mM)	Total tissue Se concentration ($\mu g/g$)	Protein Se concentration ($\mu g/g$)	Percent protein Se in total tissue Se concentration	Se tolerance (%)
Tall fescue				
0	455.3 + 74.8	243.2 + 23.3	61.4 + 4	91 + 6
1	88.1 + 12.4	27.6 + 1.5	55.0 + 17	102 + 8
5	43.0 + 4.6	20.1 + 2.8	32.0 + 6	112 + 6
Control	0.9 + 0.2	0.2 + 0.1	12.0 + 2	
White clover				
0	798.1 + 102.1	388.7 + 18.3	38.3 + 11.8	53 + 8
1	242.6 + 40.2	94.2 + 5.7	27.6 + 15.8	81 + 5
5	53.7 + 1.1	14.3 + 3.2	17.0 + 6.0	95 + 5
Control	1.0 + 0.3	1.3 + 1.0	6.0 + 1.0	

[a]Values given as mean +1 SD.
Source: Ref. 56.

and 20% growth reduction under the 2 mg Se plus 1 mmol Na_2SO_4 treatment. Plant tissue concentration was not affected by Se and sulfate treatments. White clover had a slightly but significantly higher tissue protein concentration (about 120 mg/g dry weight) than tall fescue (about 100 mg/g). Selenium concentration in nonprotein fractions was also much higher in white clover than in tall fescue ($F = 9.5$, $P < 0.01$) and higher in the shoot tissue without sulfate treatment than in the tissue with sulfate treatment ($F = 21$, $P < 0.001$). Conversely, the percent total protein Se in the total tissue Se concentration was much lower in white clover than in tall fescue (Table 12). Correlations between tissue Se concentration and protein Se concentration in both tall fescue ($r = 0.94$, $P < 0.01$) and white clover ($r = 0.98$, $P < 0.98$) are significantly and positively correlated.

Brown and Shrift [11] found that Se tolerance in *Astragalus stanleya* was due to a mechanism excluding Se from incorporation into protein. These plants may accumulate large amounts of nonprotein selenoamino acid and restrict the toxicity effects of this element [50]. Brown and Shrift [51] suggested that the replacement of cysteine by the Se analogue selenocysteine is a major cause of the toxic effects in Se-sensitive plant species. In tall fescue and white clover, there is a positive relationship between the increase of overall plant tissue Se concentration and the protein Se concentration. The increase of protein Se concentration was associated with the reduction of plant growth. These results indicated a mechanism that restricts Se uptake by the plants with a greater Se tolerance and consequently reduces the incorporation of Se into its protein instead of developing a major protein exclusion mechanism as is found in Se accumulator plants [11]. Both short-term kinetic uptake studies [41–43] and long-term Se accumulation experiments [37,52] found that an antagonistic effect exists between Se and sulfate. This is a further indication that this antagonistic effect acts at the protein assimilation level.

C. Effects of Selenium on Mineral Nutrient Uptake

It is evident that uptake and metabolism of Se in plants are influenced by various factors such as Cl^-, sulfate, and salinity [37,48,53] and by trace elements [54,55]. However, the effects of Se on mineral nutrient element uptake by plants has rarely been studied. Wu and Huang [56] demonstrated (Table 7) that under Se treatment, the tissue calcium concentration in both tall fescue and white clover was increased but the phosphorus concentration was reduced. Tissue magnesium and potassium concentrations were not affected. For micronutrient elements, white clover had considerably higher tissue Fe concentration than did tall fescue, and its concentration increased from 190 µg/g dry weight under conditions without Se treatment to 600 µg/g dry weight under 2 µg Se/L treatment. The addition of sulfate to the nutrient solution moderated the increase of tissue iron concentration under Se stress. Tall fescue had a much lower tissue iron concentration (about 70 µg/g dry wt) and was not affected by the Se and sulfate treatments. Copper, manganese, and zinc were increased only in the white clover under conditions of 2 mg Se/L treatment and severe growth inhibition (Table 7). Epstein and Burau [57] found that the effect of phosphate on sulfate absorption appeared to be dependent on the presence of other anions such as Cl^-, SO_4^{2-}, and NO^{3-}, and the above study showed an antagonistic effect between Se and phosphorus. However, more research is needed to reveal the effect of Se on P uptake. Calcium is an essential macronutrient and is very effective in detoxifying high concentrations of other mineral elements in plants [58,59]. Therefore the increase of tissue calcium concentration under Se stress may be a part of the mechanism of calcium for growth and developmental processes

Table 7 Concentrations of Nutrient Elements in Plant Tissue of Tall Fescue and White Clover Grown in Nutrient Solution Culture Supplemented with 2 mg Se/L and Different Concentrations of Sulfate[a]

Na_2SO_4 treatment (mM)	Nutrient element							
	Ca (mg/g)	K (mg/g)	Mg (mg/g)	P (mg/g)	Cu (µg/g)	Fe (µg/g)	Mn (µg/g)	Zn (µg/g)
Tall fescue								
0.0	60 a	66 a	5 a	20 b	10 a	65 a	60 a	59 a
0.5	67 a	71 a	7 a	25 a	11 a	69 a	45 b	59 a
1.0	74 a	67 a	5 a	24 a	10 a	59 a	45 b	51 a
1.5	65 a	71 a	8 a	28 a	10 a	60 a	42 a	52 a
2.0	62 a	66 a	8 a	29 a	8 a	68 a	41 a	51 a
Control	36 b	66 a	5 a	27 a	12 a	73 a	44 b	62 a
White clover								
0.0	77 a	72 a	5 a	17 c	10 a	610 a	107 a	93 a
0.5	75 a	75 a	7 a	25 bc	10 a	327 b	42 b	54 b
1.0	74 a	69 a	9 a	28 bc	9 a	210 b	45 b	51 bc
1.5	59 a	71 a	8 a	31 ab	8 a	239 b	52 b	46 bc
2.0	77 a	69 a	6 a	35 ab	9 a	239 b	42 b	44 bc
Control	38 b	74 a	5 a	39 a	10 a	190 b	44 b	41 c

[a]Mean followed by the same letter are not significantly different at the 5% level as determined by Duncan's new multiple range test.
Source: Ref. 56.

under stressful conditions. Tissue magnesium and potassium concentrations were not affected by the addition of Se to the culture solution. Orme-Johnson et al. [60], Bergersen [61], and Burger et al. [62] suggested that there is an increased requirement for iron in symbiotic legume species, but this has rarely been studied. The reason for the increase in iron uptake in white clover under Se treatment is not clear. Selenium seems to have little effect on Cu, Zn, and Mn uptake. An increased uptake of these elements was found only in white clover under conditions of severe growth inhibition.

IV. EFFECTS OF FIELD IRRIGATION PRACTICES ON SELENIUM ACCUMULATION

The task of restoring natural vegetation and the ecological habitat at Kesterson Reservoir requires application of a wide range of knowledge of soil–plant relations. One of the greatest difficulties to a controlled manipulation of Se accumulation in plants under field conditions is the insufficient knowledge of (a) the effects of various soil factors on the availability of Se and (b) how different plant species respond to these factors in terms of growth and Se accumulation. Among the different soil factors, soil moisture probably is the most critical element in determining soil Se availability and Se accumulation by plants. Effects of field irrigation on Se accumulation in forage grass and naturally established herbaceous plant species at Kesterson were studied [37,63].

A. Effects of High and Low Salinity Irrigation Water

A study of the effects of high salinity irrigation water on Se accumulation by tall fescue was conducted under field conditions at the West Side Field Station in the San Joaquin Valley, California [37]. The soil was alkaline, with a pH of 8, a sulfate concentration of about 0.4 mg/g, and no detectable soil Se. The high-salinity irrigation water had a sulfate concentration of 2816 mg/L, about 50 µg/L selenate, and 1.8 µg/L selenite; the low-salinity irrigation water had a sulfate concentration of 681 mg/L, 8 µg/L selenate, and the selenite was undetectable. Three tall fescue cultivars ('Olympic,' 'Falcon,' and 'Alta') and the field plots were irrigated twice a week through the growing season at a rate of 390 L of water for each of the six 1.5 m x 1.5 m plots. An additional 3 mg/L of Se as Na_2SeO_4 was added to both the high- and low-salinity irrigation water.

Plants irrigated with water low in salinity had tissue Se concentrations of 5 µg/g dry weight, but shoot tissue from plants irrigated with high salinity water had only 1.7 µg Se/g, even though the high-salinity water had a higher Se concentration than the low-salinity water. No apparent difference in tissue Se concentration was detected among the three tall fescue cultivars (Table 8). This result demonstrated further that tissue Se concentration of the field-grown plants is determined not only by soil Se concentration but also by the presence of sulfate. Sulfate is known to be competitive with Se uptake [64]. No genetic difference is Se accumulation was detected among the cultivars.

A significant difference in biomass production was found among the tall fescue cultivars (Table 8). 'Alta' produced greater biomass than the 'Olympic' and 'Falcon' cultivars, resulting in greater amounts of total Se uptake by 'Alta' (13 and 33 mg/m^2 of soil with high- and low-salinity irrigation water, respectively) than by the 'Olympic' and 'Falcon' cultivars (10 and 25 mg/m^2 of soil with high- and low-salinity irrigation water, respectively).

Following one season of field irrigation, bare plots receiving high-salinity water had 50 µg/g selenate and 40 µg/g selenite in the top 15 cm of soil; however, topsoil of similar plots producing tall fescue contained only 20 µg/g selenate and 15 µg/g selenite. Low-salinity irrigation water contributed only 10% as much Se as high-salinity water. Overall, bare plots had about twice as much soil Se as did the tall fescue plots.

Along the soil profile, the Se concentration decreased with increasing soil depth ($r = -0.77$, $P < 0.05$). The highest Se concentration was found in the top 15 cm of soil. Below 15 cm there was no consistent discrepancy in soil Se concentration between the bare plots and the tall fescue plots (Table 9). The lack of Se movement into the soil profile may be due to the

Table 8 Shoot Biomass Production, Tissue Se Concentration, and Total Se Uptake of Tall Fescue Grown Under Field Conditions Irrigated with High- or Low-Saline Waters and Supplemented with 2 mg Se/L as Na$_2$SeO$_4$[a]

	Low-salinity irrigation water			High-salinity irrigation water		
Cultivar	Biomass (kg/m^2)	Tissue Se (µg/g)	Total Se accumulation (mg/m^2)	Biomass (kg/m^2)	Tissue Se (µg/g)	Total Se accumulation (mg/m^2)
Olympic	4.8 + 1.0	5.05 + 1.0	24.2 + 3.4	5.32 + 0.11	1.83 + 1.0	9.7 + 3.1
Falcon	5.0 + 1.0	4.98 + 0.9	24.9 + 5.2	5.66 + 0.15	1.70 + 0.9	9.6 + 2.0
Alta	6.6 + 1.7	5.00 + 1.0	33.3 + 4.0	8.06 + 1.53	1.65 + 0.3	13.2 + 2.0

[a]Mean and 1 SD.
Source: Ref. 37.

Table 9 Soil Analysis for the Field Plots Irrigated with High- or Low Salinity Water and Supplemented with 2 mg Se/L as Na_2SeO_4[a]

Soil depth (cm)	Parameter	Low salinity irrigation		High salinity irrigation	
		Bare plot	Tall fescue plot	Bare plot	Tall fescue plot
0–15	Selenite (mg/kg)	5.50 ± 0.2	1.90 + 0.2	41.30 + 8	15.5 + 1
	Selenate (mg/kg)	4.50 + 0.2	2.90 + 0.3	52.30 + 9	20.4 + 1
	Sulfate (mg/kg)	0.14 ± 0.0	0.12 ± 0.0	19.60 ± 1	9.5 ± 1
	EC (S/m)	0.11 ± 0.0	0.13 ± 0.0	1.18 ± 0.4	1.5 ± 0.1
	pH	8.00 ± 0.1	7.90 ± 0.1	8.20 ± 0.1	8.0 ± 0
15–30	Selenite	0.80 ± 0.2	1.30 ± 0.4	12.90 ± 3	16.8 ± 3
	Selenate	0.30 ± 0.0	0.03 ± 0.0	21.70 ± 2	19.0 ± 6
	Sulfate	0.20 ± 0.0	0.20 ± 0.0	6.04 ± 0.2	7.2 ± 2
	EC	0.12 ± 0.0	0.13 ± 0.0	1.03 ± 0.4	1.38 ± 0.1
	pH	8.01 ± 0.0	8.10 ± 0.1	8.20 ± 1	8.21 ± 0
30–60	Selenite	0.20 ± 0.0	1.00 ± 0.1	8.40 ± 3	14.3 ± 3
	Selenate	0.20 ± 0.1	0.10 ± 0.0	13.90 ± 3	18.6 ± 9
	Sulfate	0.20 ± 0.0	0.30 ± 0.6	3.10 ± 0.5	3.4 ± 0.5
	EC	0.11 ± 0.0	0.15 ± 0.0	0.63 ± 0.0	0.59 ± 0.0
	pH	8.30 ± 0.1	8.20 ± 0.1	8.30 ± 0.1	8.0 ± 0.1
60–90	Selenite	0.20 ± 0.0	—[b]	10.80 ± 3	0.2 ± 0.0
	Selenate	0.40 ± 0.1	0.40 ± 0.0	9.80 ± 3	12.5 ± 2
	Sulfate	0.20 ± 0.0	0.40 ± 0.0	2.10 ± 1	12.1 ± 1
	EC	0.14 ± 0.0	0.18 ± 0.0	0.42 ± 0.1	0.46 ± 0.1
	pH	8.30 ± 0.1	8.30 ± 0.1	8.40 ± 0.0	8.3 ± 0.1
90–120	Selenite	—[b]	—[b]	0.2 ± 0.0	0.2 ± 0.0
	Selenate	0.40 ± 0.0	0.40 ± 0.0	0.90 ± 0.2	14.5 ± 4
	Sulfate	0.50 ± 0.1	0.30 ± 0.0	2.40 ± 0.0	1.0 ± 0.4
	EC	0.21 ± 0.0	0.23 ± 0.1	0.19 ± 0.0	0.33 ± 0.0
	pH	8.30 ± 0.0	8.30 ± 0.1	8.40 ± 0.0	8.40 ± 0.1

[a] Mean and 1 SD.
[b] Concentration undetectable.
Source: Ref. 31.

restriction of intensity and frequency of irrigation and the lack of rainfall during the growing season in the San Joaquin Valley.

The highest soil sulfate concentration (20 mg/g) was found in the bare plots irrigated with water high in salinity. The soil sulfate concentration of the tall fescue plots averaged 10 mg/g. In soil irrigated with high-salinity water, sulfate concentration generally decreased with increased soil depth ($r = -0.73$, $P < 0.05$). Soils irrigated with water low in salinity had a sulfate concentration of less than 0.5 mg/g, and the sulfate concentration remained similar throughout the soil profile. Electrical conductivity was positively associated with the soil sulfate concentration ($r = 0.73$, $P < 0.05$), and there was about 15 dS/m for the soil irrigated with high-salinity water and about 1 dS/m for the soil irrigated with low-salinity water. No difference in conductivity was found between the bare soil and the soil of the tall fescue plots. Soil pH averaged 8 throughout the soil profile and showed no significant difference between high- and low-salinity irrigated plots.

The field study again showed a substantial reduction of Se uptake by tall fescue due to the presence of sulfate (Table 9). Growing tall fescue for 1 year reduced Se by 50% in the top 15 cm of soil for both the high- and low-salinity treatments. A total of 40 mg Se/m^2 was added into the field plots (the Se concentration of the irrigation water was not included), and 50% of this Se was removed by tall fescue production in the low-salinity water treatment, while 25% was removed in the high-salinity treatment. Selenium might also be removed from the soil by volatilization but was not measured.

Table 10 presents the plant tissue Se concentration and the Se concentration in the top 15 cm of soil of the field plots detected in May 1989 following supplementation with 2 mg Se/L during the previous year (no supplemental Se was added in the following season). The tissue Se

Table 10 Tall Fescue Shoot Tissue and Soil Se Concentrations (μg/g Dry Weight) from Field Plots Previously Supplemented with 2 mg Se/L as Na$_2$SeO$_4$ and Subsequently Irrigated with High-Salinity or Low-Salinity Water[a]

Cultivar	Plant tissue Se Concentration		Soil (10–15 cm) Se concentration[b]
	High-salinity irrigation	Low-salinity irrigation	High-salinity irrigation
Olympic	0.70 ± 0.1	0.36 ± 0.0	1.65 ± 0.4 (selenate)
			22.31 ± 3.6 (selenate)
Falcon	0.62 ± 1	0.37 ± 0.0	1.65 ± 0.4 (selenite)
			18.44 ± 1.8 (selenate)
Altas	0.65 ± 0.1	0.40 ± 0.0	1.00 ± 0.4 (selenite)

[a]Mean and 1 SD.
[b]Selenium concentration not detectable, and therefore not reported, in soil irrigated with low-salinity water.
Source: Ref. 37.

concentration was reduced from 1.7 to a range of 0.4–0.7 µg/g dry weight. The high-salinity treatment had 20 mg/kg selenate and 1 mg/kg selenite. The Se concentration in soil receiving the low-salinity irrigation water treatment was below the level of detection. This result indicates that neither the high-salinity nor the low-salinity irrigation water caused high levels of Se accumulation by the tall fescue cultivars without continual addition of Se into the system.

B. Effects of Field Irrigation on Selenium Accumulation in Naturally Established Herbaceous Plants

The effect of field irrigation practices on Se accumulation was also studied for naturally established plant species at Kesterson [63]. A 2-acre field plot was prepared by tilling and leveling the soils in Pond 2 by the Lawrence Berkeley Laboratory, University of California (UC) as part of the cooperative effort with the UC Davis campus. This field was colonized by cattail (*Typha latifolia*) before the Kesterson Reservoir was dried out. The original pond bottom was clay loam soil, but it now had accumulations of sediments and decayed plant tissues in the topsoil horizon and had not been covered by fresh soil. Field plots 40 m x 25 m were constructed. Four different irrigation–tillage management practices were applied: (1) no irrigation, no tillage; (2) with irrigation, no tillage; (3) no irrigation, with tillage; and (4) with irrigation, with tillage. Soil tillage to 15 cm depth was conducted at 3-month intervals, and irrigation was applied biweekly throughout the dry summer months. Approximately 5 cm of water was applied per irrigation. The electrical conductivity (EC) of the irrigation water was about 0.5 dS/m, and Se was not detectable in the irrigation water. The field plots were allowed to become colonized naturally by native herbaceous plants. No fertilizer was applied throughout the experiment.

The biomass and Se accumulation of the plants established in the field plots were substantially affected by the irrigation practices. In May 1991, four plant species were found to be the predominant colonizers: *Atriplex patula* L.; *Bassia hyssopifolia* Kuntze, Rev. Gen. Pl.; *Melilotus indica* (L.), All.; and *Salsola kali* L. These species composed nearly 100% of the vegetation. *Bassia hyssopifolia*, *A. patula*, and *S. kali* are summer weeds. These plants were at the juvenile stage. *Melilotus indica* is a winter weedy legume species, and it was in blossom. *Bassia hyssopifolia* had the greatest biomass among the four species; it produced approximately 60 g dry weight per square meter in the nonirrigated and nontilled plots and 300 g dry weight/m² in the irrigated and nontilled pots, and it contributed 50% and 70% of the total biomass, respectively. Irrigation increased the plant tissue Se concentration of *B. hyssopifolia* from 25 to 50 µg/g dry weight. In terms of biomass, *M. indica* had the greatest response to irrigation. Its biomass increased from an average of 5 g dry weight/m² in nonirrigated soils to about 70 g/m² in irrigated soils. Its percentage of total biomass increased from 4% to 18% with irrigation. Among the four species, irrigation had the greatest effect on Se accumulation by *M. indica*. Its tissue Se concentration increased from 19 to 183 µg/g dry weight due to irrigation. *Atriplex patula* produced approximately 40 g/m² biomass in nonirrigated plots, but its biomass was reduced to an average of 10 g/m² in irrigated plots. Its percentage of total biomass was also reduced from 32% to 25%, but its tissue Se concentration was increased from 20 to 80 mg/g dry weight. This biomass reduction may be due to its being less competitive than the other species and/or the increase of Se toxicity due to irrigation. The biomass of *S. kali* was increased from 11 g/m² to 25 g/m² dry weight by irrigation, but the increase was not statistically significant, and it only contributed less than 10% of the total biomass. It had an average tissue Se concentration of 40 µg/g dry weight and was not significantly affected by irrigation. The field plots that were tilled at 3-month intervals essentially remained bare throughout the experiment.

Most of the summer annual weeds at Kesterson reach their maximum growth in the fall. The second biomass and Se accumulation measurements were done in September 1991. *Bassia hyssopifolia* was the only predominant plant species that still remained in the field, and it composed nearly 100% of the vegetation. Its average biomass was 8258 g/m² dry weight for the irrigated plots and 5917 g/m² for the nonirrigated plots. The plants reached a height of about 170 cm and formed nearly 100% of ground cover. In summary (for the September 1991 plant and soil sample analysis), the irrigation treatment increased the biomass and vegetative coverage from 20 g/m² dry weight and 9% ground coverage to 95 g/m² dry weight and 58% ground coverage, respectively. The total Se accumulation increased from 0.87 to 9.02 mg/m². The average tissue Se concentration of *B. hyssopifolia* in the irrigated soils was 42 µg/g dry weight; it was significantly greater than the tissue Se concentration (13 µg/g) of the plants grown in the nonirrigated soils. Total Se accumulation of *B. hyssopifolia* of the irrigated plots was approximately 347 mg/m², but for the nonirrigated plots it was only about 200 mg/m². The total Se accumulation per unit field area (m²) was much higher than in the May 1991 survey.

Total soil Se concentrations of the top 15 cm of soil measured in May 1991 were found to range from 43 to 76 mg/kg dry weight. The total soil

Se concentrations below 25 cm depth (25–40 cm) were only within a range of 8–18 mg/kg dry weight. No significant difference was detected among the four irrigation–tillage treatments. Water-extractable Se ranged from 15 to 3.4 mg/kg, which is only 4% of the total soil Se concentration, and the nontilled plots had slightly greater concentration than the tilled plots. Water-extractable soil Se concentrations below the 25 cm depth were comparable to Se concentrations of the top soil horizon.

Soil chemical characteristics measured in September 1991 indicated that total soil Se remained unchanged compared to the May 1991 soil analyses and were found to range from 43 to 76 mg/kg dry weight in spite of the irrigation and tillage practices. This fact suggests that a large portion of the Se inventory was not remobilized. Water-extractable soil Se concentrations increased slightly from the May soil analysis and ranged from 3.22 to 7.03 mg/kg for the top 15 cm soil horizon and from 1.16 to 4.26 mg/kg for the soil below 25 cm depth.

Among the four irrigation–tillage treatment categories, the irrigated field plots had lower water-extractable soil Se levels than the nonirrigated field plots. Soil EC ranged from 4 to 7 dS/m for the nonirrigated field plots and from 3 to 4 dS/m for the irrigated plots and was slightly higher than in the May 1991 soil sample. Soil surface concentrations ranged from 1626 to 3958 mg/kg and were comparable between the top 15 cm and the soil below 25 cm depth. Sulfate concentrations were significantly higher in soils of nonirrigated and nontilled field plots. Soil pH was about 7 for all the field plots under different irrigation and tillage treatments. Soil moisture content measured in July 1991, prior to the subsequent irrigation treatment, showed that the soil moisture contents of the top 15 cm of soil in the irrigated fields ranged from 8% to 10%. The soil moisture contents of the nonirrigated fields was 4%. The soil moisture increased proportionally with increase in soil depth. Soil moisture for the irrigated fields at depth of 45–65 cm was about 24%, and for the nonirrigated fields was only 15%. In both May 1991 and September 1991, soil analyses indicated that EC was positively correlated with the soil sulfate concentrations ($r = 0.89$, $P < 0.001$), and the water-extractable Se was positively correlated with the soil EC values ($r = 0.63$, $P < 0.001$). But the water-extractable Se was not correlated significantly with the total soil Se concentrations. The total soil Se concentrations were not correlated significantly with either EC or sulfate concentrations.

The presence of sulfate in the soil generally reduces Se accumulation by plants [37,44,46,52,65]. The increase of Se accumulation in the irrigated plants is at least partly due to the reduction of sulfate concentration in the root zone. However, the degree and direction of the effects were quite different among the species. Differences in Se accumulation potential have been found both between and within plant species [47,66–71]. *Melilotus indica* had the greatest increase in both tissue Se concentration and biomass under irrigation treatment. The tissue Se concentration of *M. indica* was over 200 µg/g dry weight, and its biomass increased 20-fold under irrigation treatment. Therefore, it is unlikely that the increase in Se accumulation due to irrigation is associated with phytotoxicity from Se. In addition, the field-grown plants were well nodulated. More research is needed to determine whether the nitrogen fixation symbiotic system had any influence on Se accumulation in this species.

It is known that clay content of soil has a large impact on the uptake of Se by plants, especially that of selenite, because selenite is strongly adsorbed by clay [72]. In addition, organic matter is also capable of removing Se from solution phase [73]. Gissel-Nielsen [45] found that Se uptake by red clover (*Trifolium paratense*), barley (*Hordeum vulgare*), and white mustard (*Brassica alba*) from selenate added to a muck soil having 13% organic matter was 10 times less than from some mineral soils. Soils in Pond 2 at Kesterson are clay loam with huge amounts of decayed plant materials. These factors may limit the bioavailability of Se in the soil. Selenium bioavailability may also be affected by differences in the root systems of plant species. It was reported that deep-root species can access more bioavailable Se in the deeper, more alkaline subsoil horizons [74,75]. Among the four herbaceous species, *B. hyssopifolia* and *A. patula*, which are summer weeds, have deeper root systems than the winter weed *M. indica*. But *M. indica* accumulated two to four times as much tissue Se as *B. hyssopifolia* and *A. patula*. This was due to the great majority of the Se inventory being in the top 15 cm of the soil horizon. Therefore, it is reasonable to conclude that the differences in Se accumulation among the four species due to irrigation practices are intrinsic characteristics of the plant species rather than the results of differences in root systems.

V. COLONIZATION OF NATURAL VEGETATION ON SELENIUM-CONTAMINATED SOILS AT KESTERSON

The biological cycling of Se in the environment is a complicated process, because there are different sources and physicochemical forms of Se existing in the environment [76]. In Kesterson Reservoir, the source of Se was from imported agricultural drainwaters, primarily as selenate, which was transformed into different forms of organic and inorganic species of Se. The Se-contaminated soils are now being colonized by various annual and perennial herbaceous plant species. These plants may play important roles in the bioextraction and cycling of Se in the environment.

A. Salt Grass Habitat

Simple vegetation and low species diversity are features of Kesterson Reservoir. Plant species including salt grass (*Distichlis spicata* L.), atriplex (*Atriplex patula* L.), cattails (*Typha latifolia* L.), and rushes (*Juncus effusus* var *pacificus*) were the ecologically dominant species. Prior to the conversion of the evaporation ponds to a terrestrial habitat in November 1988, two distinct types of plant species distribution existed. In wet sediment, the most striking distribution was the mosaic of cattails and rushes. Each piece of the mosaic was over 100 ft in diameter. The second distribution pattern involved salt grass. Its distribution spread over large areas of the dry land of the ponds in the Kesterson Reservoir. The soil salinity and Se concentrations of the salt grass habitat may vary 10–100-fold, indicating that there might be an unusually high degree of Se tolerance in this species and that the species may play an important role in soil Se cycling in the environment. Soil factors affecting the colonization of salt grass and the role of salt grass in bioaccumulation of Se were investigated [48].

1. Salt Grass Sampling Sites and Soil Se Conditions

Selenium concentrations in the soil's top 15 cm in the 12 evaporation ponds were found to be highest in Ponds 1 and 2, near the drainage outlets of the San Luis Drain, and the Se concentrations in the topsoil of these ponds ranged from 0.1 to more than 100 mg/kg. Soil Se concentrations decreased for the ponds further from the outlets. The topsoil Se concentrations in Ponds 6 and 7 ranged from 0.1 to 52 mg/kg dry weight.

In May 1988, four sites of salt grass habitat were chosen for plant sample collection. Ponds 2 and 6 were two of the 12 evaporation ponds at Kesterson Reservoir where the soil Se was high. A location where the soil Se was low, at Kesterson National Wildlife Refuge about 300 m from Pond 6, was used as a control site. The second control site was a vernal pool in a field near Dixon, California, about 400 km north of Kesterson NWR. The vernal pool usually becomes flooded during the winter months and is dry throughout the summer. Pure salt grass stands exist along the edge and in areas in the pool that are infrequently subject to flooding. Plant samples were collected as single pieces of rhizomes from these locations and were propagated in the greenhouse for studies of Se tolerance and Se accumulation.

The selenate concentration in the soils of the Kesterson control site was about 40 µg/kg dry weight, about twice the selenate concentration in soils at the Dixon site. Selenite concentrations ranged from 10 µg/kg for the

Kesterson control site to 3 µg/kg for the Dixon site. For Ponds 2 and 6, soil selenate concentrations were 1601 and 2758 µg/kg, respectively; selenite levels were 2101 and 360 µg/kg, respectively. These Se concentrations were 100 times higher than the soil Se concentrations of the Kesterson and Dixon control sites. In soils collected from Pond 6, the selenate concentration was much higher than the selenite concentration, but in Pond 2 the selenite concentration was about twice the selenate concentration.

Soil from the Dixon control site had a sulfate concentration of about 450 mg/kg dry weight; soil from the Kesterson control site had a much higher sulfate concentration, about 1700 mg/kg dry weight. Soil from Pond 2 had 2000 mg/kg sulfate, and that from Pond 6 had 5000 mg/kg. The soil pH values were about 7 for all the soil samples collected from the four sites. The highest EC value (33 dS/m) was found in the soil of Pond 6. The EC of the Kesterson control site was 1.9 dS/m, and the Dixon site had the lowest EC, 0.95 dS/m (Table 11).

In September 1988 and September 1989, plant species richness, biomass distribution, and plant tissue Se concentrations in a salt grass–predominant habitat in Pond 7 were examined. The above-ground plant tissue was collected from thirty-one 60 × 60 cm^2 quadrants at 5.5-m intervals along the 183-m north–south centerline of a 46-m-wide and 183-m-long field plot. The number of plant species, percentage area of vegetation coverage by each plant species, and plant tissue dry weight were recorded.

Additional soil Se analysis was conducted for soils collected from Ponds 6 and 7 over a 16-month period from May 1989 to September 1990, in order to detect possible trends of seasonal soil Se concentration changes in the salt grass habitat. Soil Se, sulfate concentrations, and EC fluctuated between seasons, tending to have minimum values during May with maximum levels in September to December. Soil pH remained relatively constant within a range of 7–8. The mean water-extractable Se concentrations for soil samples collected in March 1990 were 140 and 93 µg/kg for Ponds 6 and 7, respectively. For soil samples collected in September 1990, water-extractable Se concentrations were 285 and 50 µg/kg for Ponds 6 and 7, respectively. Soil total Se concentrations ranged from 2000 to 5000 µg/kg and remained relatively consistent. The average soil pH was about 8, and the EC about 3 dS/m. The variation of soil pH and EC between soil samples was much smaller than the variation of soil Se concentrations.

2. Biomass

For the salt grass–predominant habitat in Ponds 2 and 7, salt grass contributed an average of 80% of the groundcover for the 1988 survey and

Table 11 Chemical Character of Soils Collected in May 1988 from Locations of Salt Grass Habitats[a]

Location	Soil chemical analysis
Pond 2	
Selenate (µg/kg dry weight)	1601 ± 58
Selenite (µg/kg dry weight)	2101 ± 294
Sulfate (mg/kg dry weight)	2310 ± 500
pH	7.15 ± 0.21
EC (ds/m)	4.35 ± 0.45
Pond 6	
Selenate (µg/kg dry weight)	2758 ± 0.45
Selenite (µg/kg dry weight)	360 ± 14
Sulfate (mg/kg dry weight)	5620 ± 1630
pH	7.39 ± 0.09
EC (ds/m)	33.05 ± 1.95
Kesterson control site	
Selenate (µg/kg dry weight)	38 ± 4
Selenite (µg/kg dry weight)	10 ± 0
Sulfate (mg/kg dry weight)	1780 ± 50
pH	6.87 ± 0.04
Dixon vernal pool	
Selenate (ng/g dry weight)	19 ± 4
Selenite (ng/g dry weight)	3 ± 0
Sulfate (μg^{-1} dry weight)	450 ± 160
pH	6.70 ± 0.07
EC (ds/m)	0.95 ± 0.25

[a]Mean and 1 SD.
Source: Ref. 48.

96% for the 1989 survey. *Atriplex patula* contributed only 18% and 3% for the 1988 and 1989 surveys, respectively. Salt grass produced approximately 600 g/m^2 dry weight in both the 1988 and 1989 surveys (Table 12). *A. patula* produced only 60 and 20 g/m^2 dry weight in the two consecutive years. *A. patula* had tissue Se concentrations of about 4 µg/g, and salt grass had about 2 µg/g dry weight for both years' samples.

No significant correlations between soil Se concentration and biomass production were detected for either salt grass ($r = 0.30$, $P > 0.05$) or *Atriplex* ($r = 0.26$, $P > 0.05$). However, plant tissue Se concentration was significantly correlated with soil Se concentration: ($r = 0.53$, $P < 0.01$) for salt grass and ($r = 0.53$, $P < 0.05$) for *Atriplex*.

Table 12 Biomass Production, Species Richness, and Plant Tissue Se Concentration in Salt Grass–Predominant Habitat in Evaporation Pond 7 at Kesterson

	September 1988	September 1989
Number of samples	31	10
Species richness (percentage ground cover)[a]		
Distichlis spicata	81 ± 21	96 ± 8
Ariplex patula	18 ± 11	3 ± 1
Biomass (g/m^2)		
Distichlis spicata	740 ± 45	593 ± 125
Atriplex patula	59 ± 43	20 ± 14
Plant tissue Se concentration (μg/g)[a]		
Distichlis spicata	1.5 ± 0.3	2.7 ± 0.9
Atriplex patula	3.6 ± 1.5	4.1 ± 2.0
Soil chemical analysis[a]		
Total soil Se (μg/kg)	170 ± 72	100 ± 18
Sulfate (μg/g)	4562 ± 206	3900 ± 230
EC (ds/m)	3.6 ± 0.3	2.9 ± 0.5
pH	7.8 ± 0.2	8.0 ± 0.5

[a]Mean and 1 SD.
Source: Ref. 48.

3. Selenium Tolerance

Salt grass samples collected from both the Kesterson Se-contaminated soils and the control sites were tested for their Se tolerance in nutrient solution culture using vegetatively propagated plant materials [47]. The salt grass was found to be very sensitive to Se. Plants grown in quarter-concentration Hoagland nutrient solution amended with 0.5 mg Se/L as sodium selenate had 50% growth reduction compared to the growth in the culture solution without Se amendment, and the growth was severely inhibited by 1 mg Se/L. The Se tolerance and dry weight of salt grass collected from the four sites produced in nutrient solution culture amended with 0.5 mg/L are presented in Figure 4. The mean Se tolerance ratio of the plants collected from the Dixon site was about 50%. Plants from the other three sites had a similar mean tolerance ratio of 40%. Selenium tolerance ratios were found to be significantly different ($P < 0.05$) among clones within a site. However, no significant difference between sites was detected.

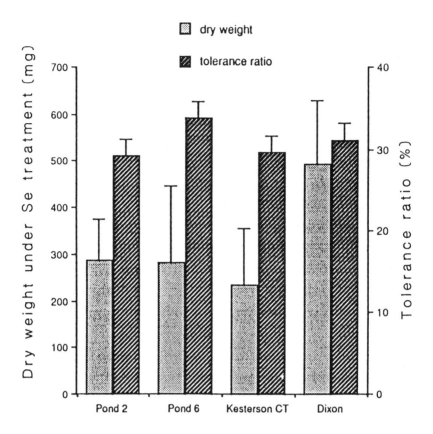

Figure 4 Dry weight and tolerance ratios of four salt grass population samples grown in nutrient solution supplemented with 2 mg Se/L as sodium selenate. (From Ref. 48.)

4. Effects of Sulfate and Chloride Salt Concentrations on Growth and Se Accumulation of Salt Grass

In addition to sulfate, chloride is also a major inorganic anion in the sediments of the Kesterson evaporation ponds. To study the possible effects of sulfate and chloride on the ability of salt grass to colonize the Se-contaminated soils at Kesterson, nutrient solution culture studies were conducted.

Under equivalent chloride and sulfate concentrations within a range of 50–100 meq/L, salt grass produced a significantly greater amount of dry weight under sulfate salt than under chloride salt. Without the addition of sulfate to the culture solution, 1 mM NaCl stimulated the growth of salt grass. The salt grass produced about 40% of normal shoot dry weight (compared to the control treatment) under 0.5 mg Se/L without the addition of salt. With the addition of 1 or 10 mM chloride salt and 0.5 mg Se/L, salt grass produced 50% of normal shoot dry weight. In the presence of 0.5 mM sulfate and 0.5 mg Se/L, salt grass produced 60% normal shoot dry weight, and in the presence of 5 mM sulfate and 0.5 mg Se/L, it produced 100% normal shoot dry weight.

Tissue Se concentration was negatively correlated with dry weight production ($r = -0.90$, $P < 0.01$). Selenium uptake was markedly inhibited by the presence of sulfate. For example, in the 1 mg Se/L treatment, the Se concentration in shoot tissue was reduced from 400 µg/g dry weight without the addition of sulfate to only about 17 µg/g dry weight with the addition of 5 mM sulfate. There was a slight increase in tissue Se concentration with the addition of chloride.

5. Selenium Volatilization

The phenomenon of Se volatilization, in which Se may be dispersed into the air through plants, could be taken advantage of to detoxify Se-contaminated soils. Certain fungi can convert selenite and selenate to dimethylselenide and/or dimethyl diselenide, volatile compounds that can readily be dispersed into the air [77]. Plants are also able to convert Se into a volatile form, but information is scant and research is needed [78,79].

The volatile Se–collecting chamber used was modified from the chamber developed by Biggar and his associates [80]. For volatile Se measurement, salt grass plants collected from Kesterson evaporation Pond 7 were established in sand culture. The plants were irrigated either with quarter concentration Hoagland nutrient solution or quarter concentration Hoagland nutrient solution plus 1 mg Se/L as Na_2SeO_4 three times a week. After 4 weeks of Se pretreatment, the Se volatilization of the salt grass was measured [47].

Plants irrigated with nutrient solution supplemented with 1 mg Se/L produced volatile Se of 65 µg/(kg·day). The addition of 0.5 mM Na_2SO_4 did not cause a reduction in volatile Se. However, with the addition of 5 mM Na_2SO_4, volatile Se production was reduced to 25 µg/(kg·day). Volatile Se released by the salt grass was found essentially to be dimethylselenide. No detectable volatile Se was found for plants under the control treatment or for sand without salt grass (Table 13). Plants grown in the greenhouse and the Kesterson native soil produced only 4 µg of volatile Se per kilogram dry weight per day. Volatile Se measured for the field-grown

Table 13 Rates of Se Volatilization of Salt Grass Grown in Sand Culture Irrigated with Quarter Concentration Hoagland Solution Supplemented with Either 2 mg Se/L or 2 mg Se/L plus 5 mM Sulfate

Selenium and salt treatment	Rate of Se volatilization[a] ($\mu g/g \cdot$ day)	Plant shoot tissue Se concentration[a] ($\mu g/g$)
Control	—[b]	—[b]
2 mg Se/L	65 ± 14	106 ± 14
2 mg Se/L + 5 mM Na$_2$SO$_4$	25 ± 5	67 ± 12
Kesterson soil	4 ± 0.1	22 ± 1
Measured in Pond 7 at Kesterson	0.3 ± 0.1	4 ± 1

[a]Mean and standard deviation.
[b]Se concentration not detectable.
Source: Ref. 48.

salt grass in Pond 7 during the summer of 1990, using an identical trap system, was about 0.3 µg/(kg·day).

B. Succession of the Grassland Communities in Reconstructed Soils at Kesterson Reservoir

Attempts were made to alleviate the deterioration of the natural habitat at Kesterson Reservoir. As a short-term solution, in the fall of 1988 the discharge of agricultural drainwater was terminated and the top 15 cm of soil in the evaporation ponds at Kesterson was covered with fill dirt in a program called the Kesterson Cleanup Action. A new effort has thus been proposed to monitor the habitat and to evaluate how a combination of soil, water, and vegetation management can clean up the Se-contaminated soils more effectively [81]. The task of restoring the ecosystem requires the application of a wide range of knowledge of soil–plant relations.

Initial vegetation-monitoring studies were conducted to survey (a) plant species diversity, biomass production, vegetation coverage, and Se accumulation of plants in the field at Kesterson, either in fill dirt or in native soil sites, after conversion of the evaporation ponds to terrestrial habitat in November 1988; (b) trends of these parameters over seasonal periods from May 1989 to September 1990; (c) the effects of soil Se on native plant species in terms of seed bank, pollen fertility, and the seed set; and (d) the potential for food chain contamination by terrestrial plants [75].

1. Soil and Vegetation

Field-grown plants and soil samples were collected in May and September 1989 and in March, May, and September 1990. In May 1989, species richness, biomass production, and plant tissue Se accumulation were studied for two sites where the topsoil was covered by fill dirt (Ponds 2 and 6) and two sites of unfilled native soil (Ponds 6 and 7). For each site, a 30 m × 15 m area (plot) was chosen to maximize the number of species sampled, since the objectives of these collections were to survey both Se uptake by plants and species richness within each area. The selected plots are thus representative of vegetation in the area.

The average water-extractable soil Se concentrations of the top 15 cm of fill dirt (May 1989 sample) in Ponds 2 and 6 were 20 and 15 µg/kg, respectively. For the native soil sites, the average soil Se concentrations in Ponds 6 and 7 were 179 and 348 µg/kg, respectively. Water-extractable Se concentrations were 132 and 653 µg/kg for Ponds 2 and 6, respectively, for the soils below 15 cm depth in the fill dirt sites and 273 and 233 µg/kg for Ponds 6 and 7 respectively, in the native soil sites. Soil EC ranged from 0.4 dS/m for the top 15 cm of fill dirt to 5.8 dS/m for the soil below 15 cm in the Pond 2 fill dirt site.

For the two fill dirt sites, the average soil sulfate concentrations in Pond 2 were 323 mg/kg dry soil for the top 15 cm of soil and 5678 mg/kg for the soil below 25 cm depth. The average soil sulfate concentrations in Pond 6 were 72 and 3329 mg/kg for the top 15 cm and below 25 cm depth, respectively. For the two native soil sites, the soil sulfate concentrations ranged from 4000 to 7000 mg/kg dry soil, and no significant difference was found between the two soil depths. Soil pH values were similar among the four sites and had an average value of 8.

Water-extractable and total Se concentrations measured for soil for seasonal periods from May 1989 to September 1990 (Figures 5a and 5b) indicated that the Se concentration of the top 15 cm of soil at the native soil sites remained relatively consistent. For the fill dirt site in Pond 6, it remained low over the various seasons. However, the soil Se concentration of the fill dirt in Pond 2 increased from 20 to 600 µg/kg from May to December 1989 but decreased in 1990. Soil EC and sulfate concentrations showed a positive correlation with the water-extractable Se concentration, and soil pH values varied slightly between 7 and 8 (Figures 6a–c).

At least 20 different plant species were found in the fill dirt sites, while only five species were found in the native soil sites (Table 14). In native soil sites, salt grass (*Distichlis spicata* L.) was the most abundant species, with the remaining species making up less than 15% of the vegetation. The biomass of the plant species in the fill dirt sites ranged from 2 to 473 g/m^2.

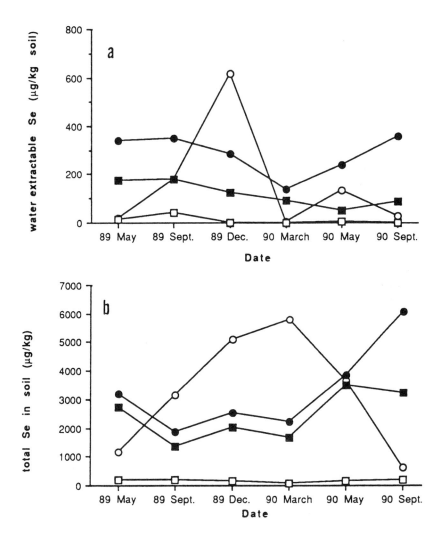

Figure 5 (a) Water-extractable and (b) total Se concentrations detected from soil samples collected from two fill dirt sites [(○) Pond 2 and (□) Pond 6] and two native soil sites [(●) Pond 6 and (■) Pond 7] in different seasons over a period of 1 year. (From Ref. 75.)

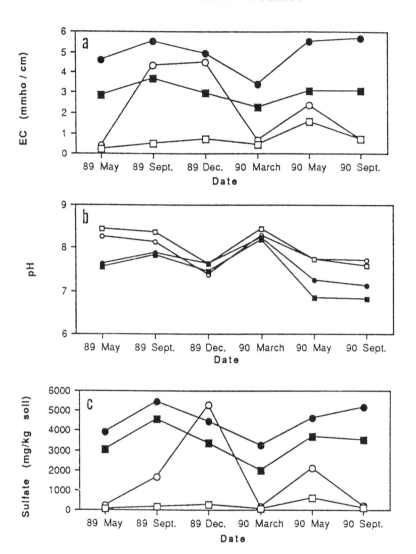

Figure 6 Soil EC (a), pH (b), and sulfate (c) concentrations detected from soil samples collected from two fill dirt sites ((○) pond 2 and (□) pond 6) and two native soil sites ((●) pond 6 and (■) pond 7) in different seasons over a period of 1 year.

Table 14 Biomass and Plant Tissue Se Concentration of Plant Species Collected from Two Ponds at Kesterson Reservoir[a]

Species	Mean Se concentration (μg/g)		Mean dry weight (g/m^2)
	Mean	SD	
Pond 2			
Atriplex canescens	12.33	1.25	91.55
Atriplex patula	12.86	3.90	70.18
Avena barbata	3.36	3.50	6.87
Avena fatua	5.00	5.02	73.00
Bassia hyssopifolia	9.31	0.77	473.35
Bromus mollis	3.31	3.21	10.70
Bromus rubens	1.57	0.42	22.11
Centaruea solstitalis	16.15	1.22	63.60
Erysium officinale	18.47	0.00	11.09
Festuca magalura	2.03	0.06	1.89
Franseria acanthicarpa	12.02	2.54	24.47
Hordeum leporinum	2.78	0.00	21.20
Matricaria matricarioides	3.36	0.00	2.97
Melilotus indica	1.06	0.78	279.06
Polypogon monspeliensis	19.01	10.50	21.82
Pond 6			
Atriplex patula	1.17	0.00	1.70
Avena barbata	3.36	0.03	27.52
Avena fatua	0.45	0.20	185.13
Bassia hyssopifolia	5.89	3.09	580.91
Bromus rubens	0.39	0.22	14.06
Erodium cicutarium	0.27	0.00	2.96
Eryisium officianale	0.88	0.00	93.85
Festuca magalura	0.24	0.21	4.81
Franseria acanthicarpa	1.66	1.31	178.70
Melilotus indica	0.20	0.16	366.24
Polygonum aviculare	0.56	0.00	8.58

[a]Collected in May 1989 where top soil was covered with fill dirt in two evaporation ponds at Kesterson.
Source: Ref. 75.

There was great variation in the vegetation coverage and the biomass of samples within the fill dirt sites. *Atriplex canescens, A. patula* L., *Avena fatua* L., *Bassia hyssopifolia* Kuntze Rev. Gen. Pl., and *Melilotus indica* (L.) All. contributed more than 60% of the biomass at the two fill dirt sites. No salt grass occurred in samples from either fill dirt site. Biomass (dry weight) of these plant species was positively correlated with vegetation coverage ($r = 0.536$, $P < 0.005$). Two years after the topsoil replacement, among the 20 plant species recorded, 14 species disappeared from the fill dirt sites, a 70% reduction of species richness in 1 year.

For the fill dirt sites, plant tissue Se concentrations ranged from an undetectable level to 19 µg/g (Table 14). Tissue Se concentrations varied depending on the sample location and plant species. Species with high tissue Se concentrations were *A. canescens, A. patula, Centaurea solstitalis* Sol. L., *Eryisimum officianale* L., *Franseria acanthicarpa* Cav. Lcon., and *Polypogon monspeliensis* (L.) Dest. Since these species all have deep root systems, this implies that plants with deep roots were able to take up Se from the subsurface soil. For the native soil sites, the plant tissue Se concentrations ranged from an undetectable level to 20 µg/g. Variation in tissue Se concentrations existed both between sites and between plant species. Generally, *A. patula* and *Polygonum aviculare* L. had higher tissue Se concentrations than salt grass.

The mean tissue Se concentrations of the plants grown in the fill dirt sites were variable between the two years. *Avena fatua* from Pond 2 had higher Se concentrations in 1989, but at the Pond 6 site Se concentration was higher in the 1990 samples (Table 15). *Bassia hyssopifolia* and *F. acanthicarpa* from both Ponds 2 and 6 had lower mean tissue Se concentrations in 1989 than in 1990, and tissue Se concentration differences varied both between years and between the two ponds. At the Pond 6 native soil site, the mean tissue Se concentrations of *A. patula* and salt grass were significantly higher in 1989 than in 1990, whereas *Frankenia grandifolia* Cham and Schl. had lower tissue Se concentrations in 1989 than in 1990. Tissue Se concentrations of salt grass at the Pond 7 native soil sites were not significantly different between the two years.

Overall, the plant species from the Pond 2 fill dirt site had the highest tissue Se concentration, followed by the Pond 6 and 7 native soil sites; the Pond 7 fill dirt soil had the lowest tissue Se concentration ($P < 0.05$).

Total plant coverage was significantly greater ($P < 0.05$) in native soil sites than in fill dirt sites for both the May 1989 and May 1990 surveys. In fill dirt sites, vegetation coverage was significantly higher for the 1989 survey than for the 1990 survey ($P < 0.05$). However, there were no significant differences in vegetation coverage between the two years for the native soil sites. Biomass in fill dirt sites was higher in 1989 than in 1990, and the difference was nearly significant ($P < 0.053$). Biomass in the native soil sites was significantly ($P < 0.01$) between the two years. The number of plant species in the fill dirt sites was significantly greater in 1989, whereas

Table 15 Mean Values of Shoot Tissue Se Concentrations (µg/g dry weight) of Species Collected in Different Seasons Over a Period of One Year from Two Evaporation Ponds With and Without Surface Soil Covered with Fill Dirt

	Pond 2				Pond 6			
	May 1989	Dec 1989	March 1990	May 1990	May 1989	Dec 1989	March 1990	May 1990
Fill dirt								
Avena fatua	3.36	—	—	—	0.02	—	—	0.45
Bassia hyssopifolia	9.31	2.23	—	14.86	5.89	2.72	—	6.09
Centaurea solstitalis	16.15	—	—	5.84	—	—	—	—
Franseria acanthicarpa	12.02	—	—	24.74	1.66	—	—	19.66
Melilotus indica	1.0	—	0.96	—	0.20	—	0.50	—
Salsola stragus	—	2.46	—	—	—	2.67	—	—
Native soil								
Atriplex patula	9.55	—	1.31	4.33	—	—	—	—
Distichlis spicata	3.25	—	1.91	4.61	3.20	—	2.10	3.22
Frankenia gradifolia	3.36	—	—	10.5	—	—	—	—

Note: —, species not present.
Source: Ref. 75.

the number of plant species in the native soil sites were not significantly different between years.

2. Effects of Soil Selenium on Establishment of Seedlings

For seed bank and seedling establishment studies, 30 cm × 30 cm quadrants of 6-cm-deep soil samples were collected from both the Pond 2 and 6 fill dirt and the Pond 6 and 7 native soil sites. Soils were contained in 15-cm-diameter and 15-cm-deep plastic pots. Seeds contained in the soil were germinated, and the plants were grown in a greenhouse [75]. The plants were irrigated with deionized water three times a week, and leaching from the pots was minimized.

Seedling density was recorded 4 weeks after the start of the experiment and was presented as the number of seedlings per square meter. When plants were flowering, pollen was collected; pollen fertility was measured by staining with acetocarmine and viewing under a microscope. Mature seeds were collected as the measure of the seed set. All plants were harvested 5 months after seed germination. The harvested tissue was oven-dried at 65°C for 72 h, and the dry weights of the various organs of each plant species were recorded for biomass partitioning. Tissue Se concentrations were measured for each plant species.

Only three plant species were found in soils collected from the fill dirt sites, while five species were found in soils collected from the native soil sites. *B. hyssopifolia* was the most abundant (density, 1650 and 4875 seedlings/m^2) in fill dirt soil. *Melilotus indica* and *Sesuvium sessile* Pres. made up nearly 80% of the overall seedling density in native soil. No significant difference was detected for overall seedling density between the fill dirt and native soil sites.

No significant differences were detected in pollen fertility and seed set for plants of *M. indica* and *A. patula* grown in the fill dirt and native soil, even though tissue Se concentrations of the plants grown from native soil were 2–10 times those of the plants grown from fill dirt (Table 16).

Selenium concentrations in the seeds of *M. indica* grown in native soil in the greenhouse ranged from 15 to 30 µg/g, while the seeds of the plants grown from fill dirt had tissue Se concentrations of less than 2 µg/g. Seed germination of *M. indica* was not significantly affected by the soil Se of the

Table 16 Plant Organ Tissue Se Concentration[a] (µg/g) Partitioning of Four Plant Species Grown in the Greenhouse from Either Fill Dirt or Native Soil Collected from the Top 15 cm Soil Horizon of the Evaporation Ponds at Kesterson

	From fill dirt		From native soil	
	Pond 2	Pond 6	Pond 2	Pond 6
Atriplex patula				
Root	14 ± 3	0.75 ± 0.01	25 ± 4	116 ± 21
Flower	8 ± 2	—	30 ± 4	—
Stem	3 ± 1	0.72 ± 0.2	10 ± 2	20 ± 6
Leaf	6 ± 1	—	31 ± 4	21 ± 6
Bassia hyssopifolia				
Root	7 ± 2	0.70 ± 0.1	—	117 ± 20
Stem	4 ± 1	0.14 ± 0.0	—	8 ± 2
Leaf	7 ± 2	0.32 ± 0.1	—	19 ± 3
Melilotus indica				
Root	5 ± 2	0.72 ± 0.1	69 ± 9	58 ± 4
Stem	2 ± 1	0.22 ± 0.0	23 ± 5	16 ± 3
Sesuvium sessile				
Root	—	—	45 ± 7	45 ± 5
Flower	—	—	24 ± 6	31 ± 4
Stem	—	—	16 ± 4	20 ± 4
Leaf	—	—	23 ± 3	26 ± 4

Note: —, species not present.
[a] Mean ±1 standard deviation.
Source: Ref. 75.

native soil (3.70 µg/g water-extractable Se). However, shoot and root growth of the seedlings was severely inhibited. Compared with the control, roots of seedlings germinated in the native soil were much thinner with fewer root hairs and lateral roots, and they had shortened radicals and swollen hypocotyls. In further experiments with nutrient solution culture supplemented with 1 mg Se/L without excessive salinity, the seedling growth of *M. indica* was also inhibited severely, especially in root elongation.

VI. DISCUSSION

The ability of a plant to colonize successfully in soils with elevated concentrations of Se and salt is a product of interactions of various soil chemical and plant biological factors, notably the type of soil salinity (such as sulfate or chloride), species of Se, and tolerance of the plants. Plants tolerant to both elevated concentrations of salts and Se are useful for Se bioaccumulation management in saline soils with excessive Se concentrations. Currently, most areas of the ponds at Kesterson are filled with topsoil in an attempt to alleviate the problem of Se contamination in the food chain. Thus far, no evidence of a detrimental effect of Se on wildlife has been found in the reconstructed Kesterson grassland habitat. However, it is evident that the uptake of Se by deep-rooted plants such as *B. hyssopifolia*, *C. solstitalis*, and *F. acanthicarpa* can certainly make the "trapped" Se bioavailable once again, and this is demonstrated by the significant positive correlation between the plant tissue Se concentration and the subsurface soil Se concentration found in Ponds 2 and 6 ($r = 0.80$, $P < 0.05$).

Kesterson is characterized by moist winters and dry summers, and many of the annual species brought into the area with the fill dirt, such as *A. fatua*, *C. solstitalis*, and *M. indica*, germinate in the fall. Rapid growth begins early in the year, flowering occurs in early spring, and seeds are ripened before the onset of the summer drought. This kind of life history ensures that the vegetative stage escapes the attention of most invertebrate herbivores. However, the plants are vulnerable to vertebrate herbivores (e.g., hares and ground squirrels) that forage throughout the winter. Since the plant species with the above-mentioned life history strategies contain low tissue Se concentrations, they provide little Se to wildlife in fill dirt sites. However, summer animals like *Atriplex* sp., *B. hyssopifolia*, and *F. acanthicarpa* overwinter as dormant seeds, then germinate in the spring and grow rapidly, to ripen their seeds in the summer or early fall. High tissue Se concentrations in this group may result in potential toxicity in animal diets. Hopefully, these plant species will be replaced by plant species with low tissue Se concentrations or species that are not favored for foraging, such as salt grass.

Although the soil Se concentrations at Kesterson had no negative effects on pollen fertility, seed set, and seed germination, seedling growth and development was impaired by the soil Se. This suggests that selection pressure may have been imposed on certain plant species, such as *M. indica*, at an early stage of the life cycle. A number of species, such as *M. indica* and *S. sessile*, may have large numbers of viable seeds buried in the soils of native soil sites. Some seeds remain viable in the soil for long periods, and these species may appear at unexpected times of succession when the soil is disturbed and soil Se and salinity are reduced.

Among vascular plant species at Kesterson, the principal life history trait of *A. canescens*, *B. hyssopifolia*, and *M. indica* is to set seed once, then die (i.e., they are monocarpic species). By contrast, salt grass has the ability to "move about" by virtue of rhizomes. Therefore, it is capable of spreading radially to form large clonal stands via vegetative reproduction. Vegetative reproduction of salt grass at Kesterson resulted in a thick mat that eliminated many other species and prevented yet others from germinating. This pattern (i.e., initial growth by certain broadleaf species such as *Atriplex* but final dominance by salt grass) partly explains why salt grass is successful and stable in the plant communities of Se- and salt-enriched soils. In addition, salt grass habitually occupies very dry, saline soils and therefore is not usually faced with competition. These conditions may be rather stable and/or predictable (what Grime [82] termed "stress-tolerant").

The Se concentrations of the soil in the evaporation ponds were much higher (3 µg/g dry weight selenate and selenite) than in the soil of the uncontaminated fields. Such high Se concentrations can severely impair the growth of salt grass when tested in nutrient solution culture. Nevertheless, no apparent symptom of Se toxicity has been noticed for the field-grown salt grass. However, there is no evidence that the salt grass growing prosperously in the evaporation ponds has evolved Se tolerance. High concentrations of sulfate were found in the soils of the evaporation ponds. Sulfate has an antagonistic effect on Se uptake and can reduce Se phytotoxicity [64]. Inland-type salt grass has been shown to be less chloride-tolerant than coastal-type salt grass [83]. Salt grass from the evaporation ponds at Kesterson may reduce its growth under 50 mM NaCl. However, the Se accumulation and growth inhibition may be reduced by the presence of a moderate concentration of NaCl (1–10 mM). Therefore, Se tolerance and its uptake by salt grass are dependent not only on soil Se concentration but also on the presence of sulfate and chloride salts. The successful colonization of salt grass in the soils with an elevated Se concentration at Kesterson apparently is attributable to the presence of high concentrations of soil sulfate. This effect helps to explain why a relatively low tissue Se concentration was found in the field-grown salt grass.

Salt grass is a Se nonaccumulator, and it is relatively sensitive to Se. It is a halophyte and accumulates less Se than other salt-tolerant species, such as *Atriplex*, existing in the same area in the evaporation ponds at Kesterson. Accumulation of high concentrations of Se by plants and their predation by insects and animals are major concerns in food chain contamination responsible for the deleterious effects on birds, fish, and mammals at Kesterson [84]. However, no predation on salt grass in the field at Kesterson has been noticed.

Salt grass habitat may transpire significant amounts of volatile Se; it may be a combination of both plant and soil microbe volatilization. The above study estimated that under greenhouse conditions without the sulfate competitive effect, salt grass produced 65 µg/g dry weight per day. Under field conditions, salt grass had a much reduced rate of volatile Se production [approximately 0.3 µg/(g·day)], which may be due partly to the unfavorable growth and soil conditions as well as high soil sulfate concentration. For a 1 m^2 salt grass field plot, it may produce 600 g of dry weight; therefore, 180 µg of volatile Se per day may be produced. Thus, a significant amount of Se may be dispersed by the salt grass habitat. Differences in intrinsic growth rate found among the salt grass populations may be important factors determining the effectiveness of Se volatilization, because the rate of volatilization per unit area in the field is dependent on the biomes production of the plants. Nevertheless, no evidence of any trend in a reduction of soil Se concentration was detected over the 1-year period in the salt grass habitat. This does not seem surprising, because the change in soil salinity, oxidation conditions, and the movement of Se inventory along the soil profile are factors that may influence the quantity and availability of soil Se in the root zone and the rate of Se volatilization of the plants. Therefore, a long-term monitoring of Se status for the salt grass habitat is needed to make predictions for the effectiveness of efforts to clean up Se-contaminated soils through the use of native plant species.

REFERENCES

1. H. G. Byers, J. T. Miller, K. T. Williams, and H. W. Lakin, 3rd Rep., USDA Tech. Bull. 601, U.S. Govt. Printing Office, Washington, D.C., 1938.

2. H. W. Lakin and H. G. Byers, 6th Rep., USDA Tech. Bull. 873, U.S. Govt. Printing Office, Washington, D.C., 1941.
3. A. L. Moxon, O. E. Olson, and W. V. Sewright, Tech. Bull. S. Dak. Agric. Exp. Sta. No. 2.
4. O. H. Muth and W. H. Allaway, *J. Am. Vet. Med. Assoc.* 142:1379 (1963).
5. D. L. Carter, W. H. Allaway, and E. E. Cary, *Agron. J.* 60:532 (1968).
6. T. S. Presser and I. Barnes, U.S. Geol. Surv. Water Resour. Invest. Rep. 84, Sacramento, Calif., 1984.
7. S. J. Deverel, R. J. Gilliom, R. Fuji, J. A. Izbick, and J. C. Field, U.S. Geol. Surv. Water Resour. Invest. Rep. 84, Sacramento, Calif., 1984.
8. H. M. Ohlendorf, D. J. Hoffman, D. J. Saiki, and T. W. Aldrich, *Sci. Total. Environ.* 52:49 (1986).
9. H. M. Ohlendorf, D. J. Hoffman, C. M. Bunck, T. W. Aldrich, and X. X. Moore, *Trans. N. Am. Wildlife Nat. Resour. Conf.* 51:330 (1986).
10. E. Marshall, *Science* 229:144 (1985).
11. T. A. Brown and A. Shrift, *Plant Physiol.* 67:1051 (1981).
12. G. Gissel-Nielsen, U. C. Gupta, M. Lamand, and T. Westermark, *Adv. Agron.* 37:397 (1984).
13. H. W. Lakin, *Geol. Soc. Am. Bull.* 83:181 (1972).
14. H. W. Lakin, in *Trace Elements in the Environment* (E. L. Kothy, Ed.), Adv. Chem. Ser. 123, Am. Chemical Society, Washington, D.C., 1973, p. 96.
15. P. J. Peterson and G. W. Butler, *Aust. J. Biol. Sci.* 15:126 (1961).
16. P. J. Peterson and G. W. Butler, *Aust. J. Bot.* 24:270 (1962).
17. K. J. Maier, C. Foe, R. S. Ogle, M. J. Williams, A. W. Knight, P. Kiffney, and L. A. Melton, in *Trace Substances in Environmental Health*, Vol. 21 (D. D. Hemphill, Ed.), Univ. Missouri Press, Columbus, 1987.
18. H. F. Mayland, L. F. James, K. E. Panter, and J. L. Sonderegger, in *Selenium in Agriculture and the Environment*, SSSA Spec. Publ. 23, Proceedings of a symposium sponsored by Am. Soc. Agron. Soil Sci. Soc. Am. in New Orleans, La., Dec. 2, 1989.
19. R. L. Mikkelson, A. L. Page, and F. T. Bingham, in *Selenium in Agriculture and the Environment*, SSSA Spec. Publ. 23, Proceedings of a symposium sponsored by the Am. Soc. Agron. Soil Sci. Am., New Orleans, Dec. 2, 1986.
20. W. H. Allaway, E. E. Cary, and C. F. Ehlig, in *Symposium: Selenium in Biomedicine* (O. H. Muth, Ed.), Avi, Westport, Conn., 1967, p. 273.
21. O. H. Muth, *J. Am. Vet. Med. Assoc.* 142:272 (1963).
22. O. E. Olson, E. I. Whitehead, and A. L. Moxon, *Cancer Res.* 3:230 (1942).
23. E. L. Patterson, R. Milstrey, and E. L. R. Stokstad, *Proc. Soc. Exp. Biol. Med.* 95:617 (1957).
24. G. N. Schrauzer, D. A. White, C. J. Schneider, and J. Field III, *Bioinorg. Chem.* 7:23 (1977).
24. K. Schwarz and C. Foltz, *J. Am. Chem. Soc.* 79:3292 (1957).
25. G. Gissel-Nielsen and B. Bisbjerg, *Plant Soil* 32:381 (1970).
26. E. B. Davies and J. H. Watkinson, *N.Z.J. Agric.* 9:641 (1969).
27. P. J. Peterson and G. W. Butler, *Nature* 212:961 (1966).

28. I. Rosenfeld and O. A. Beath, *Annu. Rev. Plant Physiol.* 20:475 (1964).
29. G. S. Smith and J. H. Watkinson, *New Phytol.* 97:557 (1984).
30. D. C. Andrian, *Trace Elements in the Terrestrial Environment*, Springer-Verlag, New York, 1986.
31. L. Wu, Z. Z. Huang, and R. G. Burau, *Crop Sci.* 28:517–522 (1988).
32. S. Mahendra and S. Narendra, *Soil Sci.* 127:264 (1979).
33. T. J. Ganje, Selenium, in *Diagnostic Criteria for Plants and Soils* (H. D. Chapman, Ed.), Quality Printing, Abilene, Tex., 1966, p. 394.
34. A. Shrift, *Annu. Rev. Plant Physiol.* 20:475 (1969).
35. K. S. Dhillon, N. S. Randhawa, and M. K. Shia, *Indian J. Dairy Sci.* 30:218 (1977).
36. C. Foe and A. W. Knight, *Proc. First Annu. Environ. Symp. Selenium in the Environment*, Fresno, Calif., June 10–12, 1985.
37. L. Wu and Z. Z. Huang, *Crop Sci.* 31:114 (1991).
38. O. R. Lunt, V. B. Youngner, and J. J. Oertli, *Agron. J.* 53:247 (1961).
39. L. J. Greub, N. P. Drolsom, and A. Rohweder, *Agron. J.* 77:76 (1985).
40. G. L. Host and N. B. Beadle, *J. Am. Soc. Hort. Sci.* 109:419 (1984).
41. J. E. Leggett and E. Epstein, *Plant Physiol.* 31:222 (1965).
42. C. J. Asher, G. W. Butler, and P. J. Peterson, *J. Exp. Bot.* 28:279 (1977).
43. M. S. Vange, K. Holmern, and P. Nissen, *Physiol. Plant.* 31:292 (1974).
44. J. E. Pratley and J. D. McFarlane, *Aust. J. Exp. Agric. Anim. Husb.* 14:533 (1974).
45. G. Gissel-Nielsen, *J. Sci. Food Agric.* 24:649–655 (1973).
46. R. L. Mikkelsen, A. L. Page, and G. H. Haghnia, *Plant Soil* 107:63 (1988).
47. L. Wu and Z. Z. Huang, *Ecotoxicol. Environ. Safety* 22:199 (1991).
48. L. Wu and Z. Z. Huang, *Ecotoxicol. Environ. Safety* 22:267 (1991).
49. J. M. Ulrich and A. Shrift, *Plant Physiol.* 43:14 (1968).
50. A. Shrift, in *Organic Selenium Compounds: Their Chemistry and Biology* (D. L. Klayman and W. H. H. Gunther, Eds.), Wiley-Interscience, New York, 1973, p. 763.
51. T. A. Brown and A. Shrift, *Plant Physiol.* 66:758 (1980).
52. R. L. Mikkelsen, G. H. Hagnia, A. L. Page, and F. T. Bingham, *J. Environ. Qual.* 17:85 (1988).
53. U. C. Gupta, K. B. McRae, and K. A. Winter, *Can. J. Soil Sci.* 62:145 (1982).
54. E. Epstein, in 1985–1986 Tech. Prog. Rep., Univ. Calif. Salinity/Drainage Task Force, Div. Agric. Res., Univ. Calif., 1986, p. 136.
55. A. L. Page, in 1985–1986 Tech. Prog. Rep., UC Salinity/Drainage Task Force, Div. Agric. Res., Univ. Calif., 1986, p. 147.
56. L. Wu and Z. Z. Huang, *J. Exp. Bot.* 43:549 (1992).
57. E. Epstein and R. G. Burau, Tech. Prog. Rep., UC Salinity/Drainage Task Force, Div. Agric. res., Univ. Calif., 1988, p. 64.
58. J. B. Hanson, in *Advance in Plant Nutrition* (P. B. Tinker and A. Lauchli, Eds.), Praeger, New York, 1984, p. 149.
59. E. A. Kirkby and D. J. Pilbean, *Plant Cell Environ.* 7:397 (1984).
60. W. H. Orme-Johnson, L. C. Davis, M. T. Henzl, B. A. Averill, N. R. Orme-Johnson, E. Munck, and R. Zimmerman, in *Recent Developments in Nitrogen*

Fixation (W. Newton, J. R. Postgate, and E. Rodriguez-Barrueco, Eds.), Academic, London, 1977, p. 131.
61. E. J. Bergersen, *Aust. J. Biol. Sci. 16*:916 (1963).
62. B. K. Burger, S. S. Yang, C. B. You, J. G. Li, G. D. Friesen, W. H. Pan, E. I. Stiegel, W. E. Newton, S. D. Conradson, and K. D. Hodgson, in *Current Perspectives in Nitrogen Fixation* (A. H. Gibson and W. E. Newton, Eds.), Wiley, New York, 1981, pp. 71–74.
63. L. Wu, A. Enberg, and K. Tangi, *Ecotoxicol. Environ. Safety 25*:237 (1993).
64. G. Ferreri and F. Renosto, *Plant Physiol. 49*:114 (1972).
65. E. E. Cary and G. Gissel-Nielsen, *Soil Sci. Soc. Am. Proc. 37*:590 (1973).
66. J. W. Hamilton and O. A. Beath, *J. Range Manage. 16*:261 (1963).
67. J. W. Hamilton and O. A. Beath, *Agron. J. 55*:528 (1963).
68. J. W. Hamilton and O. A. Beath, *J. Agric. Food Chem. 12*:371 (1964).
69. B. Bisbjerg, and G. Gissel-Nielsen, *Plant Soil 31*:287 (1969).
70. A. M. Davis, *Agron. J. 78*:727 (1968).
71. C. F. Ehlig, W. H. Allaway, E. E. Cary, and J. Kubota, *Agron. J. 60*:43 (1968).
72. G. Gissel-Nielsen, *J. Agric. Food Chem. 19*:1165 (1971).
73. M. Levesque, *Can J. Soil Sci. 54*:205–214 (1974).
74. D. L. Massey and P. J. Martin, *Alberta Agric. Agdex. 531*:1 (1975).
75. Z. Z. Huang and L. Wu, *Ecotoxicol. Environ. Safety 22*:251 (1991).
76. P. M. Haygarth, K. C. Jones, and A. F. Harrison, *Sci. Total Environ. 103*:89 (1991).
77. W. T. Frankenberger, Jr. and U. Karlson, *Soil Sci. Soc. Am. J. 53*:1435 (1989).
78. B. G. Lewis, C. M. Johnson, and T. C. Broyer, *Plant Soil 40*:107 (1974).
79. N. Terry, C. Carlson, T. K. Raab, and A. Zayed, *J. Environ. Quality 21*: (1992) (in press).
80. J. W. Biggar, G. R. Jayaweera, and D. E. Rolston, in *Hydrological, Geochemical, and Ecological Characterization of Kesterson Reservoir*, Lawrence Berkeley Lab. Annu. Rep. LBL-27993, pp. 145–161, 1990.
81. Lawrence Berkeley Laboratory and Division of Agriculture and Natural Resources (LBL and DANR), *Vegetation Sampling Progress Report*, Earth Sci. Div., Lawrence Berkeley and Div. Agric. Nat. Resour., Univ. Calif., 1989.
82. J. P. Grime, *Plant Strategies and Vegetation Processes*, Wiley, London, 1979.
83. F. W. Wrona and E. Epstein, in *Biosaline Research: A Look to the Future* (A. San Pietro, Ed.), Plenum, New York, 1982, p. 559.
84. R. L. Hothem and H. M. Ohlendorf, *Arch. Environ. Contam. Toxicol. 18*:773 (1989).

13
Vegetation Management Strategies for Remediation of Selenium-Contaminated Soils

David R. Parker and Albert L. Page

University of California
Riverside, California

I. INTRODUCTION

A. Causes of High Soil Selenium Levels

In the western United States, saline drainage waters from irrigated fields are often disposed of by transferring them to sinks for storage, leaching, and evaporation. The evaporative concentration of trace elements has led to elevated concentrations of selenium (Se) in soils and waters of the Kesterson Reservoir located on the west side of the San Joaquin Valley, California [1]. Deaths and deformities of wildfowl linked to elevated Se levels [2] led to an upsurge in research activity, much of which is summarized in this volume. Recently, attention has focused on other irrigation/drainage projects in California, Utah, Nevada, Wyoming, and other states where evidence is emerging that concentrations of Se (and perhaps other trace elements) in soils, surface waters, and aquatic biota are sufficiently high to be of concern [3,4]. Drainage water is most often discharged into aquatic environments (i.e., evaporation ponds, wetlands, streams), but the apparent sensitivity of these ecosystems to Se is causing fundamental changes in water management strategies. Examples are the closure of the San Luis Drain, which released agricultural drainage water into Kesterson, and the cessation of drainage water deliveries into other evaporation ponds in the valley, with the highly salinized sediments contained therein reverting to terrestrial ecosystems (i.e., soils).

Naturally seleniferous soils are widespread through areas of the northern Great Plains and intermountain regions of the Rocky Mountains. Most Se-rich soils in the region are derived from Cretaceous sedimentary rocks, especially shales [5,6]. The solubility of Se in these soils varies considerably with chemical conditions and weathering, so total soil Se is a poor predictor of, for example, plant tissue Se concentrations [5].

Although natural weathering of rocks and minerals is a major source of Se in aquatic and terrestrial ecosystems, human activities contribute substantially to the redistribution and cycling of Se on a global scale. The concentration of Se in coal typically exceeds that of soils, and the combustion of coal in industrial operations and in electric power generation represents the major source of anthropogenic Se. Phosphorous fertilizers from western U.S. sources, residues from the incineration of municipal refuse, and the disposal or recycling of municipal sewage sludges contribute lesser amounts of Se but can be important on a local scale. More detailed information on the distribution and behavior of Se in the environment can be found in Refs. 5 and 7.

B. Remediation and Prevention Strategies

In keeping with the emphasis of several chapters in this volume, this chapter's discussion is focused on vegetation management in the context of Se in saline drainage waters, especially those in the western San Joaquin Valley. To date, only a few research efforts have explored the possible role of higher plants in managing or attenuating Se problems. Yet three general scenarios can be envisioned wherein growing plants might play a role in minimizing or reducing the Se burden of soils, surface waters, or wetlands, particularly if useful plant species can be identified that will both accumulate and tolerate high concentrations of Se.

1. *Remediation* of previously contaminated soils and sediments by growth of Se-accumulating species with removal of the harvested plant tops from the site. Such an approach could reduce hazards to terrestrial biota, permit restoration of wetland or riparian habitats, and/or minimize the potential for Se transport to surface waters or groundwaters.
2. *Prevention* of trace element migration in irrigation drainage water. If Se-accumulating crops could be rotated into growers' fields prior to leaching of excess salinity, and if the crop could significantly reduce soluble soil Se, subsequent transfer to surface water might be reduced.

3. *Decontamination* of Se-enriched saline drainage water prior to discharge. If a salt-tolerant species could be irrigated with saline drainage water, and Se burdens reduced, then the salts could ultimately be discharged into streams, wetlands, or evaporation ponds with reduced chances of causing Se toxicity.

To date, our research has focused principally on the first scenario: the possible role of higher plants in reducing Se inventories in the highly saline and Se-enriched soils/sediments such as those found at Kesterson. Our findings, however, should have applications for the other two scenarios. In principle, plants could be harvested, removed from the site, and either dispersed on low-Se soils or used as a Se-rich animal feed additive [8]. Alternatively, the plant biomass could be incorporated into the soil to enhance microbial volatilization of methylated Se compounds [9], the enhancement arising from provision of a carbon source, more readily metabolizable Se-rich precursors, or both.

The third scenario provides much of the impetus for research into the agroforestry concept (see Chapter 10). We note, however, that unless plant uptake of Se occurs faster than evapotranspiration, the net effect of irrigating land with Se-enriched drainage water will be to increase the soil Se burden. A considerable investment in "clean" irrigation water would thus be required to achieve further plant growth and Se uptake before the accrued soil salinity could be leached and discharged. There have also been some efforts to evaluate the potential of annual herbaceous plant species to reduce soil inventories of Se (see, e.g., Ref. 8), which may fit best into the second scenario described above.

We have focused most of our studies on perennial herbaceous species as candidate crops. The rationale for this focus includes (a) rapid establishment; (b) potential persistence of the crop; and (c) ease of establishment, management, and harvest using conventional equipment. When uptake and removal of Se is the objective, these plants offer potential for a deep and extensive root system that might effectively remove Se from throughout the soil profile. Our studies have addressed only plant accumulation of Se; other research on the role of plants in the dissipation of Se as volatile gases is summarized in Chapter 14 of this volume. We also postulate that certain species of the genus *Astragalus* that are notorious accumulators of Se [10,11] represent a unique source of genetic potential for tolerance and accumulation of Se, and we have examined these species in some detail for their possible utility.

II. CHARACTERISTICS OF SELENIUM-CONTAMINATED SOILS

A. Salinity and Boron

A fundamental premise of our research has been that in the western San Joaquin Valley, high Se levels in soils and drainage waters are almost always accompanied by high levels of soluble salts and boron (B) [12]. The converse is not necessarily true; the valley is extremely variable geochemically [13], and many areas, especially some in the Tulare Basin, exhibit saline drainage waters comparatively low in Se [14,15].

In the valley, salinity levels can be exceptionally high in exposed evaporation pond sediments, in soils adjacent to evaporation ponds, and in salt-affected and unreclaimed soils. We have collected a few representative samples, mostly associated with evaporation ponds, for which 1:1 (w/w) water extracts exhibit salinities of 14–82 dS/m (Table 1). In situ porewater salinities (i.e., at field moisture contents) would likely be some three- to fivefold greater than those given in Table 1 (for comparison, the salinity of seawater is about 40 dS/m). The soluble salts in this region are dominated by Na^+ and SO_4^-, although significant quantities of Ca^{2+}, Mg^{2+}, and Cl^- are typically present as well (Table 1).

In the highly saline soils of interest, soluble B levels are also quite high, with concentrations in excess of 10 mg/L being common (Table 1). As with

Table 1 Chemical Characteristics of the Surface 15–25 cm of Some Representative Exposed Evaporation Pond Sediments (EEP), an Adjacent Berm Soil (EP-AB), and a Highly Saline, Undrained Soil (US) in the San Joaquin Valley, California[a]

Site[b]	Setting	pH	EC (dS/m)	Ca (meq/L)	Mg (meq/L)	Na (meq/L)	SO$_4$ (meq/L)	Cl (meq/L)	B (mg/L)
K-4	EEP	7.8	14.0	27	36	134	120	66	17
K-11	EEP	7.3	17.7	25	29	166	148	65	21
W-1	EEP	8.2	52.8	26	47	1270	1110	238	55
Tu-1a	EEP	8.1	76.6	26	60	1080	511	717	26
Tu-1b	EP-AB	8.7	10.6	1	<1	120	77	20	9
Tu-2	US	8.8	82.2	26	12	1320	1050	295	18

[a] All data are for 1:1 (w/w) water extracts.
[b] K-4 and K-11 are Ponds 4 and 11 at Kesterson Reservoir; W-1 is in the Westlands Water District; Tu sites are on the old Tulare Lake bed.
Source: Ref. 30 and unpublished data.

overall salinity, in situ soil solution B concentrations will often be considerably higher than the 1:1 water extract values given in Table 1.

B. Selenium and Other Trace Elements

In most soil environments, selenium is typically found in either of two oxidation states, as selenite [Se(VI)] or selenate [Se(VI)]; only highly reducing conditions lead to the formation of elemental Se(0) and selenide [Se(-II)] [5,6]. Selenate is chemically very similar to sulfate; it is absorbed only weakly by soil surfaces and is highly mobile. Alkaline, well-aerated soils favor a preponderance of the more mobile selenate [6,12], which probably accounts for elevated Se concentrations in many agricultural drainage waters. Selenite appears to be intermediate between phosphate and sulfate in that it is absorbed fairly strongly and is rather immobile. Decreasing soil pH and/or lower redox potentials favor selenite, with a concomitant decrease in soil solution concentration and mobility [6]. Most studies of soils in the western San Joaquin Valley suggest that selenate is the dominant solution species [12,16], but some limited data suggest that selenite can be a significant component under transient conditions of high soil moisture [17].

Other trace elements, including As, B, Mo, U, and V, have been observed at elevated concentrations in drainage waters of the western San Joaquin Valley [13,15,18]. Moore et al. [15] identified four trace elements of definite concern for their possible effects on wetlands and riparian habitats (arsenic, boron, molybdenum, and selenium) and recommended further research to evaluate potential impacts of silver, lithium, strontium, uranium and vanadium in drainage waters. For our purposes, these other trace elements need to be considered in the context of potential phytotoxicity, although there is presently no evidence that any trace element except boron is likely to limit plant growth. In addition, possible uses of Se-enriched vegetation (e.g., as a feed supplement) may require an assessment of any adverse effects due to other trace elements, especially molybdenum [19].

III. CRITERIA FOR PLANT GENOTYPE SELECTION

A. Salinity and Boron Tolerances

Soil salinity and boron pose serious limitations to the use of many plant species for vegetative management of Se-enriched soils. To help put the salinity levels presented in Table 1 in perspective, we refer to Maas's summary [20] of the available literature with respect to various crop species' tolerances expressed in terms of EC_e, the electrical conductivity of a saturated paste extract. For a given soil, EC_e values will be about twice as high as the 1:1 extract data given in Table 1. Maas [20] classifies plants as "tolerant" if they can yield 50% of their potential maximum yield at EC_e values of 15–21 dS/m. Clearly, then, many of the soils listed in Table 1 would not permit significant growth of any known crop species, and even the soils with lower EC will require the use of truly superior genotypes. Similarly, plants that can tolerate ca. 10 mgsB/L in a saturated paste extract are usually considered to be very boron-tolerant [20]. The soluble B levels in Table 1 suggest that only unusually tolerant genotypes will hold any promise for vegetating these soils.

B. Tendency to Accumulate Selenium

In addition to salinity and B tolerances, plant genotypes useful for direct removal of soil Se should exhibit the potential to take up significant quantities of soil Se within a reasonable time. On the other hand, if vegetation is to be used primarily as a cover and/or carbon source for volatilization, then tolerance of salinity, boron, and perhaps Se are all that may be required.

The feasibility of using vegetation to directly remove soil Se can be evaluated by considering probable above-ground biomass production and expected shoot Se concentrations. Example calculations are presented in Figure 1 for a range of assumed dry matter yields (4–16 t/ha). Based on estimates of the volume and composition of subsurface drainage water delivered to Kesterson, the total soil Se burden of the reservoir is ca. 9000 kg of Se unevenly distributed across 520 ha (Chapter 5). The average soil Se burden is thus ca. 15 kg/ha, assuming little change since drainage water deliveries ceased in 1986. If one then sets a minimum objective for removal of, for example, 1 kg Se/(ha·yr), Figure 1 would suggest a minimum "target" shoot Se concentration of ca. 100 mg/kg. Lower shoot Se levels would require unrealistically high biomass yields to achieve significant reductions in soil Se, especially since shoot Se would be expected to progressively diminish as the labile soil Se pool is reduced.

IV. SUMMARY OF SOME PERTINENT STUDIES

A. Known Selenium Accumulators

Selenium-accumulating plant species, or "indicators," were the subject of numerous studies in many parts of the western United States during the

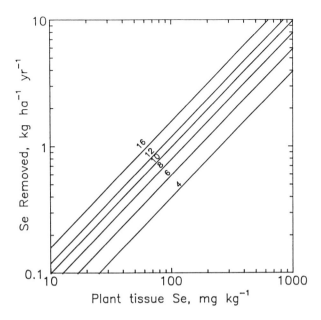

Figure 1 Calculated annual removal of soil Se in the above-ground portion of crop as a function of shoot tissue Se concentration (dry matter basis). Numbers above each line indicate the assumed biomass yield in metric tons of dry matter per hectare per year.

period 1930–1960. The principal concern at that time was the apparent poisoning of livestock (i.e., "alkali disease" and/or "blind staggers") grazing on certain indigenous range plants that colonize naturally seleniferous soils, although the exact etiology of these disorders has been recently questioned [21]. There was also some transient interest in Se indicator plants for geobotanical prospecting, especially for uranium-containing ores [22]. There are, however, considerably fewer recent published investigations of Se accumulators, and much of the older literature does not address questions we might pose today as we consider vegetative remediation of Se-contaminated soils.

A summary of the known genera that exhibit at least some tendency to accumulate Se is presented in Table 2. According to most authors, the primary Se-accumulating plants are capable of accumulating and tolerating thousands of micrograms of Se per gram of leaf tissue [5,11], although values observed in the field are often much lower, even in plants growing

Table 2 Summary of the Known Primary and Secondary Se-Accumulating Plant Genera Indigenous to North America

Genus	Common name	Family
Primary accumulators		
Astragalus	Milkvetch	Leguminosae
Stanleya	Prince's plume	Cruciferae
Haplopappus (Oonopsis section)	Goldenweed	Compositae
Machaeranthera (Xylorhiza section)	Woody aster	Compositae
Secondary accumulators		
Aster	Wild aster	Compositae
Astragalus	Milkvetch	Leguminosae
Atriplex	Saltbush	Chenopodiaceae
Castilleja	Paintbrush	Scrophulariaceae
Comandra	Toad-flax	Santalaceae
Grayia	—	Chenopodiaceae
Grindelia	Gum plant	Compositae
Gutierrezia	Snakeweed	Compositae
Machaeranthera	—	Compositae
Mentzelia	—	Loasaceae

Source: Adapted from Refs. 5, 10, and 11.

in seleniferous soils [10]. Secondary accumulators can accumulate Se to a few hundred micrograms per gram [11] and have received comparatively little attention in published research [23]. Some members of the Cruciferae (e.g., *Brassica* spp.), which typically accumulate S to high concentrations, tend to accumulate more Se than other nonaccumulators [5,8,24], and some species might be properly categorized as secondary Se accumulators [25].

Of the primary accumulators, members of the *Astragalus* genus have been by far the most studied. Only a small fraction of the known North American species are classified as accumulators [10,11], and the Se metabolism of both types of *Astragalus* has been studied in some detail. The accumulator species all seem to possess a unique pathway wherein Se is incorporated in a specialized and nontoxic amino acid, Se-methylselenocysteine, that is not observed in nonaccumulating species [11]. This unique metabolism greatly limits incorporation of Se into amino acids and proteins, which otherwise leads to metabolic dysfunction (and overall Se toxicity) and is believed to account for the Se accumulators' tolerance of very high shoot Se concentrations [5,11].

B. "Native" Vegetation

The vegetation that has naturally colonized Kesterson since drainage water deliveries ceased is dominated by salt-tolerant species of genera such as *Distichlis* and *Atriplex* [26]. These plants typically exhibit low plant Se concentrations (i.e., <10–20 µg/g) [26] that are insufficient to contribute directly to remediation efforts. Because the saline soils at Kesterson contain high concentrations of sulfate, an ion that competes with selenate for root absorption (see below), the relatively low plant concentrations of Se observed at Kesterson are not unexpected and may account for the absence of any evidence of Se phytotoxicity in the indigenous vegetation [26].

C. Other Species with Superior Salinity and Boron Tolerances

Parker et al. [27] employed solution/sand culture methods to screen a number of genotypes from the genera *Astragalus, Leucaena, Medicago, Trifolium, Elymus, Elytrigia, Festuca, Leymus, Oryzopsis, Psathyrostachys, Puccinellia,* and *Sporobolus* for tolerance to both salinity and boron. Some of the species examined were known to be salt-tolerant, but information concerning B tolerance was lacking. Electrical conductivities required to produce a 50% reduction in seed germination (EC_{50}) ranged from 5 to 30 dS/m, while B levels up to 4.0 mM had only minimal effects on germination. The 16 most promising genotypes were also tested for salinity and B tolerance during seedling growth. Lines of five species—tall wheatgrass, alkali sacaton, weeping alkaligrass, *Astragalus bisulcatus*, and *A. racemosus*—appeared most promising. All exhibited EC_{50} values >20 dS/m and were unaffected by B concentrations up to 4.0 mM during seedling growth (Table 3). The observed tolerances of the primary Se accumulators, *Astragalus bisulcatus* and *A. racemosus*, warrant particular attention in future studies of the feasibility of vegetation for remediation of Se-enriched soils [27].

Bañuelos et al. [8] presented data to suggest that wild mustard (*Brassica juncea*) is at least moderately tolerant of both salinity (chloride sources) and boron. This Cruciferae species appears capable of accumulating Se to several hundred micrograms per gram without Se phytotoxicity [8] and represents another possible option in vegetative dissipation of Se.

D. The Pivotal Role of Sulfate in Plant Uptake of Selenate

Numerous studies have demonstrated a profound inhibitory effect of sulfate on plant uptake of the analogous selenate ion [24]. Short-term

Table 3 Selected Plant Genotypes Exhibiting Superior Salt and Boron Tolerance During Seedling Growth[a]

Common name	Scientific name	EC_{50} (dS/m)	$[B]_{50}$ (mM)
Alkali sacaton	*Sporobolus airoides* (Torr.) Torr.	29.6	>4.0
Tall wheatgrass	*Elytrigia pontica* (Podb.) Holub.	27.2	>4.0
Weeping alkaligrass	*Puccinellia distans* (L.) Parlat.	25.3	>4.0
Two-groove milkvetch	*Astragalus bisulcatus* (Hook.) Gray	20.6	>4.0
—	*Astragalus racemosus* Pursh.	26.0	>4.0
Indian ricegrass	*Oryzopsis hymenoides* (Roem & Schult.)	12.1	>3.0

[a]The EC_{50} and $[B]_{50}$ values indicate the electrical conductivity and boron concentration, respectively, required to reduce shoot yields by 50%. Indian ricegrass, a grass with only moderate tolerance of salinity and boron, is included for comparison.
Source: Ref. 27.

excised-root experiments (e.g., Ref. 28) have suggested that selenate and sulfate transport is facilitated by a single membrane carrier with an approximately equal affinity for both anions. But results of longer term studies clearly indicate that a simple competition between SeO_4^{2-} and SO_4^{2-} often fails to explain observed patterns of uptake [25,29]. Primary Se accumulators (e.g., some *Astragalus* species) have been particularly puzzling because they can accumulate Se to very high concentrations even in gypsiferous soils high in soluble SO_4^{2-} [10].

In a recent study, Bell et al. [25] grew a primary Se accumulator (*Astragalus bisulcatus*) and alfalfa (a nonaccumulator) in identical nutrient solutions with varied SeO_4^{2-} (0.002–0.08 mM) and SO_4^{2-} (0.05–15.5 mM) concentrations, taking pains to ensure that solutions of SeO_4^{2-} and SO_4^{2-} were held constant throughout the 3–5-week growth period. Alfalfa shoot Se concentrations ranged from 4 to 154 µg/g, while the identical treatments resulted in shoot levels of 175–1200 µg/g in *A. bisulcatus*. Consistent with other reports [24], uptake of SeO_4^{2-} by alfalfa was profoundly inhibited by increases in solution SO_4^{2-}, but Se uptake by *A. bisulcatus* was much less affected. Although *A. bisulcatus* accumulated sulfur to levels typical of the S-accumulating Cruciferae, this did not explain its much higher shoot Se levels.

By comparing plant molar Se/S ratios (whole-plant basis) with nutrient solution SeO_4^{2-}/SO_4^{2-} ratios, Bell et al. [25] demonstrated that while alfalfa discriminates against Se, *A. bisulcatus* exhibits preferential uptake of Se relative to S (Figure 2). Given completely nondiscriminate

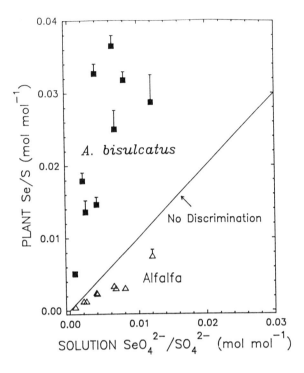

Figure 2 Plant molar Se/S ratio versus the SeO_4^{2-}/SO_4^{2-} ratio in nutrient solutions used to grow *Astragalus bisulcatus* (a primary Se accumulator) or alafalfa (a nonaccumulator). Plant ratios were computed on the basis of whole plant (roots plus shoots) uptake of Se and S. Solid line (slope = 1) denotes no plant discrimination between the analogues SeO_4^{2-} and SO_4^{2-}. (From Ref. 25.)

transport and assimilation of SeO_4^{2-} and SO_4^{2-}, the two molar ratios should be equal—the solid line in Figure 2. Thus, tissue Se in *A. bisculatus* was some 2.5- to 9.0-fold greater than expected, assuming nondiscriminate uptake of SeO_4^{2-} and SO_4^{2-}, while tissue Se in alfalfa was about one-third to two-thirds of that expected under the same assumption. Bell et al. [25] also conducted a survey of previously published results and concluded that primary Se accumulators are uniquely able to accumulate SeO_4^{2-} in the face of competition from SO_4^{2-}, probably as a consequence of their specialized Se metabolism, which is not found in other plant species. Field studies have also indicated that when grown side by side on a seleniferous soil, *A. bisulcatus* contains much more Se relative to S than do nonaccumulating species [10].

E. Results with Kesterson Soil

Retana et al. [30] recently conducted a greenhouse study using columns of soil collected from Pond 4 at Kesterson (see Table 1 for some surface horizon characteristics). Preliminary attempts using the Pond 4 topsoil failed to obtain any germination or growth of five candidate genotypes due to excessive salinity. The results indicated that more sophisticated management of these soils is required. By applying just enough water (7.5 cm) to leach salts out of the seed zone (but not out of the soil profile), surface soil salinity was sufficiently reduced to permit germination, emergence, and seedling growth [30]. By temporarily reducing the irrigation regime, the soils were allowed to gradually dry, and at termination the distribution of salts and Se in the profile was only slightly altered from the initial conditions [30].

Two salt-tolerant grasses (tall wheatgrass and alkali sacaton) grew relatively well, and the sum of the dry matter yields of three cuttings was equivalent to 9.5–14 t/ha (Table 4). The persistence of these two genotypes, even with gradual drying of the columns and increases in surface-soil EC, suggest an ability to acclimate to very high soil salinity. Indian ricegrass became severely necrotic after just one modest cutting (Table 4) and shows little promise as a component of management strategies due to its lower salt tolerance. The growth and vigor of the two *Astragalus* species was only moderate (Table 4), indicating that use of these species might be facilitated by improved salinity and/or B tolerance. The relative performance of these five genotypes was quite consistent with their relative salinity/B tolerances during sand-culture screenings (Tables 3 and 4).

Table 4 Summary of Shoot Dry Matter Yields, Shoot Se Concentrations, and Computed Equivalent of Biomass and Se Removed for Five Plant Genotypes[a]

		Dry matter		Shoot Se	
Species	No. of cuttings	Total g/column	Equiv. kg/ha	Mean µg/g	Equiv. kg/ha
Alkali sacaton	3	11.1	13,820	9	0.12
Tall wheatgrass	3	7.6	9,510	12	0.11
Indian ricegrass	1	0.8	950	19	0.02
Astragalus bisulcatus	2	3.8	4,680	596	2.79
Astragalus racemosus	2	4.4	5,470	670	3.66

[a]Plants were grown for 7 months in a greenhouse study using 1-m-long columns of Se-contaminated soil from Pond 4 at Kesterson Reservoir.
Source: Adapted from Ref. 30.

None of the grasses accumulated Se to very high tissue concentrations (Table 4), and several harvests over a growing season would remove only about 100 g Se/ha, even with the vigorous growth of alkali sacaton. It is thus unlikely that nonaccumulators such as these grasses would be capable of significantly reducing soil Se inventories via harvest and removal. Alkali sacaton, however, seems capable of providing a substantial carbon source that might enhance microbial volatilization of Se [9]. The low tissue Se concentrations in the grasses are consistent with observations that sulfate profoundly inhibits plant uptake of selenate and that nonaccumulators can partially discriminate against Se and thus limit uptake [25]. In contrast, the Se-accumulating *Astragalus* species accrued Se to 500–600 mg/kg in their shoots and reduced the Se inventory in the Pond 4 soil by 3–4 kg/ha (Table 4). These results corroborate the solution culture studies of Bell et al. [25] in that, despite inhibition of selenate uptake by sulfate, the effect is much less profound with primary Se accumulators than with nonaccumulating species.

V. OUTLOOK AND SUMMARY

We have identified at least four plant genotypes that are sufficiently tolerant of salinity and boron to be grown successfully in a highly contaminated evaporation pond sediment such as those found at Kesterson Reservoir, as long as the seed zone is leached to reduce salinity prior to planting and during the early stages of growth. Of the two grass species examined, alkali sacaton is particularly promising as a vegetative cover and/or carbon source to enhance microbial volatilization, but repeated harvest and removal will not substantially diminish the soil Se inventory. Due to the high soluble sulfate levels in Kesterson soils, Se concentrations in nonaccumulator plant species do not seem to reach particularly high levels. Potential problems with terrestrial food chain transfer and Se toxicity to wildlife may be minimal.

Two Se-accumulating *Astragalus* species are also quite tolerant of salinity and B and seem uniquely capable of accumulating Se to high concentrations in the presence of high soil sulfate levels. Harvest of these species and their removal from the site could contribute significantly to Se dissipation strategies. The practicality of using Se-accumulating *Astragalus* species in remediation efforts, however, may be limited, at least currently. These species have not been genetically improved, and their general agronomic traits (i.e., overall vigor, disease and pest resistance, responsiveness to fertilization, etc.) are neither superior nor well-characterized. Moreover, seed is not commercially available, and the

USDA Plant Introduction Center at Pullman, Washington has only a very limited supply of seed for a few genotypes. Any large-scale project to evaluate or utilize *Astragalus* spp. for Se dissipation will consequently require that seed stocks be increased. Genetic improvement would be highly desirable, but it remains unclear whether there is sufficient impetus to justify the expense, effort, and time required to breed superior strains. A clarification of the breadth and scope of Se contamination problems derived from irrigation/drainage projects across the western United States [3,4] should assist in establishing future research priorities.

ACKNOWLEDGMENTS

We are indebted to Dr. Paul Bell, Jaime Retana, and David Thomason for their contributions to much of the research summarized in this chapter.

REFERENCES

1. J. Letey, C. Roberts, M. Penberth, and C. Vasek, *An Agricultural Dilemma: Drainage Water and Toxics Disposal in the San Joaquin Valley*, Spec. Pub. 3319, Agric. Exp. Sta. Univ. California, Oakland, 1986.
2. H. M. Ohlendorf, in *Selenium in Agriculture and the Environment* (L. W. Jacobs, Ed.), Soil Sci. Soc. Am. Spec. Pub. No. 23, Madison, Wisc., 1989, p. 133.
3. National Research Council, *Irrigation-Induced Water Quality Problems: What Can Be Learned from the San Joaquin Valley experience?*, Nat. Acad. Sci. Press, Washington, D.C., 1989.
4. J. P. Deason, *J. Irrig. Drainage Eng.* 115:9 (1989).
5. H. F. Mayland, L. F. James, K. E. Panter, and J. L. Sonderegger, in *Selenium in Agriculture and the Environment* (L. W. Jacobs, Ed.), Soil Sci. Soc. Am. Spec. Pub. 23, Madison, Wisc., 1989, p. 15.
6. R. H. Neal, in *Heavy Metals in Soils* (B. J. Alloway, Ed.), Blackie, London, 1990, p. 237.
7. B. W. Mosher and R. A. Duce, in *Occurrence and Distribution of Selenium* (M. Ihnat, Ed.), CRC Press, Boca Raton, Fla., 1989, p. 295.
8. G. S. Bañuelos, D. W. Meek, and G. J. Hoffman, *Plant Soil* 127:201 (1990).
9. W. T. Frankenberger, Jr., and U. Karlson, *Soil Sci. Soc. Am. J.* 53:1435 (1989).
10. I. Rosenfeld and O. A. Beath, *Selenium: Geobotany, Biochemistry, Toxicity, and Nutrition*, Academic, New York, 1964.
11. T. A. Brown and A. Shrift, *Biol. Rev.* 57:59 (1982).
12. T. S. Presser and H. M. Ohlendorf, *Environ. Manage.* 11:805 (1987).
13. R. A. Schroeder, D. U. Palawski, and J. P. Skorupa, U.S. Geol. Surv. Water Resour. Invest. Rep. 88-4001, 1988.
14. T. S. Presser and I. Barnes, U.S. Geol. Surv. Water Resour. Invest. Rep. 85-4220, 1985.

15. S. B. Moore, J. Winckel, S. J. Detwiler, S. A. Klasing, P. A. Gau, N. R. Kanim, B. E. Kesser, A. B. DeBevec, K. Beardsley, and L. K. Puckett, *Fish and Wildlife Resources and Agricultural Drainage in the San Joaquin Valley California*, 2 volumes, San Joaquin Valley Drainage Program, Sacramento, Calif., 1990.
16. J. L. Fio, R. Fujii, and S. J. Deverel, *Soil Sci. Soc. Am. J.* 55:1313 (1991).
18. G. R. Bradford, D. Bakhtar, and D. Westcot, *J. Environ. Qual.* 19:105 (1990).
17. Lawrence Berkeley Laboratory, *Hydrological, Geochemical, and Ecological Characterization of Kesterson Reservoir*, Annu. Rep. LBL-24250, Berkeley, Calif., 1987.
19. N. Albasel and P. F. Pratt, *J. Environ. Qual.* 18:259 (1989).
20. E. V. Maas, *Appl. Agric. Res.* 1:12 (1986).
21. L. F. James, K. E. Panter, H. F. Mayland, M. R. Miller, and D. C. Baker, in *Selenium in Agriculture and the Environment* (L. W. Jacobs, Ed.), Soil Sci. Soc. Am. Spec. Pub. 23, Madison, Wisc., 1989, p. 123.
22. H. L. Cannon, *Am. J. Sci.* 250:735 (1952).
23. A. Shrift, in *Organic Selenium Compounds: Their Chemistry and Biology* (D. L. Klayman and W. H. H. Gunther, Eds.), Wiley, New York, 1973, p. 763.
24. R. L. Mikkelsen, A. L. Page, and F. T. Bingham, in *Selenium in Agriculture and the Environment* (L. W. Jacobs, Ed.), Soil Sci. Soc. Am. Spec. Pub. 23, Madison, Wisc., 1989, p. 65.
25. P. F. Bell, D. R. Parker, and A. L. Page, *Soil Sci. Soc. Am. J.* 56:1818 (1992).
26. Z. Z. Huang and L. Wu, *Ecotoxicol. Environ. Safety* 22:251 (1991).
27. D. R. Parker, A. L. Page, and D. N. Thomason, *J. Environ. Qual.* 20:157 (1991).
28. J. E. Leggett and E. Epstein, *Plant Physiol.* 31:222 (1956).
29. A. M. Hurd-Karrer, *Am. J. Bot.* 25:666 (1938).
30. J. Retana, D. R. Parker, C. Amrhein, and A. L. Page, *J. Environ. Qual.* 22:805 (1993).

14
Selenium Volatilization by Plants

Norman Terry and Adel M. Zayed

*University of California
Berkeley, California*

I. INTRODUCTION

Recent interest in the volatilization of selenium (Se) is related to the buildup of Se in soils, which potentially poses risks to the health of humans and wildlife. Selenium may accumulate to toxic levels in many areas of the world. In the Kesterson Reservoir of the San Joaquin Valley of California, Se has accumulated to such high levels that it has been responsible for death and deformities in the wildfowl population. There is concern that the Se content of some crops might increase to levels high enough to jeopardize human health [1]. Thus, the agricultural management of Se-contaminated soils is of major importance.

One way to remove Se from such soils is by biological volatilization. Several trace elements, including Se, arsenic, lead, and mercury, are known to be capable of being biomethylated, producing organic molecules that are more volatile than the original forms. The volatilization of Se from soils is known to be a biological process because no volatilization occurs when soils are autoclaved or sterilized to eliminate microbiological activity [2,3]. The biological volatilization of Se may be carried out by microorganisms (e.g., bacteria, actinomycetes, fungi, and yeast) as well as by plants. Ross [4] estimated that as much as 10,000 metric tons of Se may be emitted to the atmosphere annually in the northern hemisphere alone, and more than one-fourth of it originates from soils and

343

plants. The volatile Se compounds reported to evolve from soil include dimethylselenide (DMSe), dimethyl diselenide (DMDSe), and dimethyl selenone [3].

Microorganisms play an important role in the volatilization of Se from soils [5,6]. However, plants also make an important contribution to biological Se volatilization. The addition of plants to soil increases the rate of Se volatilization above that found for the soil alone. This was shown by Zieve and Petersen [7], who found greater volatilization of ^{75}Se from soil after barley was planted. Duckart et al. [8] obtained relative volatilization rates, expressed as a percentage of that of fallow soil, as high as 225% for *Astragalus bisulcatus*.

The first indication that Se could be volatilized from plants came from the research of Beath et al. [9]. They found that Se compounds were released from plant materials (*A. bisulcatus*) during storage and that these volatile Se compounds could be trapped by passing air from the drying plant material through concentrated sulfuric acid. Later it was suggested by some workers that the distinctive, unpleasant garliclike odor produced by plants growing in soils or nutrient solution containing substantial amounts of Se arose from the release of volatile Se compounds [10].

The release of volatile Se compounds from intact higher plants was first shown by Lewis et al. [11] in work with Se accumulator plants (e.g., *A. racemosus*). Subsequently, they showed that volatilization of Se may also occur in nonaccumulators such as alfalfa. Evans et al. [12] were the first to identify a volatile Se compound released by plants; they found that the Se accumulator *A. racemosus* produced DMDSe. Lewis [13] later showed that live cabbage leaves produce DMSe, the compound most typically produced by nonaccumulator plants, and that they do not produce DMDSe. Lewis [13] demonstrated that Se volatilization from cabbage leaves did not appear to be microbial but was due directly to metabolism within leaf tissues. She suggested that DMSe may be produced by enzymatic cleavage of Se-methyl selenomethionine selenonium salt.

It is now known that most of the Se produced by the plant is volatilized from the root [14]. Initially, however, it was thought that Se was volatilized mostly from the shoot. This view was held by Lewis et al. [11], who found that considerably less Se was volatilized after shoots of alfalfa were removed. Later, Asher et al. [15] reported that during oven drying roots released as much of the volatile Se compounds or more than the tops. They suggested that Lewis et al. [11] observed only small amounts of Se volatilized from living roots because a large proportion of the volatile Se, being water-soluble, was retained in the nutrient solution.

The goal of this chapter is to review the physiology of Se as it relates to our knowledge of Se volatilization by plants and to discuss its possible significance in the remedial management of Se pollution.

II. POSSIBLE ROLE OF SELENIUM AS A PLANT MICRONUTRIENT

Although it is well established that Se is an essential nutrient for the normal functioning of animals and humans, it is less clear whether Se is essential for plants. There is some evidence that it may be essential to the growth of Se-accumulating plants such as *A. racemosus*. Species of *Astragalus* that are not Se accumulators, for example, *A. crassicarpus*, appear not to have a Se requirement [16]. The evidence that Se serves as a micronutrient in Se-accumulating plants is summarized by Shrift [17] as follows: (a) Se-accumulating plants are found only in seleniferous soils; (b) they accumulate Se in very high concentrations (i.e., milligrams of Se per gram dry weight), while nonaccumulators contain only a few micrograms of Se per gram dry weight; (c) the addition of small amounts of selenate or selenite to the culture medium stimulates the growth of accumulator species and inhibits the growth of nonaccumulator species; (d) the assimilation pathways of selenite, selenate, and selenomethionine differ substantially between accumulators and nonaccumulators; and (e) Se is metabolized differently from the essential element sulfur in higher plants, microorganisms, and animals.

There is an alternative point of view about the beneficial effects of Se addition on plant growth. Broyer et al. [18] found no effect of Se addition on the growth and yield of alfalfa and subterranean clover when 0.02–0.2 µg/l selenite was added to highly purified nutrient solution; when the Se level was increased to 2µg/l, growth was reduced. Later, Broyer et al. [19,20] suggested that the results of Trelease and Trelease [16] showing that Se improves growth may have been due to an amelioration of phosphate toxicity by selenite. Broyer and his colleagues showed that plants (Se accumulators or nonaccumulators) supplied with low levels of phosphate did not exhibit a growth stimulation with increased Se supply and that selenite depressed the uptake of phosphate. Clearly, the plant requirement for Se has yet to be established.

In recent experiments with both the marine diatom *Thalassiosira pseudonana* [21] and the green alga *Chlamydomonas reinhardtii* [22], Se-dependent glutathione peroxidase (GSH-Px) activity has been identified and partially purified. If such activity turns out to be conserved in evolution up to the angiosperms, then the need for Se in plant metabolism must remain an open question, since the mammalian form of this enzyme accounts for a large proportion of the dietary requirement for this trace element [23].

III. SELENIUM UPTAKE, ASSIMILATION, AND TOXICITY IN PLANTS

Selenium is taken up by plants as selenate, selenite, and organic selenium. (Most of the soluble Se in soils and subsurface drainage waters is selenate.) There is evidence that the uptake of selenate is driven metabolically whereas the uptake of selenite may have a passive component. Ulrich and Shrift [24,25] demonstrated that excised roots of *A. Crotalariae* (Se accumulator) and *A. lentiginosus* (Se nonaccumulator) take up selenate actively, whereas some uptake of selenite is energy-independent. Asher et al [26] showed that when [^{75}Se]selenate was added to the nutrient solution, ^{75}Se concentrations in the xylem exudate of excised tomato roots were 6–13 times higher than in the external solution. With selenite, ^{75}Se concentrations in the exudate were always lower than in the external solution. Plants may also take up organic Se. Hamilton and Beath [27] extracted organic Se compounds from leaves of the Se accumulator species of *Astragalus* and found that other plants could absorb these compounds through their roots. Recent work by Abrams et al. [28] showed that wheat plants can take up selenomethionine supplied in nutrient solution and that this absorption is a metabolically driven process.

With respect to selenate, its absorption and translocation in plants is believed to resemble closely the uptake and movement of sulfate [29]. According to Leggett and Epstein [30], selenate is absorbed by the same carrier in root cell membranes as sulfate. Ferrari and Renosto [31] demonstrated that the uptake mechanism for selenate by barley roots is identical to that of sulfate. Sulfate ions have been shown to be antagonistic to the uptake of selenate ions [32–37], an observation consistent with the idea that sulfate and selenate compete for the same carrier on the root membrane.

Competition between sulfate and selenate may also depend on the concentrations of the two ions. Using broccoli as a test plant, Zayed and Terry [38] showed that when plants were supplied with 20 µM Se as Na_2SeO_4, increasing the supply of sulfate in the culture solution from 0.25 to 1 mM had no effect on the concentrations of Se in plant tissue, but that with further increase in sulfate from 1 to 5 or 10 mM, tissue Se concentrations were markedly decreased. These results indicate that the effect of sulfate supply on Se uptake by broccoli plants varied with sulfate concentration in the nutrient solution. When sulfate levels are low, there may also be a synergistic (rather than competitive) effect between selenate and sulfate uptake; increasing the concentration of selenate results in an increase in shoot sulfur concentration [36,37,39]. Sulfate levels may also affect the transport of Se from roots to shoots; Singh et al. [40] showed that in the absence of sulfur, Se tends to accumulate in the roots of *Brassica juncea* and in the presence of sulfur more Se is translocated to the shoot.

Once selenate has entered the plant, it is almost certainly metabolized by the enzymes of the sulfur assimilation pathway [29,41]. This is because Se has the ability to mimic S, thereby forming Se analogues of S compounds that are substrates for the S-assimilation enzymes. There is substantial evidence that Se analogues of S compounds compete for enzymes in the S-assimilation pathway. Burnell [42] showed that ATP-sulfurylase, which normally activates sulfate to form adenosine 5'-phosphosulfate (APS), which in turn is reduced nonenzymatically to sulfite, had a greater affinity for selenate than for sulfate and that the two ions compete for the same site on the enzyme. Ng and Anderson [43] concluded that there was a competitive inhibition between sulfide and selenide for the enzyme cysteine synthase; selenide inhibited the synthesis of cysteine, and sulfide inhibited the synthesis of selenocysteine. Dawson and Anderson [44] found that cystathionine γ-synthase exhibited a greater affinity for selenocysteine than for cysteine. McCluskey et al. [45] demonstrated that the enzyme β-cystathionase (which catalyzes the formation of homocysteine from cystathionine, the immediate precursor of methionine) metabolizes cystathionine and selenocystathionine with similar affinities.

The toxicity of Se to plants is almost certainly related to the competitive interactions between S compounds and their Se analogues. This view is supported by the fact that the inhibition of plant growth caused by the addition of Se compounds (e.g., selenate) can be overcome by supplying their S analogues (e.g., sulfate) [37,40]. The primary cause of Se toxicity is likely to be the replacement of S by Se in the amino acids of proteins, thereby disrupting catalytic activity. A significant loss of enzyme function can be expected to result from a synthesis of polypeptides in which some or all of the S-containing amino acids have been replaced by their selenium equivalents [29]. Consistent with this hypothesis is that Se accumulators (compared to nonaccumulators) limit the incorporation of endogenous selenoamino acids into protein as a mechanism of avoiding Se toxicity [46].

The replacement of S by Se in the S-containing amino acids cysteine and methionine can have especially significant effects on the catalytic properties of enzymes. Cysteine plays a critical structural role in proteins because of its involvement in the formation of disulfide bonds between residues in adjacent stretches of the same or different polypeptide chains. Because the Se atom is larger than that of sulfur (the radius of Se^{2+} is 0.5 Å and that of S^{2+} is 0.37 Å [47]), the bond between two Se atoms will be longer and therefore weaker than the equivalent disulfide bond. Huber and Criddle [48] showed that the selenol group of selenocysteine is completely ionized at physiological pH values (7.0–7.5), while the sulfhydryl group of cysteine is 80% protonated at the same pH. Such differences in atomic size and in ionization properties make it likely that replacement of cysteine by selenocysteine would interfere with the formation of disulfide bridges between adjacent polypeptide chains. As a result, selenoproteins would have a slightly different tertiary structure, resulting in a conformational change that could have a large effect on catalytic activity [29].

Both selenocysteine and selenomethionine have been identified in the proteins of plants supplied with Se [49,50]. Replacement of S by Se in methionine may also result in a slowing of protein synthesis. Eustice et al. [51] compared selenomethionine with methionine as substrates for in vitro aminoacylation, ribosome binding, and peptide bond formation in wheat germ preparations. Their results showed that selenomethionine paralleled methionine in most steps of the translation process; however, in the initiation process of peptide bond formation, selenomethionyl-$tRNA_i$ was less effective as a substrate than methionyl-$tRNA_i$. This effect on the protein synthesis process could be partly responsible for Se toxicity. Substitution of selenomethionine for methionine does not always prove detrimental; more than one-half of the 150 methionine residues present in the bacterial enzyme β-galactosidase could be replaced by the Se analogue without deleterious effects on enzyme activity [48].

It seems likely, therefore, that the major cause of Se toxicity is its interference in sulfur metabolism (as proposed by Brown and Shrift [29]), especially during the formation of catalytically active proteins. Additionally, Se toxicity might also disrupt the methylation function of methionine [29]. Methionine serves as a methyl donor through its reaction with ATP to form S-adenosylmethionine (SAM). SAM transfers its methyl group to a variety of metabolites. Replacement of Se for S in methionine could affect the production of SAM and therefore the efficiency of methylation.

Selenium toxicity is markedly attenuated by increasing the external sulfate supply. This was illustrated in a study of the effects of sulfate level on the growth of broccoli supplied with 20 µM selenate [38]. Zayed and Terry [38] found that with each increase in sulfate level in the nutrient solution, plant growth was enhanced along with a stepwise decrease in the Se/S ratio in plant tissues. They concluded that the increase in plant growth with increase in sulfate supply is most likely caused by an increase in competition between sulfate and other S compounds in plant tissue, with selenate and other Se compounds as substrates for the enzymes of the S-assimilation pathway. This view is supported by the research of Chow et al. [52], who showed that an increase in selenate supply to the Se accumulator *A. bisulcatus* resulted in an increase in the concentration of methyl selenocysteine at the expense of methylcysteine and that increasing the supply of sulfate had the opposite effect.

IV. FACTORS AFFECTING PLANT VOLATILIZATION OF SELENIUM

A. Plant Species

Unfortunately, there are only a few studies available showing the effectiveness of different crop or range plants in volatilizing Se. One of the first comparative studies was done by Lewis [13], who extracted cell materials from leaves and compared the rate of ^{75}Se volatilization of these leaf extracts for six different crops. These included five cruciferous species, mustard (*B. juncea* Czern L.), turnip (*B. rapa*), cabbage (*B. oleracea* var. *Capitata*), Chinese cabbage (*B. pekinensis*), radish (*Rhaphanus sativus*), and a composite, sunflower (*Helianthus annuus*). She indicated that the proportion of volatile Se produced did not appear to be affected appreciably by the plant species. This might arise from the fact that most of the plant species used in her comparison were cruciferous species that accumulate appreciable amounts of Se in their tissues.

It is now evident that the uptake and volatilization of Se by agricultural crops is dependent on plant species. Terry et al. [53] measured volatilization rates for 15 crop species grown under standardized environmental conditions. The plants were cultured in growth chambers in quarter-strength Hoagland's nutrient solution with 20 μM Se supplied as sodium selenate. Under these conditions, they found the rate of Se volatilization to be dependent on plant species (Table 1). The three best volatilizers were rice, broccoli, and cabbage. These crop plants volatilized Se at 200–350 μg/(m^2·day) on a leaf area basis and at 1500–2500 μg/(kg·day) on a plant dry weight basis. Other species of plants volatilized at much lower rates. For example, sugar beet, bean, lettuce, and onion volatilized Se at rates less than 15 μg/(m^2·day) on an area basis and less than 250 μg/(kg·day) on a dry weight basis. Other plant species—carrot, barley, alfalfa, tomato, cucumber, cotton, eggplant, and maize—exhibited intermediate values of 30–100 μg/(m^2·day) or 300–750 μg/(kg·day). Although this research has identified those crops that appear to be superior volatilizers of Se, more

Table 1 Rates of Se Volatilization for Crop Plant Species Grown in Quarter-Strength Hoagland's Solution with Se Supplied Continuously at 20 µM (as Sodium Selenate)

	Rate of Se volatilization	
Plant species	Per leaf area [µg Se/(m^2·day)]	Per plant dry weight [µg Se/(kg·day)]
Rice	340	1500
Broccoli	273	2393
Cabbage	221	2309
Carrot	102	548
Barley	102	486
Alfalfa	72	300
Tomato	70	742
Cucumber	57	752
Cotton	48	499
Eggplant	44	462
Maize	32	370
Sugar beet	22	246
Bean	14	217
Lettuce	11	179
Onion	ND	229

ND = not determined.

research is still required to obtain Se volatilization rates under field conditions for both cultivated and noncultivated plants.

In an attempt to compare rates of Se volatilization from plants growing in soil, Duckart et al. [8] measured volatilization rates for five different plant species (i.e., tomato, alfalfa, broccoli, tall fescue, and *A. bisulcatus*) grown in a greenhouse in pots filled with Panoche fine sandy loam soil containing 16.5 µg extractable Se/kg soil and amended with 500 µgSe(VI)/kg soil as sodium selenate. Their data show large differences among plant species, suggesting a significant role for plants in Se volatilization from soils. *A. bisulcatus* and broccoli showed the highest rates of Se volatilization, both on a leaf area basis and on a soil dry weight basis, followed by tomato, tall fescue, and alfalfa. These results are in agreement with results obtained by Terry et al. [53] for plants cultured hydroponically, indicating the importance of the solution culture experiments in identifying those crops that appear to be superior volatilizers of Se.

B. Chemical Form of Selenium

There is evidence that the rate of Se volatilization by plants may depend on which chemical form of Se is available to the plant. The availability of Se to plants is governed by the chemical form of the element in the soil and by the pH and redox potential in the root–soil environment [54]. The forms of Se that are most likely to accumulate in plant tissues are those that are mobile in the soil–water–plant system—selenate, selenite, and the soluble forms of organic Se. These forms will be transported readily through the soil and into the plant, where they accumulate; a small proportion may then be volatilized by various plant tissues. Selenium may be more readily volatilized if it is supplied as selenite than if it is supplied as selenate. Asher et al. [15] measured the amount of volatile Se compounds released from oven-drying shoots or roots of *Medicago sativa* plants supplied with 15 µM selenite or selenate for 6 weeks. Roots of the selenite-treated plants released more than 11 times as much volatile Se as those of the selenate-treated plants. In another study by Lewis et al. [55], cabbage leaves from plants cultured in nutrient solutions to which selenite was added released 10–16 times as much volatile Se as leaves from selenate-cultured plants.

In some unpublished research from our laboratory, we measured the rate of Se volatilization from broccoli plants cultured in nutrient solution containing 20 µM Se in the form of selenate, selenite, or selenomethionine. Our results showed that when broccoli plants were supplied with selenomethionine, the roots volatilized Se 13 times faster and the shoots volatilized Se 3 times faster than the roots and shoots of selenate-supplied plants, respectively. When we supplied plants with selenite, Se was volatilized 4 times faster in roots and 60% faster in shoots than in roots and shoots of the selenate-grown plants. From the above results, it seems likely that as selenate is reduced to selenite or selenomethionine it becomes more readily available for volatilization by plant metabolic processes. This could be attributed to the fact that selenite and selenomethionine will require less reducing power (NADPH) to be reduced to the volatile forms of Se than will selenate.

C. Presence of Other Ions

The addition of Na_2SO_4, NaCl, and $CaCl_2$ to soils has been shown to affect the volatilization of Se by microbes [56]. In plants, the presence of sulfate and orthophosphate is known to affect Se uptake [57]. Regarding volatilization of Se by plants, there are no data available to show the influence of other ions except the sulfate study by Zayed and Terry [38] referred to earlier. Increasing the concentration of sulfate in the culture solution caused a progressive decrease in the rate of Se volatilization (Figure 1). With increase in the sulfate concentration from 0.25 to 10 mM, the rate of volatilization decreased from 97 to 14 µg Se/(m^2·day) [38]. The decrease in Se volatilization rate with increased sulfate was correlated with an increase in the S/Se ratio in plant tissues (as discussed above). This suggests that sulfate outcompeted selenate for the active sites of enzymes responsible for the conversion of inorganic Se to volatile forms.

Other research from our laboratory indicates that nitrate supply has an inhibiting effect on Se volatilization similar to that of sulfate [Zayed and Terry, unpublished data]. With increase in the NO$_3$-N supply in the culture solution from 1.5 to 7.5 to 15 mM, the rate of Se volatilization by broccoli roots decreased from 1.79 to 1.12 to 0.78 mg/(kg·day) on a dry weight basis, respectively. Rates of Se volatilization in shoots were less affected. Nitrate ions may inhibit Se uptake and accumulation in rape plants [58], forage crops [59,60], and excised barley roots [61]. Since nitrate and selenate ions are not chemically similar, it is unlikely that the reduction of plant Se uptake and volatilization by nitrate are through competitive inhibition, but how nitrate exerts its inhibitory action is unknown.

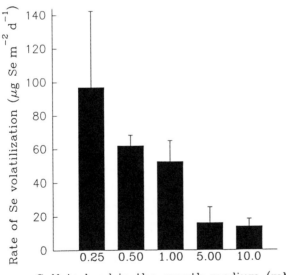

Figure 1 Effect of sulfate supply on the rate of Se volatilization by broccoli plants. Plants were supplied with 20 µM Se as sodium selenate in half-strength Hoagland's solution. (From Ref. 38.)

D. Temperature, pH, and Light

The suggestion that Se volatilization by plants is an enzymatically mediated process led Lewis et al. [55] to carry out several experiments to determine the effects of light, temperature, and pH on the rate of Se volatilization by plants. Using cabbage leaf homogenates, they demonstrated that the optimal temperature for Se volatilization was close to 40°C (Figure 2) and the optimal pH was around 7.8 (Figure 3). It is worth noting here that Mazelis et al. [62] reported a similar pH dependence and pH optimum for the analogous evolution of dimethyl sulfide from S-methyl methionine sulfonium salt in the presence of a bacterial enzyme fraction.

In a greenhouse experiment with alfalfa, Lewis et al. [11] demonstrated that, after a lag of about 2 h, the rate of Se volatilization increased with increase in irradiance level. They suggested that this increase was not due to a direct effect of light but to enhanced stomatal opening and/or to heating of plant tissues such that transpiration rates and enzyme reactions would be accelerated, resulting in greater production of volatile Se. Later,

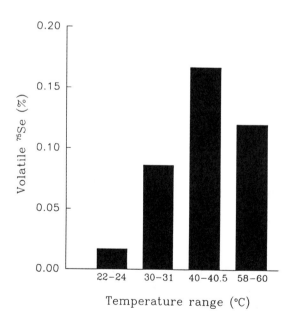

Figure 2 Influence of temperature on Se volatilization from cabbage leaf homogenates. (Adapted from Ref. 55.)

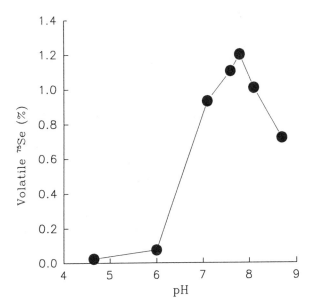

Figure 3 Influence of pH on Se volatilization from cabbage leaf homogenates. (Based on data of Ref. 55.)

Lewis et al. [55] found no measurable effect of light per se on the rate of Se volatilization from cabbage leaf homogenates under dark and light treatments.

E. Roots, Shoots, and Shoot Removal

Comparisons of root with shoot Se volatilization rates show that roots volatilize Se at much faster rates than shoots. In broccoli, Zayed and Terry [14] found that roots volatilized a large proportion (87.5%) of the Se volatilized by the whole plant even though the roots made up only 12.5% of the total plant dry weight. Per unit dry weight of tissue, roots volatilized 7–20 times faster than shoots. These greater volatilization rates in roots compared to shoots do not, however, reflect the amount of Se accumulated in the various plant tissues. The results obtained by Zayed and Terry [14] indicate that the Se concentrations in broccoli shoot tissues were more than two- to five-fold those of root tissues. Similar findings were obtained for five other species—rice, cabbage, cauliflower, Chinese mustard, and wild brown mustard (*B. juncea* Czern L.) (Table 2). Thus, it would appear

Table 2 Selenium Volatilization Ratios (Dry Weight Basis) of Root/Shoot and Detopped Root/Intact Root of Six Plant Species Grown in Half-Strength Hoagland's Solution Containing 20 µM Se as Selenate

Species	Se volatilization ratio	
	Root/shoot	Detopped root/intact root
Rice	2.2	3.68
Broccoli	14.3	2.90
Cabbage	14.7	5.15
Cauliflower	18.6	1.58
Chinese mustard	10.1	2.18
Wild brown mustard	3.1	4.34

that the root is the primary site of Se volatilization in plants despite its smaller mass and lower Se concentration.

When the shoot of the broccoli plant is removed, there is a significant increase in the rate of Se volatilization by the detopped root (Figure 4). Shoot removal increased the Se volatilization rate in roots in the following 24 h by 1.5–5-fold. Increasing the level of sulfate decreased the rate of volatilization by both intact and detopped roots but did not affect the extent to which detopping increased root volatilization. This effect of detopping was not confined to broccoli: five other crop plant species exhibited the same effect (Table 2). In broccoli, Se volatilization increased progressively for 72 h after shoot removal, attaining rates that were 20–30 times the rate of volatilization of the intact root (Figure 5).

The reason for the enhanced rates of Se volatilization following shoot removal is not known. We discuss the possibility (below) that the volatilization process may involve microbes in the rhizosphere. If this is true, then one explanation for the enhancement of root volatilization is that shoot removal results in a leakage of reduced carbon compounds into the rhizosphere, thereby accelerating the production of volatile Se by rhizosphere microorganisms. Studies on S volatilization by plants in response to mechanical injury suggest an alternative explanation. These studies show that plants with mechanically injured roots evolve greater amounts of volatile S compounds than noninjured plants [63]. Since Se compounds appear to be metabolized via the S-assimilation pathway, it is possible that the mechanical injury resulting from shoot removal might also enhance Se volatilization.

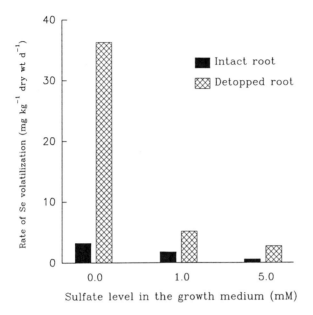

Figure 4 Selenium volatilization by broccoli roots as affected by shoot removal (detopping) and sulfate supply. (Adapted from Ref. 14.)

F. Microbes

The greater efficiency of roots than shoots in volatilizing Se may be due solely to plant factors or it might be that roots have microbes in their rhizosphere (or within the root itself) that contribute directly or indirectly to Se volatilization. Although we have little firm knowledge yet about the potential role(s) of microbes in the volatilization of Se by plants, it is perhaps worthwhile to review some of the possibilities. Volatile Se compounds collected from plants might come from the physiological activities of various biological agents. These are (a) microorganisms in the root medium (i.e., soil or nutrient solution); (b) microbes in the rhizosphere or some other location within the root; (c) metabolism of Se by root tissue itself, with or without the help of microbes acting indirectly (production of intermediates, for example); (d) microbes in the shoot; and (e) metabolism of Se in shoots, with or without the help of microbes acting indirectly (in the roots, for example). Furthermore, if microbes are involved, the question arises as to whether these microbes are bacteria or fungi. These possibilities and any evidence for or against them are presented below.

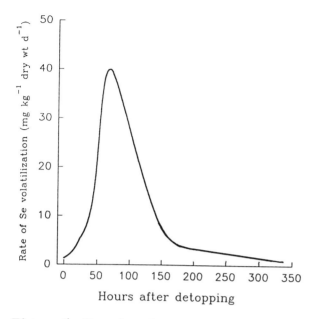

Figure 5 Time dependence of the enhancement in Se volatilization rate of broccoli roots following shoot removal (detopping). (Adapted from Ref. 14.)

1. Microbial Volatilization in the Nutrient Solution

Several types of microorganisms, including fungi, bacteria, and yeast, are capable of causing Se volatilization from soils and sediments [2,3,6,56,64]. The extent to which any of these organisms could exist and volatilize Se in hydroponic culture is uncertain. However, we have observed that when plants are removed from the nutrient solution, the remaining solution still has low but measurable rates (10% of total Se volatilized by plant plus the nutrient solution) of Se volatilization [14]. In a water column study, Cooke and Bruland [65] demonstrated that Se can be methylated and volatilized in natural waters (in the absence of plants). They detected three dissolved methylated Se compounds (DMSe, DMDSe, and an unidentified alkyl dimethylselenonium ion: $DMSe^+$-R) in natural waters from three sites in the San Joaquin and Imperial Valleys of California. They estimated the amount of DMSe (the major volatile form in their study, comprising 99.8% of the total volatile Se) to be 6%, possibly reaching up to 12–22% of the total Se present in waters collected from Kesterson Reservoir. Although these data provide evidence for Se volatilization in natural waters, they do not prove that DMSe was released by microbes. In fact, the above study provided data indicating that the nonvolatile DMSe$^+$-R can be transformed into volatile DMSe at natural pH, probably via hydrolysis, providing a mechanism for the extracellular transformation of dissolved DMSe$^+$-R into dimethylselenide. Furthermore, Cooke and Bruland [65] were not able to observe any detectable levels of dimethyl selenone and/or methyl methane selenonate, which have been previously proposed to be intermediates in the microbial biosynthesis of DMSe [5], in their water columns. However, these results did not rule out the direct volatilization of Se by microbes in natural waters. Alternatively, they proposed that if DMSe can be produced by microbial biosynthesis, an alternative microbial pathway analogous to that proposed for higher plants [13] involving a DMSe$^+$-R intermediate may be responsible. It seems, therefore, that the issue of the microbial volatilization of Se in natural waters and nutrient solutions remains unresolved.

2. Microbial Volatilization in the Root or Rhizosphere

When prokaryotic antibiotics were added to the nutrient solution, the total rate of Se volatilization by root and solution was decreased by 80–90% [14]. The inhibition of volatilization by the antibiotics could have been caused by the action of antibiotics on bacteria volatilizing Se in the rhizosphere or in the root itself. The effect of antibiotics on bacteria volatilizing in the nutrient solution would be minimal because the amount of Se volatilized is small (see above) and because removal of bacteria from the nutrient solution by filtration had no effect on the rate of Se volatilization by plant roots. Thus, the inhibitory action of antibiotics suggests that bacteria may have been volatilizing Se directly or that bacteria may assist plants in some way to volatilize Se (e.g., in facilitating the chemical reduction of Se or Se transport, etc.).

3. Volatilization by Root Tissue

There is no direct evidence yet that Se is volatilized from root tissue without the aid of microbes. However, there is evidence that root tissue previously supplied with Se volatilizes Se at high rates even when the roots are placed in a Se-free nutrient solution. This is consistent with the view that the site of volatilization is inside the root (where microbial activity is probably low) rather than in the rhizosphere. Zayed and Terry [14] found that detopped roots of plants enriched with Se and placed in a Se-free nutrient solution volatilized Se 4 times as fast as detopped roots of plants grown without Se and placed in Se-containing nutrient solution. These results suggest that volatile Se is produced from the metabolism of Se present inside the root cells rather than from microbes in the rhizosphere.

4. Volatilization by Shoot Tissue—Microbial or Plant?

Lewis [13] obtained evidence suggesting that volatilization from shoot tissue was independent of microbes. She measured Se volatilization from homogenates of surface-sterilized cabbage leaves and found that sterile homogenates volatilized as well as unsterile homogenates. She concluded that microbes did not contribute to the leaf volatilization of Se.

Shoots do not volatilize Se as fast as roots [14]. The volatilization rate of shoots can be increased, however, by supplying Se in chemically reduced forms such as selenite or selenomethionine. Although Se volatilization in shoots may proceed axenically, it is conceivable that microbes associated with the root may play a role in Se volatilization either by enhancing Se uptake or by chemically reducing selenate to forms more easily processed by the S-assimilation pathway.

5. Plant/Microbe Volatilization—Fungal or Bacterial?

If some biological agent other than the plant itself contributes to a plant's production of volatile Se compounds, it is likely to be bacterial rather than fungal. In experiments involving the use of antifungal or antibacterial agents, Zayed and Terry [14] found that while prokaryotic antibiotics such as penicillin and chlortetracycline inhibited Se volatilization of detopped roots, the eukaryotic antibiotic cycloheximide had no effect. As mentioned above, the most likely explanation for the large reduction in Se volatilization by detopped root with the application of prokaryotic antibiotics is that antibiotics may have inhibited the direct volatilization of Se by bacteria or the indirect bacterial contribution to Se volatilization by plants, that is, promotion of Se uptake and reduction.

If bacteria are contributing to a plant's volatilization of Se, where in the plant are the bacteria active? In addition to being in the nutrient solution, bacteria may be present in the rhizosphere or inside the root itself. According to Brock et al. [66], the rhizosphere is a zone where microbial activity is usually high. The bacterial count is almost always higher in the rhizosphere than it is in regions of the soil devoid of roots, and often many times higher. Brock et al. [66] stated that there is some evidence that the rhizosphere microorganisms benefit the plant by promoting the absorption of nutrients.

There is also evidence that microorganisms have the capacity to penetrate root cells and reach the interior of the plant; clear examples are the penetration of the bacteria of the genus *Rhizobium* to the roots of legume plants to form root nodules and the penetration of the fungal endomycorrhiza to root cells of many plants. Huang [67] indicated that bacteria are capable of penetrating both cuticle and cell wall. In fact, the presence of bacteria in the cytoplasm and vacuoles of host plant cells is well documented [68]. As discussed by Huang [67], microbes penetrate plants through wounds, through openings of the epidermis created by lateral root emergence, or through epidermal cell walls. Interestingly, he pointed out that colonization of the root surface and possible breaching of the more resistant surface cells of the plant can be facilitated by injury. Clearly, more research is required to uncover the physiological and biochemical basis of Se volatilization by plants and the extent to which soil microorganisms might be involved in this process.

V. POSSIBLE MECHANISMS OF PLANT VOLATILIZATION

Assuming, as the evidence from Lewis [13] suggests, that Se may be volatilized by plant metabolic processes, what is the biochemical pathway for the conversion of selenate to volatile Se? There would appear to be two potential pathways depending on whether the plant is or is not a Se accumulator (Figure 6). Because of the chemical similarity between Se and S, selenate is most probably metabolized via the S-assimilation pathway, and the tentative schemes outlined in Figure 6 are developed by analogy with that pathway. Certain enzymatic steps of the pathway have been shown to function as predicted using Se analogues of the appropriate S compounds. These steps are identified below.

In our treatment of this topic, we have focused on the mechanism of volatilization by Se nonaccumulators. This may be considered in terms of five main processes: (a) reduction of selenate to selenite, (b) reduction of selenite to selenide, (c) conversion of selenide to selenocysteine, (d) conversion of selenocysteine to selenomethionine, and (e) conversion of selenomethionine to the volatile Se compound DMSe.

A. Reduction of Selenate to Selenite

The first step in the metabolism of selenate is thought to be the formation of an activated form of selenate, APSe (adenosine 5'-phosphoselenate), by combination of selenate with ATP [42]. ATP sulfurylase may catalyze this reaction [42]. The reduction of APSe to selenite may proceed by nonenzymatic reactions similar to those proposed for APS (adenosine 5'-phosphosulfate) reduction, but no information exists concerning the details of this step for Se.

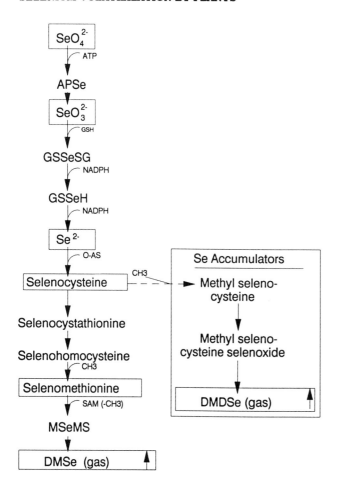

Figure 6 Possible mechanisms of Se volatilization by plants. (Compiled from Refs. 13, 42, 44, and 69.)

B. Reduction of Selenite to Selenide

Ng and Anderson [69] proposed the following steps. Selenite is reduced nonenzymatically to selenotrisulfide (GSSeSG) using reduced glutathione. GSSeSG is then reduced to selenide in two steps by the enzyme glutathione reductase: GSSeSG is reduced using NADPH to GSSeH (selenoglutathione); this is then reduced to selenide (again using NADPH).

C. Conversion of Selenide to Selenocysteine

Ng and Anderson [43] suggested that inorganic selenide is converted to selenocysteine by cysteine synthase, which combines selenide with O-acetylserine. Although this is an S-metabolism enzyme, it has been shown to have a greater affinity for selenide than sulfide.

D. Conversion of Selenocysteine to Selenomethionine

Burnell [42] and Dawson and Anderson [44] proposed the next three steps leading to the production of selenomethionine (shown in Figure 6). Cystathionine γ-synthase catalyzes the combination of selenocysteine with O-phosphorylhomoserine to form selenocystathionine. Cystathionine β-lyase splits selenocystathionine at its β linkage into selenohomocysteine, pyruvate, and NH_4^+. Selenohomocysteine is then methylated to selenomethionine, possibly by a methyltransferase enzyme.

E. Conversion of Selenomethionine to Dimethylselenide

Lewis [13], by analogy with the S-assimilation pathway, proposed that selenomethionine is methylated, most probably by S-adenosyl methionine (SAM), to form methyl selenomethionine selenonium salt (MSeMS). This step may be catalyzed by S-methyltransferase. Lewis et al. [55] concluded that the final step in the volatilization pathway is likely to be the cleavage of MSeMS to DMSe and homoserine (the enzyme for this might be S-methylmethionine hydrolase). Recent studies of the expression and localization of SAM synthetase (the enzyme that catalyzes the only step leading to the biosynthesis of SAM) in *Arabidopsis thaliana* showed that mRNA levels were 10–20 times greater in roots and stems than leaves, correlating well with the higher enzyme activity in the stems [70]. The higher rates of Se volatilization in roots compared to leaves may be related to the larger amounts of SAM synthetase in roots than leaves.

In Se-accumulating plants, the process is thought to be the same as for nonaccumulators up to the formation of selenocysteine [42]. At this point, selenocysteine is then methylated twice to form the volatile Se compound DMDSe (Figure 6). Brown and Shrift [29] suggested that the methylation of selenocysteine in accumulator plants is probably by the transmethylation of selenocysteine to methyl selenocysteine. An alternative view, not shown in Figure 6, is that methyl selenocysteine might be produced by the hydrolytic cleavage of selenocystathionine in an enzymatic reaction [13]. The final two steps shown in Figure 6 were proposed by Lewis [13]: methyl selenocysteine is converted enzymatically to methyl selenocysteine selenoxide, which is then acid hydrolyzed to DMDSe, the form of Se commonly volatilized by Se accumulators. There is no experimental evidence that these last two reactions occur; they were proposed by analogy with the S-assimilation pathway.

VI. PLANT VOLATILIZATION AND VEGETATIVE MANAGEMENT

Plants volatilize Se in substantial amounts, up to 16.2 g Se per acre per day, according to data of Terry et al. [53]. Plants are effective in removing Se from deeper regions of the soil, and root penetration can be extensive [71]. Selenium removed by plant roots may be volatilized to the atmosphere, or the Se accumulated in shoots can be harvested and removed [72]. Clearly, plant volatilization should be considered a potential bioremediation process in the vegetative management of Se. The effectiveness of plants was demonstrated in an experiment by Duckart et al. [8]. They calculated the time required for an amount of Se equal to one-half of the Se(VI) added to the soil to be removed from the soil by both plant uptake and volatilization. This time was 461 days for the soil alone (without plants) and ranged from 5.9 to 12 days when different plants were added to the soil. Similarly, actual field measurements of Se volatilization conducted in Kesterson Reservoir of the San Joaquin Valley of California by Biggar and Jayaweera [73] showed that Se volatilization was 19.6 times greater from soil with barley and 6.9 greater from soil with the native salt grass than from soil alone.

If plant volatilization is to be used as part of a vegetative management scheme, it is important to know how efficiently different plants may volatilize Se from the crop or soil surface. One can calculate maximum potential rates of Se volatilization from crop surfaces using data obtained for hydroponically grown plants cultured in growth chambers under optimum environmental conditions. Although actual measurements for crop volatilization in the field may be lower than the maximum potential rates obtainable under ideal growing conditions, the latter do provide an estimate of the upper limit of Se volatilization for different crops. Our data show that rice may volatilize Se at a rate of 340 $\mu g/m^2$ leaf area per day [53]. Rice has been shown to have a maximum leaf area index (LAI, total leaf area per unit area of soil surface) of 11.8 [74]. Thus, rice at maximum LAI could potentially volatilize Se at rates of 4.01 mg/day per square meter of soil surface. Broccoli and cabbage have been shown to have LAIs of 5 or more [75]. Assuming that broccoli volatilizes at 273 $\mu g/(m^2 \cdot day)$ and cabbage at 221 $\mu g/(m^2 \cdot day)$, with an LAI of 5 in each case, this yields rates of 1.3 and 1.1 mg/m^2 soil surface per day for broccoli and cabbage, respectively. It must be stressed that these estimates should be interpreted with caution since they are extrapolated from values obtained under growth chamber conditions, which may be considerably different from those found in the field. For instance, Wu and Huang [76], using the highly saline, Se-rich Kesterson soil, estimated that under field conditions a 1-m^2 salt grass plot may produce 180 µg of volatile Se per day. One reason for the lower estimates of Wu and Huang [76], which were derived from actual field measurements, might be that the high level of sulfate salinity found in Kesterson soils decreased the rate of Se volatilization; as mentioned earlier, high sulfate salinity substantially diminishes the rate of Se volatilization [38]. Alternatively, the difference between these latter two estimates is probably due to the variation in the effectiveness of the different plant species in volatilizing Se. It is clear that more research is needed to obtain Se volatilization rates under field conditions for growing different plant species.

Other vegetative management schemes might be used in tandem with biological volatilization. Plant material high in Se might be disposed of as a feed for animals in Se-deficient areas [77] or disked into the soil to be degraded subsequently and volatilized by soil microorganisms [78]. The idea of blending Se-concentrating forage with forage from Se-deficient areas has received support from research carried out by Bañuelos and coworkers [72,78]. They identified wild mustards (e.g., "wild brown mustard," *B. juncea* Czern L.) as possible Se-amendment forages [72]. These "rapid cycling" *Brassica* species go from seed to flower in as little as 19–35 days [79]. Brassicas may also be valuable in cropping-based bioremediation schemes due to their high Se-volatilization rates and response to partial shoot removal [14].

VII. FUTURE OUTLOOK

The discoveries that roots are the primary site of Se volatilization in plants and that shoot removal stimulates root volatilization even more offer some exciting opportunities for unraveling the mechanism of Se volatilization. What are the respective contributions of plants and microbes to these enhanced rates of volatilization? If a substantial portion of the enhanced volatilization is from the plant, this will permit us to determine which enzymes and which plant parts are responsible for the enhanced volatilization rates, for example, the conversion of selenate to selenite, selenite to selenide, selenide to selenocysteine, selenocysteine to selenocystathionine, or MSeMS to DMSe. If a substantial portion of the enhanced volatilization (in roots and on detopping) is microbial, we may determine the factors responsible (e.g., the extent to which the enhancement is due to increased leakage of root exudates) or identify the microbe(s) that contribute to volatilization so that it may be possible to artificially infect plants with that microbe and thereby substantially increase the volatilization rate.

Elucidation of the critical rate-limiting steps will enable us to develop ways of significantly enhancing Se volatilization by plants in the field. If it can be shown that certain key enzymes are rate-limiting for Se volatilization, we may be able to use modern biotechnological approaches such as genetic engineering or the application of chemical modifiers to achieve greater rates of Se volatilization by plants. Furthermore, the demonstration that shoot removal significantly increases Se volatilization by plants suggests an immediate agronomic way to remove Se from Se-contaminated soils. For example, since total shoot removal stimulates Se volatilization, it may be possible to increase Se volatilization by partial shoot removal. Bañuelos and Meek [72] showed that plant clipping significantly increased the accumulation of Se in the total harvested shoot tissue of wild brown mustard. Such clipping practices may well have the additional benefits of increasing Se volatilization, especially in the roots.

REFERENCES

1. L. Valoppi and K. Tanji, in *Selenium Contents in Animal and Human Food Crops Grown in California* (K. K. Tanji, L. Valoppi, and C. Woodring, Eds.), Coop. Ext., Univ. California, Div. Agric. Nat. Resour., Publ. 3330, 1988, p. 97.
2. J. W. Doran and M. Alexander, *Soil Sci. Soc. Am. J.* 40:687 (1977).
3. D. C. Reamer and W. H. Zoller, *Science* 208:500 (1980).
4. H. B. Ross, in *Atmospheric Selenium*, Report CM-66, Dept. Meteorology, Univ. Stockholm, Int. Meteorological Inst. Stockholm, 1984, pp. 1–19.
5. J. W. Doran, *Adv. Microb. Ecol.* 6:1 (1982).
6. U. Karlson and W. T. Frankenberger, Jr., *Soil Sci. Soc. Am. J.* 53:749 (1989).
7. R. Zieve and P. J. Peterson, *Sci. Total Environ.* 32:197 (1984).
8. E. C. Duckart, L. J. Waldron, and H. E. Donner, *Soil Sci.* 53:94 (1992).
9. O. A. Beath, H. F. Eppson, and C. S. Gillbert, *Wyoming Agr. Sta. Bull.* 206:1 (1935).
10. I. Rosenfeld and O. A. Beath, *Selenium: Geobotany, Biochemistry, Toxicity, and Nutrition*, Academic, New York, 1964.
11. B. G. Lewis, C. M. Johnson, and C. C. Delwiche, *J. Agric. Food Chem.* 14:638 (1966).
12. C. S. Evans, C. J. Asher, and C. M. Johnson, *Aust. J. Biol. Sci.* 21:13 (1968).
13. B. G. Lewis, Ph.D. Thesis, Univ. California, Berkeley, 1971.
14. A. M. Zayed and N. Terry, *J. Plant Physiol.* in press (1993).
15. C. J. Asher, C. S. Evans, and C. M. Johnson, *Aust. J. Biol. Sci.* 20:737 (1967).
16. S. F. Trelease and H. M. Trelease, *Am. J. Bot.* 26:530 (1939).
17. A. Shrift, *Ann. Rev. Plant Physiol.* 20:475 (1969).
18. T. C. Broyer, D. C. Lee, and C. J. Asher, *Plant Physiol.* 41:1425 (1966).

19. T. C. Broyer, C. M. Johnson, and R. P. Huston, *Plant Soil* 36:635 (1972).
20. T. C. Broyer, C. M. Johnson, and R. P. Huston, *Plant Soil* 36:651 (1972).
21. N. M. Price and P. J. Harrison, *Plant Physiol.* 86:192 (1988).
22. S. Shigeoka, T. Takeda, and T. Hanaoka, *Biochem. J.* 275:623 (1991).
23. T. C. Stadtman, *J. Biol. Chem.* 266:16257 (1991).
24. J. M. Ulrich and A. Shrift, *Plant Physiol.* 43:14 (1968).
25. A. Shrift and J. M. Ulrich, *Plant Physiol.* 44:803 (1969).
26. C. J. Asher, G. W. Butler, and P. J. Peterson, *J. Exp. Bot.* 28:279 (1977).
27. J. W. Hamilton and O. A. Beath, *Agron. J.* 55:528 (1963).
28. M. M. Abrams, C. Shennan, R. J. Zasoski, and R. G. Burau, *Agron. J.* 82:1127 (1990).
29. T. A. Brown and A. Shrift, *Biol. Rev.* 57:59 (1982).
30. J. E. Leggett and E. Epstein, *Plant Physiol.* 31:222 (1956).
31. G. Ferrari and F. Renosto, *Plant Physiol.* 49:114 (1972).
32. A. Hurd-Karrer, *Am. J. Bot.* 25:666 (1938).
33. E. B. Davies and J. H. Watkinson, *N.Z. J. Agric. res.* 9:641 (1966).
34. J. E. Pratley and J. D. McFarlance, *Aust. J. Exp. Agric. Anim. Husb.* 14:533 (1974).
35. R. L. Mikkelsen, A. L. Page, and G. H. Haghnia, *Plant Soil* 107:63 (1988).
36. R. L. Mikkelsen, G. H. Haghania, A. L. Page, and F. T. Bingham, *J. Environ. Qual.* 17:85 (1988).
37. R. L. Mikkelsen and H. F. Wan, *Plant Soil* 121:151 (1990).
38. A. M. Zayed and N. Terry, *J. Plant Physiol.* 140:646 (1992).
39. G. S. Smith and J. H. Watkinson, *New Phytol.* 97:557 (1984).
40. M. Singh, N. Singh, and D. K. Bhandari, *Soil Sci.* 129:238 (1980).
41. J. W. Anderson and A. R. Scarf, in *Metals and Micronutrients: Uptake and Utilization by Plants* (D. A. Robb and W. S. Pierpoint, Eds.), Academic, New York, 1983, p. 241.
42. J. N. Burnell, *Plant Physiol.* 67:316 (1981).
43. B. H. Ng and J. W. Anderson, *Phytochemistry* 17:2069 (1978).
44. J. C. Dawson and J. W. Anderson, *Phytochemistry* 27:3453 (1988).
45. T. J. McCluskey, A. R. Scarf, and J. W. Anderson, *Phytochemistry* 25:2063 (1986).
46. T. A. Brown and A. Shrift, *Plant Physiol.* 67:1051 (1981).
47. O. Kennard, in *CRC Handbook of Chemistry and Physics,* 60th ed. (R. C. Weast, Ed.), CRC Press, Boca Raton, 1979, p. F217.
48. R. E. Huber and R. S. Criddle, *Biochim. Biophys. Acta* 141:587 (1967).
49. G. W. Butler and P. J. Peterson, *Aust. J. Biol. Sci.* 20:77 (1967).
50. T. A. Brown and A. Shrift, *Plant Physiol.* 66:758 (1980).
51. D. C. Eustice, F. J. Kull, and A. Shrift, *Plant Physiol.* 67:1054 (1981).
52. C. M. Chow, S. N. Nigam, and W. B. McConnell, *Phytochemistry* 10:2693 (1971).
53. N. Terry, C. Carlson, T. K. Raab, and A. Zayed, *J. Environ. Qual.* 21:341 (1992).
54. C. M. Johnson, in *Trace Elements in Soil–Plant–Animal Systems* (D. J. D. Nicholas, and A. R. Egan, Eds.), Academic, New York, 1975, p. 165.
55. B. G. Lewis, C. M. Johnson, and T. C. Broyer, *Plant Soil* 40:107 (1974).
56. U. Karlson and W. T. Frankenberger, Jr., *Soil Sci.* 149:56 (1990).

57. R. L. Mikkelsen, A. L. Page, and F. T. Bingham, in *Selenium in Agriculture and the Environment* (L. W. Jacobs, Ed.), SSSA Spec. Pub. 23, Madison, Wisc., 1989, p. 65.
58. M. B. Ali, *Nucleus* 7:126 (1970).
59. W. F. Raymond, *Adv. Agron.* 21:1 (1969).
60. G. Gissel-Nielsen, *Z. Pflanzenernaehr. Bodenkd.* 138:97 (1975).
61. T. El Kobia, *Agronchimica* 19:5 (1975).
62. M. Mazelis, B. Levin, and N. Mallinson, *Biochim. Biophys. Acta* 105:106 (1965).
63. L. G. Wilson, R. A. Bressan, and P. Filner, *Plant Physiol.* 61:184 (1978).
64. F. Challenger, D. B. Lisle, and P. B. Dransfield, *J. Chem. Soc.* 1954:1760 (1954).
65. T. D. Cooke and K. W. Bruland, *Environ. Sci. Technol.* 21:1214 (1987).
66. T. D. Brock, D. W. Smith, and M. T. Madigan, Microbial symbiosis, in *Biology of Microorganisms*, 4th ed., Prentice-Hall, Englewood Cliffs, N.J., 1984, pp. 472–481.
67. J. Huang, *Annu. Rev. Phytopathol.* 24:141 (1986).
68. P. Y. Huang, J. Huang, and R. N. Goodman, *Physiol. Plant Pathol.* 6:283 (1975).
69. B. H. Ng and J. W. Anderson, *Phytochemistry* 18:573 (1979).
70. J. Peleman, W. Boerjan, G. Engler, J. Seurinck, J. Botterman, T. Alliotte, M. V. Montagu, and D. Inze, *Plant Cell* 1:81 (1989).
71. E. Epstein, *Mineral Nutrition of Plants: Principles and Perspectives*, Wiley, New York, 1972.
72. G. S. Bañuelos and D. W. Meek, *J. Environ. Qual.* 19:772 (1990).
73. J. W. Biggar and G. R. Jayaweera, *Soil Sci.* 155:31 (1993).
74. H. F. Schnier, M. Dingkuhn, S. K. De Datta, K. Mengel, and J. E. Faronilo, *Crop Sci.* 30:1276 (1990).
75. R. Fordham and A. G. Biggs, *Principles of Vegetable Crop Production*, Collins, London, 1985.
76. L. Wu and Z. Z Huang, *Ecotoxicol. Environ. Safety* 22:267 (1991).
77. C. Ross, *Calif. Farmer* 274:68 (1991).
78. G. S. Bañuelos, R. Mead, and S. Akohoue, *J. Plant Nutr.* 14:701 (1991).
79. P. H. Williams and C. B. Hill, *Science* 232:1385 (1986).

15
Microbial Volatilization of Selenium from Soils and Sediments

W. T. Frankenberger, Jr.

*University of California
Riverside, California*

Ulrich Karlson

*National Environmental Research Institute
Roskilde, Denmark*

I. INTRODUCTION

Selenium was discovered in 1817 by Berzelius and Gahn when they were working with selenium-bearing pyrites. Selenium (Se) is classified as a metalloid, since it has properties of both a metal and a nonmetal. It is markedly similar to sulfur in its chemistry, with its primary oxidation states being VI, IV, 0, and -II. Elemental Se (Se^0) exhibits a zero valence state and is commonly associated with sulfur compounds such as selenium sulfide (Se_2S_2) and polysulfides. Selenate (SeO_4^{2-}) and selenite (SeO_3^{2-}) are common ions in natural waters and soils. Reduced Se compounds include volatile methylated species such as dimethylselenide (DMSe, $[CH_3]_2Se$), dimethyl diselenide (DMDSe, $[CH_3]_2Se_2$), dimethyl selenone ($[CH_3]_2SeO_2$), and selenium substituted amino acids, including selenomethionine, selenocysteine, and selenocystine. Inorganic reduced Se forms include mineral selenides and hydrogen selenides (H_2Se).

Among the elements, Se ranks seventieth in order of abundance and is widely distributed in the earth's crust at low concentrations [1,2]. Recently, public attention has been drawn to Se because in several areas of the world it has been discovered to be an environmental threat. Selenium is a widespread contaminant in the United States in areas in Arizona, California, Colorado, Montana, Nevada, New Mexico [3], South Dakota, Texas [4], Utah, and Wyoming [5]. This chapter focuses on efforts to develop a bioremediation approach to deselenify the soil environment.

II. CYCLING OF SELENIUM

Most soils contain between 0.1 and 2 mg Se/kg [6,7], but elevated concentrations of Se are associated with soils of marine sedimentary parent material, coal and petroleum by-products (including fly ash waste), metal refining operations, and mine tailings [8]. The Se concentration in soil depends on the parent material, climate, topography, age of soil, and agricultural or industrial utilization. The rate and extent to which Se is mobilized depends on its chemical speciation and partitioning in soils and sediments.

III. FORMS OF SELENIUM

The concentration and speciation of Se in soil depends on the pH, redox potential, solubility, complexing ability of soluble and solid ligands, biological interactions, and reaction kinetics [9,10]. Under acidic reducing conditions in soils, which may be waterlogged and rich in organic matter, elemental Se and selenides are the predominant species. Since metal selenides, Se sulfides, and elemental Se are insoluble, they are less biologically available for uptake. At a high redox potential, in well-aerated alkaline soils, the highly soluble SeO_4^{2-} is the predominant species, and at neutral pH SeO_3^{2-} occurs in approximately equal concentrations [6]. Selenite ions form stable adsorption complexes or coprecipitants with sesquioxides [11]. Selenate is stable in oxidized environments and is the form in which Se is most readily taken up by plants [12,13]. In most soils, SeO_3^{2-} and SeO_4^{2-} are the predominant Se species.

IV. MICROBIAL TRANSFORMATIONS OF SELENIUM

A. Reduction

There is a diverse group of microorganisms in soils, sediments, and waters capable of reducing Se. Oremland et al. [14] demonstrated in situ (anoxic sediments) and in pure culture the anaerobic bacterial conversion of seleno-oxyanions to Se^0. Their results indicate that dissimilatory SeO_4^{2-} and SeO_3^{2-} reduction to Se^0 may be an important biological transformation for reduction of Se oxyanions in anoxic sediments. Macy and Rech [15] were able to isolate two anaerobic bacteria: one that reduced SeO_3^{2-} to Se^0, and another that reduced SeO_4^{2-} to SeO_3^{2-}. Macy isolated *Thauera selenatis* from a seleniferous soil that reduces SeO_4^{2-} to SeO_3^{2-} via selenate reductase and SeO_3^{2-} to Se^0 via nitrite reductase (see Chapter 17).

B. Oxidation

Although there are several reports of Se reduction, there are few well-documented cases of Se oxidation. Reduced forms of Se such as elemental Se [16–19] and copper selenide [20] have been reported to be oxidized by laboratory bacterial cultures and soils. *Aspergillus niger* has also been reported to oxidize SeO_3^{2-} to SeO_4^{2-} [21].

C. Volatilization

Volatilization through methylation is thought to be a protective mechanism used by microorganisms to avoid Se toxicity in seleniferous environments. In effect, the process permanently removes Se from soil and water under aerobic conditions. The predominant groups of Se-methylating organisms isolated from soils and sediments are bacteria and fungi [22–26]. In water bodies, bacteria are thought to play a dominant role (Table 1) [27]. Biomethylation of toxic Se species, including SeO_3^{2-}, SeO_4^{2-}, Se^0, and various organoselenium compounds, into a less toxic, volatile form, DMSe, is apparently a widespread transformation in seleniferous environments [25,28,29]. Dimethylselenide is the major metabolite of Se volatilization, although other Se compounds such as DMDSe, dimethyl selenone ($[CH_3]_2SeO_2$), methane selenol (CH_3SeH), and dimethyl selenenyl sulfide (CH_3SeSCH_3) may also be produced [28,30–32].

V. FATE OF GASEOUS SELENIUM

Once Se is methylated into a volatile species, it is released into the atmosphere, diluted, and dispersed by air currents away from the contaminated source. DMSe reacts with OH and NO_3 radicals and ozone (O_3) within a few hours to yield products that are yet unknown [33]. However, it is likely that these oxidized products may be scavenged onto aerosols or sorbed onto particulates that have a relatively long residence time (7–9 days) in the atmosphere [23] and can travel considerable distances [7]. At the present time, the fate of DMSe in the atmosphere is subject to much debate among Se researchers.

Table 1 Microorganisms That Volatilize Selenium

Microorganisms	Source	Se substrate	Se conc. (ppm)	Aerobic	Anaerobic	Se product	Ref.
Bacteria							
Aeromonas sp.	Lake sediment	SeO_3	5	+	−	$(CH_3)_2Se$	28
Flavobacterium sp.						$(CH_3)_2Se_2$	
Pseudomonas sp.						Unknown volatile Se	
Corynebacterium sp.	Seleniferous soil	SeO_3, SeO_4, Se^0	—	+	−	$(CH_3)_2Se$	54
Pseudomonas fluorescens	Evaporation pond sediment	SeO_4	0.8	−	+	$(CH_3)_2Se$	30
						$(CH_3)_2Se_2$	
Unidentified species	Evaporation pond sediment	Mainly SeO_4	1.2	+	−	$(CH_3)_2Se$	Up[a]
Fungi							
Cephalosporium sp.	Garden soil	SeO_3	457	+	−	$(CH_3)_2Se$	23
Fusarium sp.		SeO_4	418				
Penicillium sp.							
Scopulariopsis sp.							
Scopulariopsis brevicaulis	Unspecified	SeO_3, SeO_4	15	+	−	$(CH_3)_2Se$	55
Schizopyllum commune	Wood	SeO_4	366	+	−	$(CH_3)_2Se$	56
Aspergillus niger	Unspecified	SeO_4	—	+	−	$(CH_3)_2Se$	57
Candida humicola	Sewage	$SeO_3, SeO_4,$ $^{75}SeO_3$	46	+	−	$(CH_3)_2Se$	58
			418	+			
Acremonium falciforme	Evaporation pond sediment	SeO_3, SeO_4	100	+	−	$(CH_3)_2Se$	26
Penicillium citrinum							
Ulocladium tuberculatum							
Acremonium falciforme	Evaporation pond sediment	SeO_4	0.79	+	−	$(CH_3)_2Se$	30
Penicillium citrinum						$(CH_3)_2Se_2$	
Penicillium sp.	Sewage	SeO_3	457	+	−	$(CH_3)_2Se$	59
Alternaria alternata	Evaporation pond water	SeO_3, SeO_4	1	+	−	$(CH_3)_2Se$	46
			100	+			

VI. TOXICITY OF METHYLATED PRODUCTS

Dimethylselenide is 500–700 times less toxic to rats than aqueous SeO_3^{2-} and SeO_4^{2-} ions [34–37]. Recently, an acute toxicity study was conducted by O. G. Raabe and M. A. Al-Bayati (reported in Ref. 36) on the inhalation of DMSe by rats. This study consisted of exposing 85 adult rats to four concentrations of DMSe (0, 1607, 4499, and 8034 ppm) for 1 hr. Not a single animal was killed by gaseous DMSe. After exposure, the animals were observed for a 1-week period for clinical abnormalities, and all appeared normal. The exposed and control rats were killed, and their major tissues and organs were examined. The effect of DMSe was one of irritation rather than injury. There was a slight increase in lung weight after 1 day of exposure, a small injury to the spleen at the highest concentration tested, and elevated Se levels in the lungs and serum. Within 7 days, all affected organs exhibited complete recovery. The half-life of DMSe in animals appears to be short, and the compound is eliminated mainly via the lung. The data indicated that the inhaled DMSe vapor is nontoxic to the rat at concentrations up to 8034 ppm or 34,000 mg/m^3 [36].

VII. FACTORS ENHANCING VOLATILIZATION OF SELENIUM

There are a number of factors that affect the overall efficiency of Se methylation into a volatile gas as a removal process. To effectively use this novel biotechnology to bioremediate seleniferous soils, dewatered sediments, and water, it is important to consider the factors that affect volatilization of Se in both soil and water.

A. Microorganisms

Selenium volatilization is microbially mediated (Table 1). Sterilization of seleniferous soil and water by autoclaving completely eliminates this reaction [22,27,31,38,39]. The addition of the bactericide chloramphenicol to soil reduces, but does not eliminate, Se volatilization, indicating that both bacteria and fungi are important in this process [40]. Adding a fungal inoculant of 2.8×10^7 cells of *Candida humicola* per gram of soil caused Se evolution to double [40]. Selenium-methylating fungi have been isolated from seleniferous soils and agricultural drainage waters of the western San Joaquin Valley in California. These microbes are able to withstand extreme osmotic stress as produced by fluctuating saline conditions [27,39]. There is often no need to add a microbial inoculum to seleniferous soils or water to promote volatilization of Se.

B. Nutrients

Soil alkylselenide production is often carbon-limited. In general, the rate of Se evolution from soils, sediments, and water increases with the addition of certain organic amendments (Tables 2 and 3). It is possible to achieve more than a 10-fold increase in volatile Se evolution with the addition of organic amendments to soil. Short-term studies of naturally seleniferous sediments conducted in our laboratory indicated that Se volatilization is accelerated through the addition of saccharides, amino acids, and especially proteins [26]. The best treatments for accelerating volatilization of Se from seleniferous soils are gluten, casein, pectin, and citrus peel [41–43]. In some soils, N may be a limiting factor. The optimum C/N ratio for volatilization of Se in soil was found to be 20:1 [26]. In aqueous environments, the addition of monosaccharides, polysaccharides, and acidic saccharides, alcohols, amino acids, fats, and oils had little effect on volatilization, while proteins such as casein and albumin dramatically stimulated DMSe production [27,44].

C. Selenium Concentration

Although the Se volatilization capacity of a soil is dependent on Se concentration [26], it is the level of available or water-soluble Se that controls the process. Zieve and Peterson [40] were able to correlate a decrease in water-soluble Se with decreasing methylation rates as volatilization proceeded. The Se concentration of soil not only affects the volatilization capacity but can also change the ratio of volatile Se species evolved. Reamer and Zoller [31] reported that the relative abundance of volatile Se species evolved from SeO_3^{2-} upon the addition of sewage sludge was dependent on the Se concentration. The major volatile Se species evolved with an initial SeO_3^{2-} concentration between 1 and 10 mg/kg was DMSe, while at a concentration $\geq$100 mg/kg the portion of DMSe was decreased markedly and the relative concentrations of DMDSe and dimethylselenone were increased.

D. Selenium Species

Selenium can be transformed into volatile species from both organic and inorganic Se (Table 4). Doran and Alexander [45] found that the evolution of gaseous Se was an order of magnitude higher in the presence of trimethylselenonium chloride (TMSe), selenomethionine, and selenocysteine compared with SeO_3^{2-}, SeO_4^{2-}, and Se^0. Volatilization of these organic Se substrates varied from 7 to 87%. Frankenberger and Karlson [42] ranked the following organic Se substrates as best-to-least suitable in terms of promoting Se volatilization from seleniferous soils: selenomethionine > selenocysteine = selenoguanosine = selenoinosine > selenoethionine = selenopurine > selenourea. However, *Alternaria alternata* was found to volatilize various Se species in pure culture in the following order of magnitude: SeO_4^{2-} > SeO_3^{2-} >> selenoinosine > selenomethionine > selenopurine > selenium sulfite [46]. When organoselenium compounds in the form of plant materials are added to soil, volatilization is less enhanced than when noncomplexed selenoamino acids or inorganic Se species serve as substrates. Elemental Se is poorly methylated compared with other inorganic Se species, probably as a result of its low solubility [31].

According to Doran's hypothesis [25], the pathway of Se biomethylation requires reduction to the Se^{2-} species and subsequent methylation to form DMSe. It should therefore be more favorable energetically to methylate SeO_3^{2-} rather than SeO_4^{2-}. Experimental evidence confirms this hypothesis; in addition, because the availability of SeO_3^{2-} can be limited by its ability to bind to iron oxides and clays, particularly at lower pH values, the microbial uptake and therefore biomethylation of this compound may be limited in some matrices. Karlson and Frankenberger [26] found that, without a carbon amendment, volatilization rates were an order of magnitude higher when SeO_3^{2-} was provided as a substrate than when SeO_4^{2-} was added. The addition of an organic amendment largely canceled out this difference, possibly because there was energy available for the additional reduction step from SeO_4^{2-} to SeO_3^{2-}. Barkes and Fleming [23] found that of 11 pure isolates of soil fungi, all were capable of producing DMSe from SeO_3^{2-}, while only six were capable of producing DMSe from SeO_4^{2-}. In contrast, Ganje and Whitehead [38] and Thompson-Eagle and Frankenberger [47] discovered that seleniferous shale and evaporation pond water, respectively, methylated SeO_4^{2-} more efficiently than SeO_3^{2-} regardless of the presence of organic matter.

E. Temperature

Selenium volatilization is temperature-dependent. The maximum release of DMSe from lake sediments occurred at 20°C [28], in California evaporation pond water at 35°C [27], from a loamy soil at 20°C [40], and from a California sandy-textured seleniferous soil at 35°C [42]. However, in each case, maximum DMSe emission occurred at the maximum temperature tested. Therefore, the optimum temperature for Se volatilization may not have been reached. In field measurements, volatilization of Se was reported to fluctuate seasonally, with rates being greater in the spring and summer months than in the fall and winter months. There was also a

Table 2 Enhancement of Selenium Volatilization from Soils and Dewatered Sediments by Organic Amendments

Matrix	Organic amendment	Amendment appl rate (g/kg)	Native Se conc (mg/kg)	Se spike conc (mg/kg)	Species	% Se volatilized Unamended	% Se volatilized Amended	Incubation time (days)	Ref.
Soils									
Lima loam	Wheat	20	0.9			0	0.67	25	45
	Wheat	20	0.9	50	Se(IV)	0.4	3.1		
Sansac clay	Wheat	20	36.0			0.01	0.03		
	Wheat	20	36.0	50	Se(IV)	0.16	2.24		
Pierre Formation, South Dakota									
0–12"	Wheat	50	6.6			0.05	0.05	25	22
12–24"	Wheat	50	6.9			0.08	0.46		
24–36"	Wheat	50	9.1			0.05	0.62		
Muck soil	Wheat	25	0.42	0.04	^{75}Se(IV)	0.68^{75}Se	0.51^{75}Se	60	50
Clay loam	Wheat	25	0.35	0.04	^{75}Se(IV)	1.41^{75}Se	2.32^{75}Se		
Sandy loam	Wheat	25	0.14	0.04	^{75}Se(IV)	2.12^{75}Se	0.70^{75}Se		
Los Banos clay loam	Galacturonic acid	2 g C	0.22	100.0	^{75}Se(IV)	2.90^{75}Se	6.87^{75}Se	13	26
	Pectin	2 g C	0.22	100.0	^{75}Se(IV)	2.90^{75}Se	8.77^{75}Se		
	Cellulose	2 g C	0.22	100.0	^{75}Se(IV)	2.90^{75}Se	3.27^{75}Se		
	Sewage sludge	2 g C	0.22	100.0	^{75}Se(IV)	2.90^{75}Se	2.50^{75}Se		
	Corn	2 g C	0.22	100.0	^{75}Se(IV)	2.90^{75}Se	6.53^{75}Se		
	Cowpea	2 g C	0.22	100.0	^{75}Se(IV)	2.90^{75}Se	7.35^{75}Se		
	Manure	2 g C	0.22	100.0	^{75}Se(IV)	2.90^{75}Se	3.53^{75}Se		

MICROBIAL VOLATILIZATION OF SE FROM SOILS

Soil	Amendment				Se form				
S. Dakota clay	Glucose	10					0	45	29
Seleniferous shale	Oat straw	10	0.4	12.3	^{75}Se(IV)	^{75}0.41Se	5.69 ^{75}Se	60	38
	Astragalus	10	0.4+ 0.6	12.8	^{75}Se(IV) + org Se		70.3 ^{75}Se		
Seleniferous sediments (dewatered)									
Kesterson pond	Casein	7.5	7.5			2.40	3.47	140	41
	Gluten	7.5	7.5				7.73		
	Cattle manure	7.5	7.5				3.33		
	Orange peel	7.5	7.5				7.47		
San Luis Drain	Casein	7.5	17.1			2.11	4.85	140	41
	Gluten	7.5	17.1				7.73		
	Cattle manure	7.5	17.1				3.33		
	Orange peel	7.5	17.1				7.47		
Kesterson Pond 4	Casein	2 g C	60.7			0.56	9.0	5	26
	Albumin	2 g C	60.7				10.2		
	Gluten	2 g C	60.7				1.40		
	Methionine	2 g C	60.7				1.83		
Kesterson salt grass vegetation	Serine	20	7.3			6.85	6.85	35	60
	Methionine	20	7.3				31.5		
	Cysteine	20	7.3				28.7		

Table 3 Enhancement of Selenium Volatilization from Evaporation Pond Water (Sumner Peck Ranch) by Organic Amendments

Organic amendment	Amendment appl rate (g C/L)	Native Se conc (mg/L)	% Se volatilized Unamended	% Se volatilized Amended	Incubation time (days)
Glucose	2	1.2	0.81	1.21	21
Maltose	2			1.21	
Casein	2			17.5	
Methionine	2	1.1	0.09	4.06	15
Serine					
Albumin	20	1.1	0.45	66.8	43
Gluten	2			13.9	
Casein	2			64.4	

Source: Ref. 27.

diurnal peak of volatile Se emission from midday to midafternoon that correlated with soil temperature [36].

F. Moisture

Air-drying the soil severely inhibits Se volatilization [40], while water-saturating it to a 1:1 or 1:3 soil/water paste causes anaerobiosis, which also decreases production of volatile Se [42]. Field studies have shown that Se emission rates are much lower at dry sites than in corresponding damp or wet conditions [34,36,42,48]. Maximum Se volatilization from seleniferous dewatered sediments occurs at 70% of the water-holding capacity (field-moist soil) [42,49], whereas in a heavy clay soil, between 18 and 25% moisture was optimal [22]. In a loam soil, Zieve and Peterson [40] found that 28% was optimal for volatilization, while 16 and 40% moisture regimes gave rise to considerably less volatile Se. There are some indications that fluctuations in the soil water content can stimulate Se volatilization because sequential drying and rewetting cycles promote microbial activity [50].

G. pH

pH affects not only the biologically mediated biomethylation process but also the solubility and availability of Se. The optimum pH for Se biomethylation and subsequent volatilization from a seleniferous

Kesterson sediment (pH 7.7) was 8.0 [39]. The addition of lime to a sandy soil increased the pH from 6 to 7 and increased Se volatilization 1.2-fold [50].

H. Aeration

Many studies show that greater quantities of volatile Se are evolved under aerobic conditions than under anaerobic conditions. Saturation of a Pierce formation soil with nitrogen gas almost completely eliminated Se evolution during a 26-day incubation period [22]. Francis et al. [29] found that glucose- and Na_2SeO_3-amended seleniferous clay soils evolved trace quantities of DMSe under argon. Under air, the same soils volatilized 83- and 64-fold as much DMSe, respectively. Soil, sewage sludge, and seleniferous pond water samples exposed to air produced larger quantities of volatile Se than corresponding samples exposed to N_2 [27,31]. However, substantial evolution of volatile Se from seleniferous soils has occurred under anaerobiosis [45,51].

VIII. FACTORS INHIBITORY TO VOLATILIZATION OF SELENIUM

A. Heavy Metals

There are very few studies on the effects of heavy metals and metalloids on biomethylation of Se. Karlson and Frankenberger [26] found that the addition of 5 mmol/kg of molybdenum (Mo), mercury (Hg), chromium (Cr), and lead (Pb) to seleniferous soils greatly inhibited Se volatilization, whereas arsenic (As), boron (B), and manganese (Mn) had little effect. The addition of cobalt (Co), zinc (Zn), and nickel (Ni) to seleniferous sediments stimulated volatilization of Se. Karlson and Frankenberger [51] postulated that Zn and Ni may inhibit the utilization of a readily available organic source by the nonmethylating microbial population, thus making more C available to the Se-methylating microbiota.

B. Nitrate and Nitrite

The presence of high levels of NO_3^- and NO_2^- inhibits Se methylation and subsequent volatilization. Volatilization of Se in evaporation pond water was inhibited by NO_3^- and NO_2^- ions at concentrations of 0.1 M [44]. The application of KNO_3 to seleniferous soil in combination with galacturonic acid inhibited methylation by 11.8% when added to yield a C/N ratio of 5 [26]. Nitrates have also been found to inhibit anaerobic Se transformations [14].

Table 4 Effect of Selenium Speciation on Volatilization from Soils, Sediments, Sludge, and Water

Matrix	Native Se conc (mg/kg)	Se spike conc (mg/kg)	Species	% Se volatilized	Organic amendment	Incubation time (days)	Ref.
Soils							
Lima loam	0.9	50	(IV)	3.08	−	25	45
	0.9	0.6	(org)	2.0	+	25	
	0.9	50	(org+IV)	2.08	+	25	
	0.9	250	Se^0	0.01	−	17	
	0.9	50	(IV)	0.59	−	17	
	0.9	50	(VI)	0.10	−	17	
	0.9	5.0	Se-cysteine	7.0	+	32	
	0.9	5.0	Se-methionine	28.0	+	32	
	0.9	5.0	TMSe	87.0	+	32	
Sansarc clay	36.0	50	(IV)	2.2	+	25	45
	36.0	0.6	org	0.19	+	25	
	36.0	50	(org+IV)	1.05	+	25	
Seleniferous shale	0.4	12.3	^{75}Se(IV)	5.69^{75}	+	60	38
	0.4	3.0	^{75}Se(IV)	100.00^{75}	+	60	
	0.4+0.6 org	12.8	^{75}Se(IV)	70.31^{75}	+	60	
	0.4+0.6 org	12.8	^{75}Se(IV)	53.33^{75}	+	60	

Los Banos clay loam	0.22	5	$^{75}Se(IV)$	10.4^{75}	−	29	26
	0.22	20	$^{75}Se(IV)$	9.0^{75}	−	29	
	0.22	100	$^{75}Se(IV)$	4.8^{75}	−	29	
	0.22	5	$^{75}Se(VI)$	4.0^{75}	−	29	
	0.22	25	$^{75}Se(VI)$	2.3^{75}	−	29	
	0.22	150	$^{75}Se(VI)$	1.5^{75}	−	29	
	0.22	5	$^{75}Se(VI)$	43.1^{75}	+	29	
	0.22	25	$^{75}Se(VI)$	51.1^{75}	+	29	
	0.22	5	$^{75}Se(IV)$	41.6^{75}	+	29	
	0.22	25	$^{75}Se(IV)$	31.0^{75}	+	29	
Dewatered seleniferous sediments							
Kesterson Pond 4	60.7	219	Se-cysteine	0.2	−	5	42
		219	Se-ethionine	0.1	−	5	
		263	Se-methionine	0.6	−	5	
		132	Se-guanosine	0.2	−	5	
		263	Se-purine	0.1	−	5	
		1317	Se-urea	0.01	−	5	
		132	Se-inosine	0.2	−	5	

IX. VOLATILIZATION OF SELENIUM IN THE FIELD

Microbial volatilization of Se is being considered as a bioremediation technique to remove toxic levels of Se at Kesterson Reservoir. A field investigation was initiated in July 1987 with the goals of identifying the most effective practices for accelerating Se volatilization and to obtain information necessary for the determination of time constraints and other factors affecting this technology [34,49]. The field study was conducted on cattail-enriched sediments in Pond 4 at Kesterson Reservoir, which represented one of the more contaminated areas, containing Se concentrations ranging from 10 to 209 mg/kg (median 39 mg/kg). The treatments consisted of the application of water alone or of water with cattail straw, cattle manure, citrus (orange) peel, and protein sources (casein and gluten). Some plots were also treated with fertilizers such as ammonium nitrate and zinc sulfate.

All plots were sampled for gaseous Se with an inverted box and an alkali peroxide trap (Figure 1). Seasonal variation of gaseous Se emission was evident, with the highest emissions recorded in the late spring and summer months. The volatilization rates were correlated with

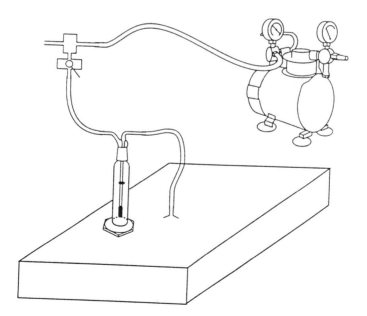

Figure 1 Apparatus used to monitor alkylselenide production in the field.

temperature (Figure 2). Less volatile Se was released in the fall and winter months. The greatest emission of gaseous Se with all treatments occurred at the initiation of the project when the Se inventory was high. As time passed, the Se inventory available for volatilization decreased each season with warmer temperatures. The emission flux of gaseous Se varied with each of the treatments [34]. Irrigation with tillage alone resulted in an average volatile Se emission of 16 µg Se/(m^2·h) over a 2-year field study. Citrus peel + N + Zn provided the most stimulatory soil treatment for Se volatilization. The highest emission rate recorded with this treatment was 808 µg Se/(m^2·h), which was approximately 42-fold higher than the background level. The application of casein (a milk protein) also promoted Se volatilization with an average emission rate of 50 µg Se/(m^2·h).

Volatilization of Se can be enhanced in the field with an available carbon source, aeration, moisture, and high temperatures. Rototilling promotes volatilization as long as the soil remains moist. Frequent tillage is needed to support the aerobic methylating organisms, to enhance soil porosity (this facilitates the diffusion of the alkylselenide gas), and to break any crust that may form as a result of sprinkler irrigation. Irrigation with wetting and drying cycles releases organic-bound Se to the methylating organisms. Water should be applied only to moisten the upper few

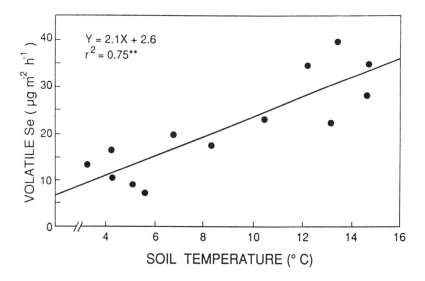

Figure 2 Linear regression analysis of soil temperature on alkylselenide production in the field.

inches of the soil; otherwise, water-soluble Se may migrate out of the surface layer, thus making it unavailable for volatilization.

X. ATMOSPHERIC DISSIPATION OF SELENIUM

Submicrometer particulate Se is the dominant phase in the atmosphere, with the remaining 25% being in the vapor phase [52]. The global distribution of Se is relatively uniform, with concentrations ranging from 5 ng/m^3 in urban areas to 0.05–0.1 ng/m^3 in remote marine and continental areas [52]. Little is known about the speciation of Se in the atmosphere because of problems associated with sampling, fractionation, and measuring Se concentrations on the border of detectability. Jiang et al. [53] identified three different organoselenium species (DMSe, DMDSe, and dimethylselenone) in the atmospheric vapor phase within the immediate vicinity of a sewage treatment plant, a coal-fired plant, a smelter, and some lakes. DMSe has also been detected in the vicinity of sewage digestion tanks [31], moist seleniferous soils [34,36,42], and seleniferous pond waters [27].

It has been suggested that a substantial portion of particulate Se in remote marine regions may result from the gas-to-particle conversion of biologically produced Se gaseous species [52]. The first attempt to determine the fate of a volatile Se compound, DMSe, in the atmosphere was by Atkinson et al. [33], who found that DMSe reacted with OH and NO$_3$ radicals in O$_3$. The rate constants at room temperature and atmospheric pressure with respect to the gas-phase reaction with OH and NO$_3$ radicals and O$_3$ and DMSe lifetimes are reported in Table 5. The rate constants are also shown for dimethylsulfide (DMS) for comparison. No evidence of photolysis of DMSe was observed. A comparison between the two alkylated compounds (DMS and DMSe) in terms of their fate in the atmosphere indicates that DMSe is slightly more reactive than DMS.

XI. CONCLUSIONS

Over the past five years, a large body of information has been collected that demonstrates the range of environmental factors influencing Se volatilization rates. The challenge for future efforts in this area will be to control and enhance these processes to develop cost-effective remediation schemes. Future studies on aquatic Se volatilization will focus on the development of a seleniferous water treatment process that can be used on-site in treatment of incoming drainage water.

Table 5 Rate Constants for the Gas-Phase Reaction of DMSe and DMS with OH and NO$_3$ Radicals and with O$_3$

Reactant species	Rate constant [cm^3/(molecule · s)]		(CH$_3$)$_2$Se lifetime
	(CH$_3$)$_2$Se	(CH$_3$)$_2$S	
OH	$(6.78 \pm 1.7) \times 10^{-11}$	6.3×10^{-12}	2.7 h[a]
NO$_3$	1.4×10^{-11} [b]	1.0×10^{-12}	5 min[c]
O$_3$	$(6.8 \pm 0.72) \times 10^{-17}$	$>1 \times 10^{-18}$	5.8 h[d]

[a] At a 12-h average daytime OH radical concentration of 1.5×10^6 molecules/cm^3.
[b] Extrapolation to zero NO$_3$ concentration.
[c] At a 12-h average nighttime NO$_3$ radical concentration of 2.4×10^8 molecules/cm^3 (10 parts per trillion).
[d] At an average O$_3$ concentration of 7×10^{11} molecules/cm^3 (30 ppt).
Source: Ref. 33.

REFERENCES

1. M. L. Berrow and A. M. Ure, in *Occurrence and Distribution of Selenium* (M. Ihnat, Ed.), CRC Press, Boca Raton, Fla., 1989, pp. 295–325.
2. R. G. Crystal, in *Organic Selenium Compounds: Their Chemistry and Biology* (D. L. Klayman and W. H. H. Günther, Eds.), Wiley-Interscience, New York, 1973, pp. 13–27.
3. K. T. Williams, H. W. Lakin, and H. G. Byers, *U.S. Dept. Agric. Tech. Bull.* 702:1 (1940).
4. P. J. Clark, R. A. Zingarro, K. J. Irgolic, and A. N. McGinley, *Int. J. Environ. Anal. Chem.* 7:295 (1980).
5. D. Y. Boon, in *Selenium in Agriculture and the Environment* (L. W. Jacobs, Ed.), Spec. Pub. 23, Soil Sci. Soc. Am., New Orleans, La., 1989, pp. 107–131.
6. M. A. Elrashidi, D. C. Adriano, and W. L. Lindsay, in *Selenium in Agriculture and the Environment* (L. W. Jacobs, Ed.), Spec. Pub. 23, Soil Sci. Soc. Am., Madison, Wisc., 1989, pp. 51–63.
7. H. F. Mayland, L. F. James, K. E. Panter, and J. L. Sonderegger, in *Selenium in Agriculture and the Environment* (L. W. Jacobs, Ed.), Spec. Pub. 23, Soil Sci. Soc. Am., Madison, Wisc., 1989, pp. 15–50.
8. K. A. Gruebel, J. A. Davis, and J. O. Leckie, *Soil Sci. Soc. Am. J.* 52:390 (1988).

9. N. J. Barrow and B. R. Whelan, *J. Soil Sci.* 40:29 (1989).
10. J. M. McNeal and L. S. Balistrieri, in *Selenium in Agriculture and the Environment* (L. W. Jacobs, Ed.), Spec. Pub. 23, Soil Sci. Soc. Am., Madison, Wisc., 1989, pp. 1–13.
11. D. E. Ullrey, in *Selenium in Biology and Medicine* (J. E. Spallholz et al., Eds.), AVI, Westport, Conn., 1981, pp. 176–191.
12. G. Gissel-Nielsen and B. Bisbjerg, *Plant Soil* 32:382 (1970).
14. R. S. Oremland, J. T. Hollibaugh, A. S. Maest, T. S. Presser, L. G. Miller, and C. W. Culbertson, *Appl. Environ. Microbiol.* 55:2333 (1989).
15. J. M. Macy and S. Rech, The selenate reductase and nitrate reductase of a selenate- and nitrate-respiring *Pseudomonas* sp. are different enzymes, 90th Annu. Meeting Am. Soc. Microbiol., Anaheim, Calif., 1990, p. 241.
16. B. Bisbjerg, Studies on Selenium in Plants and Soil, Danish Atomic Energy Commission, Risø Rep. 200, Copenhagen, 1972.
17. H. R. Geering, E. E. Cary, L. H. P. Jones, and W. H. Allaway, *Soil Sci. Soc. Am. Proc.* 32:35 (1968).
18. J. G. Lipman and S. A. Waksman, *Science* 57:60 (1923).
19. S. U. Sarathchandra and J. H. Watkinson, *Science* 211:600 (1981).
20. A. E. Torma and F. Habashi, *Can J. Microbiol.* 18:1780 (1972).
21. M. L. Bird, F. Challenger, P. T. Charlton, and J. O. Smith, *Biochem. J.* 43:78 (1948).
22. G. M. Abu-Eirreish, E. I. Whitehead, and O. E. Olson, *Soil Sci.* 106:415 (1968).
23. L. Barkes and R. W. Fleming, *Bull. Environ. Contam. Toxicol.* 12:308 (1974).
24. F. Challenger, *Chem. Rev.* 36:315 (1945).
25. J. W. Doran, in *Advances in Microbial Ecology*, Vol. 6 (K. C. Marshall, Ed.), Plenum, New York, 1982, pp. 1–31.
26. U. Karlson and W. T. Frankenberger, Jr., *Sol Sci. Soc. Am. J.* 52:1640 (1988).
27. E. T. Thompson-Eagle and W. T. Frankenberger, Jr., *J. Environ. Qual.* 19:125 (1990).
28. Y. K. Chau, P. T. S. Wong, B. A. Silverberg, P. L. Luxon, and G. A. Bengert, *Science* 192:1130 (1976).
29. A. J. Francis, J. M. Duxbury, and M. Alexander, *Appl. Microbiol.* 28:248 (1974).
30. T. G. Chasteen, G. M. Silver, J. W. Birks, and R. Fall, *Chromatographia* 30:181 (1990).
31. D. C. Reamer and W. H. Zoller, *Science* 208:500 (1980).
32. A. Shrift, in *Organic Selenium Compounds: Their Chemistry and Biology* (D. L. Klayman and W. H. H. Günther, Eds.), Wiley, New York, 1973, pp. 763–814.
33. R. Atkinson, S. M. Aschmann, D. Hasegawa, E. T. Thompson-Eagle, and W. T. Frankenberger, Jr., *Environ. Sci. Technol.* 24:1326 (1990).
34. W. T. Frankenberger, Jr., *Dissipation of Soil Selenium by Microbial Volatilization at Kesterson Reservoir*, U.S. Dept. of the Interior, Bureau of Reclamation, Contract No. 7-FC-20-05240, 1989.
35. K. P. McConnell and O. W. Portman, *Proc. Soc. Exp. Biol. Med.* 79:230 (1952).
36. W. T. Frankenberger, Jr. and U. Karlson, *Dissipation of Soil Selenium by Microbial Volatilization at Kesterson Reservoir*, U.S. Dept. of the Interior, Bureau of Reclamation, Contract No. 7-FC-20-05240, 1988.

37. K. W. Franke and A. L. Moxon, *J. Pharmacol. Exp. Ther.* 58:454 (1936).
38. T. J. Ganje and E. I. Whitehead, *Proc. S. Dakota Acad. Sci.* 37:85 (1958).
39. U. Karlson and W. T. Frankenberger, Jr., *Soil Sci. Soc. Am. J.* 53:749 (1989).
40. R. Zieve and P. J. Peterson, *Sci. Tot. Environ.* 19:277 (1981).
41. S. J. Calderone, W. T. Frankenberger, Jr., D. R. Parker, and U. Karlson, *Soil Biol. Biochem.* 22:615 (1990).
42. W. T. Frankenberger, Jr. and U. Karlson, *Soil Sci. Soc. Am. J.* 53:1435 (1989).
43. U. Karlson and W. T. Frankenberger, Jr., *Sci. Total Environ.* 92:41 (1990).
44. E. T. Thompson-Eagle and W. T. Frankenberger, Jr., *Water Res.* 25:231 (1991).
45. J. W. Doran and M. Alexander, *Soil Sci. Soc. Am. J.* 41:70 (1977).
46. E. T. Thompson-Eagle, W. T. Frankenberger, Jr., and U. Karlson, *Appl. Environ. Microbiol.* 55:1406 (1989).
47. E. T. Thompson-Eagle and W. T. Frankenberger, Jr., *Environ. Toxicol. Chem.* 9:1453 (1990).
48. W. T. Frankenberger, Jr. and U. Karlson, Land treatment to detoxify soil of selenium, U.S. Patent 4,861,482 (1989).
49. W. T. Frankenberger, Jr., Dissipation of Soil Selenium by Microbial Volatilization at Kesterson Reservoir, U.S. Dept. of the Interior, Bureau of Reclamation, Contract No. 7-FC-20-05240, 1989.
49. W. T. Frankenberger, Jr., U. Karlson, and K. E. Longley, Microbial Volatilization of Selenium from Sediments of Agricultural Evaporation Ponds, State Water Resources Control Board, Interagency Agreement No. 7-125-250-1, 1990.
50. A. A. Hamdy and G. Gissel-Nielsen, *Z. Pflanzenernaehr. Bodenk.* 6:671 (1976).
51. U. Karlson and W. T. Frankenberger, Jr., in *Metal Ions in Biological Systems*, Vol. 29 (H. Sigel and A. Sigel, Eds.), Marcel Dekker, New York, 1993, pp. 185–227.
52. B. W. Mosher and R. A. Duce, in *Occurrence and Distribution of Selenium* (M. Ihnat, Ed.), CRC Press, Boca Raton, Fla., 1989, pp. 295–325.
53. S. Jiang, H. Robberecht, and F. Adams, *Atmos. Environ.* 17:111 (1983).
54. J. W. Doran and M. Alexander, *Abstr. Annu. Meeting Am. Soc. Microbiol.* N22:188 (1975).
55. F. Challenger and H. E. North, *J. Chem. Soc.* 1934:68 (1934).
56. F. Challenger and P. T. Charlton, *J. Chem. Soc.* 1947:424 (1947).
57. F. Challenger, *Adv. Enzymol.* 12:429 (1951).
58. D. P. Cox and M. Alexander, *Microb. Ecol.* 1:136 (1974).
59. R. W. Fleming and M. Alexander, *Appl. Microbiol.* 24:424 (1972).
60. O. Weres, A. Jaouni, and L. Tsao, *Appl. Geochem.* 4:543 (1989).

16
Biogeochemical Transformations of Selenium in Anoxic Environments

Ronald S. Oremland

U.S. Geological Survey
Menlo Park, California

I. INTRODUCTION

Selenium is a micronutrient involved in diverse biochemical reactions [1] and has long been known to be required in the diet of animals and for the growth of plants. Indeed, a deficiency of dietary selenium in cattle results in deleterious effects, including anemia and death [2,3]. Selenium is required for the growth of several species of phytoplankton [4–6] and bacteria [7,8]. In contrast with its widespread use as a cofactor in biochemical systems, when absorbed at high concentrations selenium poses a threat to the health of animals [9,10], and selenium oxyanions (e.g., selenate and selenite) inhibit the growth of a variety of microorganisms [11–15].

Anthropogenic sources of selenium to the environment are diverse and include fly ash from fossil fuel combustion, petroleum refining, mine drainage, and domestic household sources such as dandruff shampoo. Ironically, it is selenium that is already abundant in the environment in the form of naturally occurring seleniferous salts that have caused the most problems. Thus, the leaching of selenium oxyanions caused by agricultural activity on irrigated seleniferous soils, such as those of the western

---

This chapter is dedicated to the memory of my colleague, Dr. Ivan Barnes.

San Joaquin Valley of California, have commanded the most attention and have become the focus of debate in the scientific community, among water management professionals, in the agricultural industry, and at public forums. Consequently, attention was drawn to the fact that although much was known concerning the toxicity of selenium to individual species, there was little information available with regard to the geochemical properties of selenium, its biologically mediated geochemical reactions (biogeochemistry), or the bioavailability of its various chemical forms. For the past decade, therefore, research has been focused on the transport of selenium in the environment and on how these properties can be exploited to devise treatment technologies and enlightened management practices.

Microorganisms are known to have biochemical interactions with selenium that can affect its chemical speciation and complexation [16] and hence may be of significance in affecting its mobility in nature. Indeed, a biogeochemical redox selenium "cycle" analogous that of sulfur or nitrogen was first proposed three decades ago [17]. Because many of the environments that receive seleniferous waste can be characterized as anoxic—for example, subsurface saturated soils or organic-rich marsh sediment—biochemically mediated transformations of selenium carried out by anaerobic bacteria have been a logical area of investigation. This chapter summarizes the work conducted in this area, taking special note of the efforts made over the past decade.

II. ASSIMILATORY REDUCTION AND TOXICITY OF SELENIUM OXYANIONS

A discussion of the uptake and toxicity of selenium oxyanions is relevant to understanding its chemical speciation and biogeochemical cycling in nature [18]. Selenium and sulfur are neighbors in the periodic table as Group VIA elements and hence share a great similarity of chemical and biochemical properties. The toxicity of selenium oxyanions appears related to their mimicry of sulfur oxyanions, and a number of investigations have documented this aspect of the sulfur–selenium antagonism [12,13,19,20]. In many microbes, selenate is transported into the cells by sulfate permeases, while in some species, selenite transport is distinct from the one for selenate/sulfate or sulfite [21–23]. Selenate and selenite, upon entering cells, undergo assimilatory reduction to the level of selenide, at which point they may be incorporated into protein [24,25] or released as alkylselenides (see Section III). Hence, reduced organosulfur compounds, such as methionine, can also alleviate selenium toxicity [11,26].

The effects of various sulfur compounds in relieving toxicity of selenate in an aerobic marine bacterium are shown in Figure 1. In a defined mineral medium, growth was dependent upon the provision of sulfur sources, either organic or inorganic (Figure 1A). With addition of 100 μM selenate, a lag was evident, as was inhibition (Figure 1B). However, inhibition was relieved, in order of increasing effectiveness, by sulfite, sulfate, and cysteine. At 300 μM selenate, growth occurred only in the presence of cysteine, suggesting that cysteine represses assimilatory sulfate reductase (or permease) and hence the toxicity of selenate. Selenite at 300 μM inhibited growth in mineral medium, regardless of what sulfur compounds were present (not shown), suggesting that uptake of selenite was not controlled by a sulfate/sulfite transport system. However, when the isolate was grown in a rich peptone–trypticase broth, neither selenate nor selenite exhibited any inhibitory influence upon growth, even when present at 10-fold higher (3 mM) concentrations (data not shown). Similarly, no inhibition was reported for 100 μM selenate or selenite when *Pseudomonas marina* was grown in a rich broth [27], and neither *Butyrivibrio fibrisolvens* nor *Bacteroides ruminicola* transported selenate or sulfate when grown in a trypticase broth [28]. Hence, the composition of the medium is a critical factor when conducting uptake and inhibition studies of this type since rich undefined broths will likely repress the permeases and assimilatory reductases involved in the metabolism of both sulfur and selenium oxyanions.

One common observation of these types of studies has been the precipitation of red elemental selenium (Se^0) when cells were exposed to selenite (Figure 2). Precipitation from selenite (but not selenate) has been observed in several obligate anaerobic bacteria [22,28], aerobic bacteria [16,27], fungi [29–31], and algae [32]. Hence, the ability to precipitate extracellular Se^0 from selenite is a common feature of many diverse microorganisms. The reasons for this phenomenon are unclear, although McCready et al. [15] postulated that it is a detoxification mechanism. However, considering the fact that it also occurs under culture conditions when selenite is not toxic (i.e., in rich broth-type media), this explanation seems unlikely. Selenite is readily reduced to the elemental state by chemical reductants such as sulfide or hydroxylamine [33] or biochemically by systems such as glutathione reductase [34,35]. Hence, this precipitation reaction, which has been associated with bacterial dissimilatory selenate reduction (see Sections IV and V), has great environmental significance.

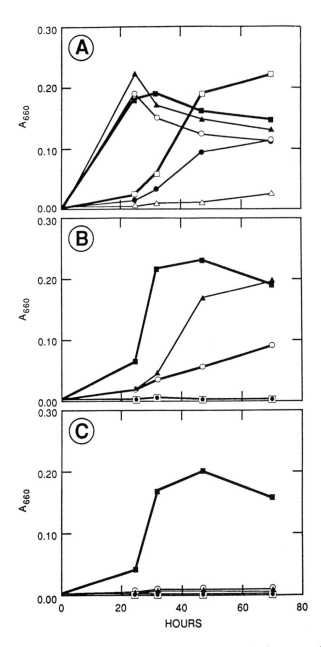

Figure 1 Effect of some sulfur sources and selenate on the growth of a marine bacterium in a glucose–mineral salts medium (A) without selenate, (B) with 100 μM selenate, and (C) with 300 μM selenate. (▲) 100 μM sulfate; (○) 100 μM sulfite; (□) 100 μM mercaptoethanesulfonate; (■) 100 μM cysteine; (●) 200 μM methionine (Δ) without added sulfur. (From J. T. Hollibaugh and R. S. Oremland, unpublished data.)

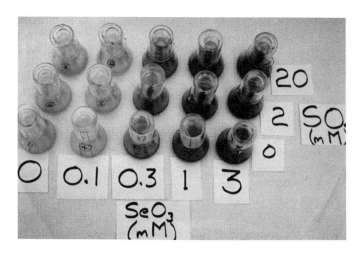

Figure 2 Precipitation of Se^0 by an aerobic marine bacterium grown in a rich broth across a gradient of selenite and sulfate concentrations. (From J. T. Hollibaugh and R. S. Oremland, unpublished observations.) (See color plate.)

III. METHYLATION AND DEMETHYLATION REACTIONS

A number of microorganisms, including both bacteria and fungi, are capable of forming volatile, alkylated selenium compounds like dimethylselenide (DMSe) and dimethyl diselenide (DMDSe) from inorganic forms of selenium [36–39]. This transformation first requires assimilatory reduction of extracellular selenate, selenite, or elemental selenium to the levels of an organoselenide compound. The biochemistry of this methylation reaction has not been elucidated, and the precursor molecules have as yet to be identified. Proposed precursor compounds include selenomethionine [40] or perhaps dimethylselenopropionate (DMSeP) (see below), compounds that are analogues of methionine and dimethylsulfoniopropionate (DMSP), which give rise to dimethylsulfide upon catabolism. Selenium biomethylation is of interest because it represents a potential mechanism for physical loss from selenium-contaminated environments, and it is believed that DMSe is less toxic than dissolved selenium oxyanions.

Biomethylation products (primarily DMSe) have been detected after amendment of soils or sewage sludge with inorganic selenium compounds [41,42]. In general, formation of DMSe in such systems is more extensive under aerobic conditions as opposed to anaerobic ones. A number of investigations have been conducted to determine mechanisms for optimizing selenium loss from soils by amendment with organic substrates or inorganic nutrients [43,44]. In wastewater evaporation ponds this loss appears to be primarily a bacterial process [45]. Dimethylselenide was detected at high concentrations (up to ca. 0.9 µM) in aerobic pond water from the Kesterson National Wildlife Refuge [40]. An outward flux to the atmosphere of 6.6 µmol/(m^2·day) of DMSe was calculated, which represented a significant loss from the total dissolved selenium pool. However, DMSe was not detected in either the aerobic or anoxic water column of Saanich Inlet [46]. The presence of hydrogen selenide (H_2Se) in the anoxic waters may have been caused by demethylation of DMSe (see below).

Any discussion of selenium biomethylation must be tempered by a consideration of concomitant demethylation reactions. Dimethylselenide is a molecular analogue of dimethylsulfide (DMS), which is widespread in nature [47,48] as a volatile degradation product DMSP, an internal compatible solute involved in cellular osmoregulation [49]. Dimethylsulfide is degraded by both aerobic [50,51] and anaerobic bacteria. In the case of anaerobes, both sulfate reducers and methanogens attack the compounds [52], as do purple sulfur bacteria [53], and it can serve as a substrate to support the growth of obligate methyltrophic methanogens [52,54]. The degradation products of methanogenic attack of DMS include CH_4, CO_2, and H_2S [52], while aerobic attack yields CO_2 and sulfate, and phototrophic attack, dimethylsulfoxide. Experiments with [^{14}C]DMSe using anaerobic sediments indicated that this analogue is also metabolized by the same organisms and pathways as for DMS, yielding $^{14}CH_4$, $^{14}CO_2$, and H_2Se [55]. Addition of DMS to anaerobic slurries completely inhibited metabolism of [^{14}C]DMSe (Figure 3). These results represented a conceptual departure from previous models based upon cell-free preparations of methanogens, which indicated a capacity for methylation of selenium, arsenic, and tellurium [56]. The demethylation of both [^{14}C]DMSe and [^{14}C]DMS was studied in parallel with sediments from Kesterson National Wildlife Refuge. Anaerobic degradation of the DMSe occurred immediately, whereas DMS exhibited a short (ca. 2 h) lag (Figure 4). These results indicated that a significant sink for DMSe exists in aquatic systems exposed to high levels of selenium. The absolute magnitude of this sink is not well known; however, it is probably significant, especially if aerobic bacteria are involved as they are in DMS degradation [57].

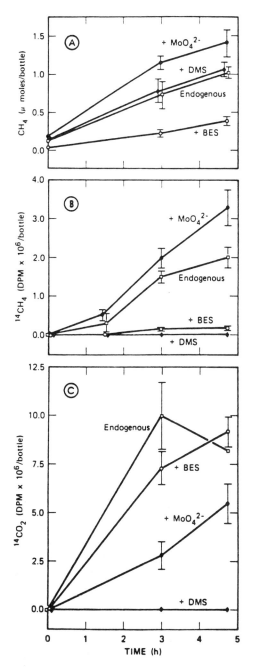

Figure 3 Production of (A) CH_4, (B) $^{14}CH_4$, and (C) $^{14}CO_2$ in anaerobic San Francisco Bay sediment slurries incubated with [^{14}C]DMSe. (From Ref. 54, reproduced with permission of *Applied & Environmental Microbiology*.)

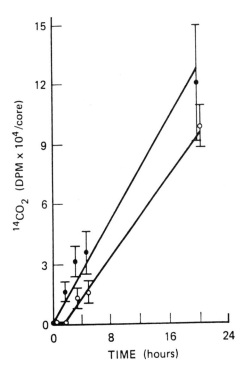

Figure 4 Formation of $^{14}CO_2$ from (●) [^{14}C]DMSe or (○) [^{14}C]DMS in sediments from the Kesterson National Wildlife Refuge. (From Ref. 54, with permission of *Applied & Environmental Microbiology*.)

One important aspect of this anaerobic demethylation reaction is the formation of toxic and reactive H_2Se from the less toxic DMSe [58]. Speciation studies are not usually conducted in assessments of volatile selenium flux [44] and hence cannot determine whether H_2Se or DMSe is the major effluxive molecule. Although H_2Se, like H_2S, undergoes rapid chemical (and possibly bacterial) oxidation under oxic conditions, it can exist for finite periods in an aerobic environment. Thus, it is possible that a major component of soil selenium volatilization efforts could be a short-lived but toxic gas. Recently, anaerobic demethylation of monomethylmercury has been shown to be carried out by methanogens and sulfate reducers [59], thereby extending the significance of this reaction to another toxic element. Because this demethylation produced CO_2 in addition to CH_4, it proceeded by oxidative pathways used in substrate metabolism rather

Chapter 6

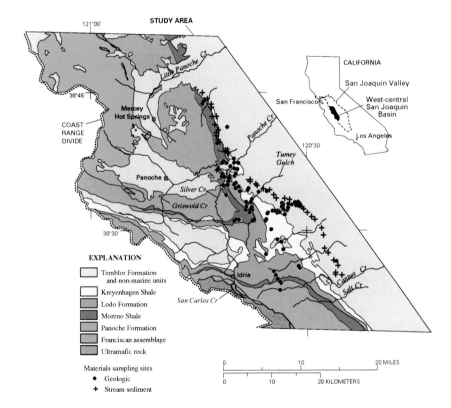

Figure 1 Surficial geology and areal distribution of sampling sites for geologic materials in the study area. (Compiled from Refs. 9 and 16–19 by the Geographic Information Service of the U.S. Geological Survey.)

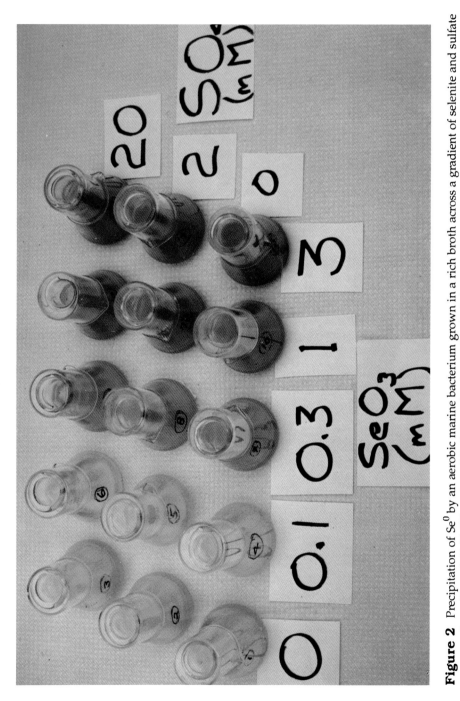

Figure 2 Precipitation of Se^0 by an aerobic marine bacterium grown in a rich broth across a gradient of selenite and sulfate concentrations. (From J. T. Hollibaugh and R. S. Oremland, unpublished observations.)

Chapter 17

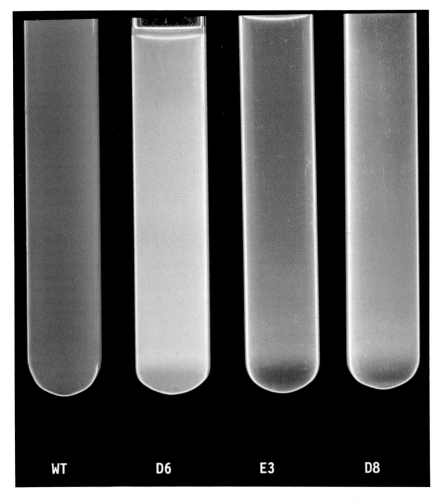

Figure 6 Growth of wild-type (WT), NIR⁻ mutants (D6, E3, D8) in minimal medium plus nitrate (20 mM), selenate (20 mM), and acetate (20 mM). Incubation time: 48 h.

than by the previously described lyases. Aerobic demethylation of DMSe will yield selenate, thereby retaining selenium in the system. The reactions involved in methylation and demethylation are summarized in Figure 5.

IV. DISSIMILATORY REDUCTION OF SELENATE

The possibility that microorganisms involved in the sulfur cycle could also achieve reductive transformations of traces of selenium was investigated by Zehr and Oremland [60]. The rationale was that since the sulfate/selenate ratio in selenium-impacted environments was typically about 10,000:1 [61], perhaps some selenate could cycle through the reductive systems operative for sulfate. Sulfate-reducing bacteria are inhibited by oxyanions of group VI, including molybdate, tungstate, chromate, and selenate [62–64]. These anions form unstable analogues of adenosine-5'-phosphosulfate, the product of the first reaction of the sulfate reduction pathway involving ATP-sulfurylase. The reaction requires the "investment" of a molecule of ATP in order to activate the thermodynamically unfavorable reduction of sulfate:

$$SO_4^{2-} + ATP \xrightarrow{\text{ATP sulfurylase}} APS + \text{pyrophosphate}$$

Subsequent reactions, starting with APS reductase, achieve a sequential biochemical reduction of sulfate coupled with the synthesis of more ATP than was initially invested. In the case of group VIB analogues of sulfate (e.g., molybdate, tungstate, and chromate), ATP is not regenerated and cellular ATP drops below critical levels [65]. However, $APSeO_4^{2-}$ is relatively stable compared to the molybdate or tungstate complexes [66] and is less effective at diminishing the ATP pools of sulfate reducers [65]. Hence, the possibility was indicated for reduction of trace concentrations (e.g., picomolar–nanomolar) of selenate to selenide via the pathway for sulfate reduction. This reduction was shown to occur with washed cell suspensions of *Desulfovibrio desulfuricans*, which reduced trace quantities (ca. 70 picomolar) or [^{75}Se]selenate to [^{75}Se]selenide (Figure 6). The reduction was inhibited by 1 mM selenate or tungstate (not shown) and by sulfate concentrations greater than 1 mM (Figure 6). However, although activity could be expressed with manipulated cells, only an insignificant fraction (<0.01%) of [^{75}Se]selenate was reduced to [^{75}Se]selenide by anaerobic sediments (Figure 7). Thus, even when sediments were manipulated to achieve an intensified activity of sulfate reducers (e.g., substrate amendments with H_2 or lactate) and when sulfate concentrations were lowered (<4 mM) to avoid competition with selenate, the reduction to

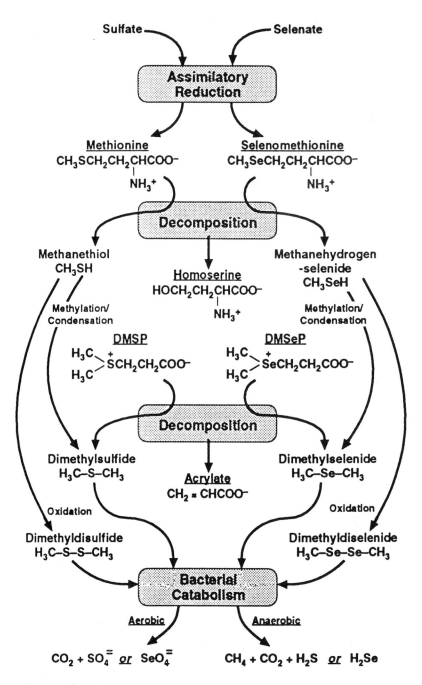

Figure 5 Pathways and possible precursors for the formation and destruction of volatile alkylated sulfur and selenium compounds.

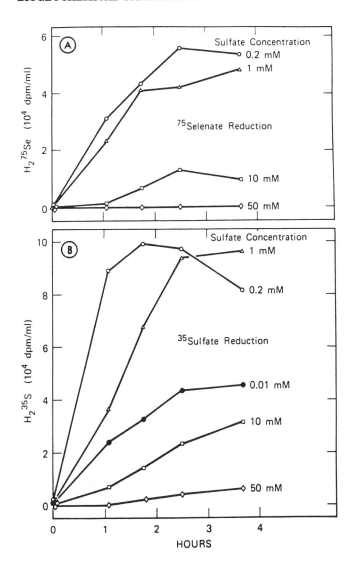

Figure 6 Reduction of (A) [^{75}Se]selenate and (B) [^{35}S]sulfate by cell suspensions of *D. desulfuricans* incubated with different quantities of sulfate. (From Ref. 60, with permission of *Applied & Environmental Microbiology*.)

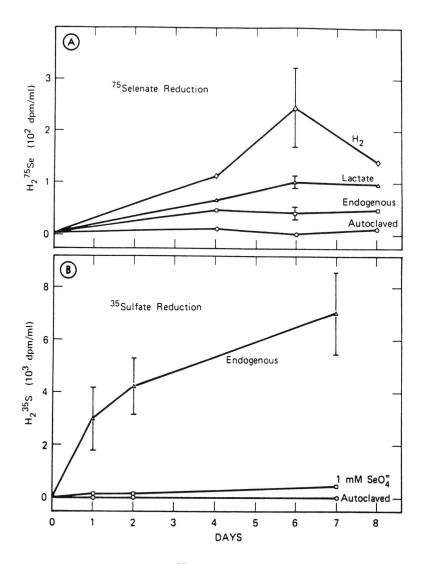

Figure 7 Reduction of (A) [^{75}Se]selenate and (B) [^{35}S]sulfate in anaerobic sediment slurries incubated with various amendments. (From Ref. 60, with permission of *Applied & Environmental Microbiology*.)

selenide was still trivial. The reason for the discrepancy between the cell suspension and the sediment experiments was the sequestration of the selenate by its reduction to Se^0, which was achieved by a previously undescribed bacterial respiration [59].

The effects of this novel dissimilatory reduction could be observed in sediment porewater profiles of selenate and sulfate taken from an agricultural wastewater evaporation pond (Figure 8). Whereas sulfate was consumed at depth by bacterial reduction, selenium oxyanions were consumed by a process occurring near the sediment surface and in the presence of extremely high concentrations (ca. 320 mM) of sulfate. In the laboratory, anaerobic sediment slurries consumed millimolar quantities of selenate, and this was enhanced by provision of electron donors such as acetate, hydrogen, or lactate. Alternative competing electron acceptors, for

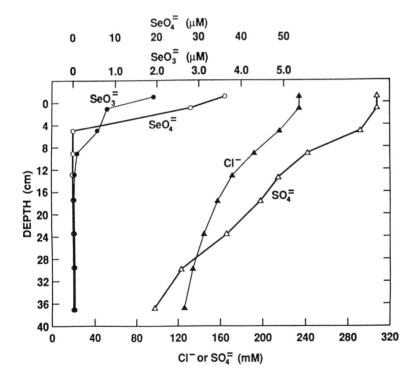

Figure 8 Interstitial porewater profiles of selenate, selenite, sulfate, and chloride taken from an agricultural evaporation pond in the San Joaquin Valley. (From Ref. 61, with permission of *Applied & Environmental Microbiology*.)

example, nitrate or MnO_2, were inhibitory, but FeOOH and sulfate had no effect (Figure 9). The reduction proceeded with selenite as a transient intermediate, followed by the accumulation of a red precipitate. No volatile gases (e.g., DMSe) or selenide precipitates (e.g., FeSe) were detected. The product of this reduction was elemental selenium (Se^0). The reduction was dependent upon direct bacterial activity since autoclaved sediments, even when supplemented with sulfide and held under a hydrogen atmosphere, did not achieve a chemical reduction of selenate. Significantly, the bacterial reaction was inhibited by tungstate ions but not by molybdate. Selenate reduction was shown to be caused by bacterial respiration, and an anaerobic bacterium (strain SES-1) was isolated from San Francisco Bay sediment that was capable of growth in mineral medium using acetate as an electron donor with selenate as the electron acceptor (see Section V).

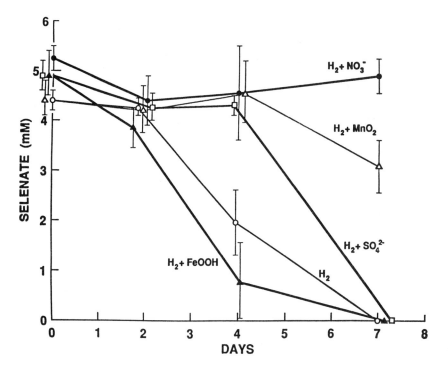

Figure 9 Reductive removal of selenate from anaerobic sediment slurries incubated under hydrogen with various electron acceptors. (From Ref. 61, with permission of *Applied & Environmental Microbiology*.)

The significance of dissimilatory selenate reduction (DSeR) in nature was assessed by the development of an in situ radioassay that measures the precipitation of $^{75}Se^0$ into sediment from injected [^{75}Se]selenate [68]. Adsorption of [^{75}Se]selenite produced during DSeR did not pose difficulties. The assay was applied to the sediments from an agricultural evaporation pond in the San Joaquin Valley and demonstrated that along with denitrification, DSeR was confined to the surficial (top few centimeters) sediments while sulfate reduction occurred at greater depth (Figure 10). The potential for DSeR and denitrification extended down the length of the core. The areal rate of DeSR was calculated to be 0.3 mmol/(m^2·day), which was about a factor of 3 lower than denitrification and about a factor of 30 lower than that for sulfate reduction. Nonetheless, the turnover time for selenate in the water column was 82 days, which was markedly more rapid than that for nitrate (2009 days) or sulfate (49,197 days).

These results indicate that biological immobilization of selenate by dissimilatory reduction to Se^0 is a practical approach for remediation. The in situ radioassay was also used to measure DSeR in the sediments of three drainage systems in western Nevada, systems with much lower levels of dissolved selenium oxyanions (13–456 nM) than those of the evaporation pond (ca. 40 µM). In general, activity declined with depth, and areal rates of DSeR were 14, 38, and 155 µmol/(m^2·day). The reduction of selenate to Se^0, presumably due to DSeR, was also reported in subsurface samples from the Panoche Fan [69], Kesterson Wildlife Refuge pond sediments [70], alluvial soils of the San Joaquin Valley [71], and salt marsh sediments [72].

The distribution and characteristics of DSeR were investigated in a survey of potential DSeR activity made with 11 different sediment types [73]. Potential DSeR differs from that of in situ DSeR in that, in addition to the radioisotope, unlabeled selenate (ca. 20 µM) is also injected. Hence, an easy comparison can be made of sediments to determine how rapidly DSeR proceeds in each sample. The 11 sediment types were collected from markedly different environments, some of which were Se-contaminated and some of which were pristine. Salinities and pH varied over orders of magnitude. Environments included muds from freshwater alpine glacial lakes, estuaries, alkaline-hypersaline lakes (e.g., Mono Lake with a pH of 9.8 and a salinity of ca. 90 g/L), and dense brines. Only in the case of a sample taken from a roadside salina (salinity = 320 g/L; pH 9.6) was incorporation indistinguishable from autoclaved controls (Figure 11). These results indicated that DSeR activity is widespread in nature, being present in environments that can be considered pristine with respect to contamination by the element. Other findings were the demonstration of

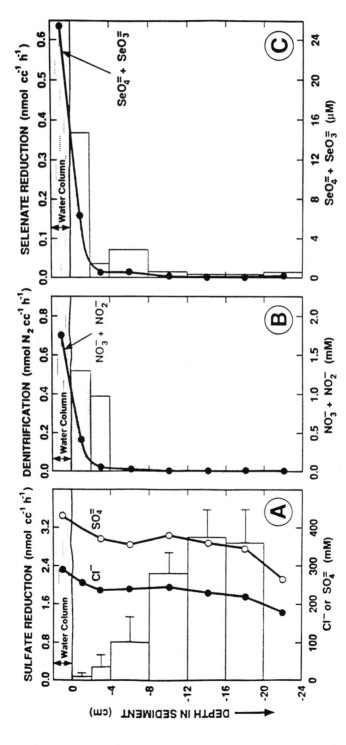

Figure 10 Profiles of in situ activity (bars) of (A) sulfate reduction; (B) denitrification; and (C) DSeR in a core taken from an evaporation pond in the San Joaquin Valley. (From Ref. 68, with permission of *Environmental Science & Technology*.)

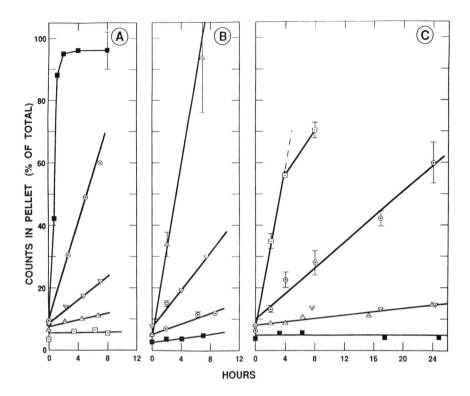

Figure 11 Potential DSeR activity in sediments taken from (A) (■) Massie Slough, (○) Searsville Lake, (▽) San Francisco Bay, (▲) Leslie Salt Pond, (□) Massie Slough (autoclaved control); (B) (▲) Big Soda Lake, littoral; (▽) Hunter Drain, (○) Big Soda Lake, pelagic, (■) Mono Lake; (C) (□) Lead Lake, (○) June Lake, (▲) Roadside Salina, experimental, (▽) Roadside Salina, autoclaved, and (■) June Lake, autoclaved control. (From Ref. 73, with permission of *Applied & Environmental Microbiology*.)

Michaelis-Menten kinetics (Figure 12), with apparent K_s values well above the ambient levels of selenate. Hence, these sediments have the capacity to receive additional selenate without saturating the rate of DSeR activity. In addition, there was a significant correlation ($r = 0.81$) between potential DSeR and potential denitrification, indicating possible involvement of nitrate-respiring microorganisms in DSeR. In a comparison of saline versus freshwater drainage sediments, nitrate or nitrite additions caused substantial inhibition of DSeR in both environments, but they differed in their responses to added tungstate, sulfate, and molybdate (Table 1).

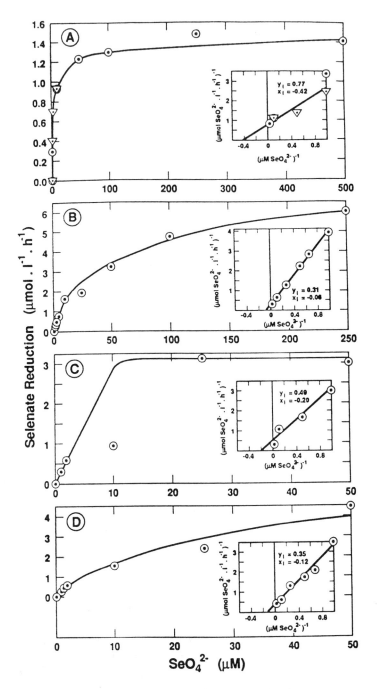

Figure 12 Michaelis-Menten kinetics of DSeR in sediments from (A) Hunter Drain (K_s = 8 µM), (B) Searsville Lake (K_s = 20 µM), (C) Lead Lake (K_s = 16 µM), and (D) Big Soda Lake, littoral zone (K_s = 34 µM). For details see Steinberg and Oremland [73].

Table 1 Percent Inhibition of DSeR in Hunter Drain and Massie Slough Sediments by Nitrate, Nitrite, and Group VI Oxyanions Added at Final Concentrations of 20 µmol/mL of Sediment

Inhibitor	Percent inhibition of selenate reduction[a]	
	Hunter Drain	Massie Slough
None[b]	0	0
Sulfate	7.5	47.0
Nitrate	70.9	96.9
Nitrite	71.9	98.7
Molybdate	52.8	70.0
Tungstate	38.7	84.1

[a]Selenate concentrations were 20 µM.
[b]Uninhibited rates of DSeR for Hunter Drain and Massie Slough were 0.199 and 11.72 µmol/(L·h), respectively.
Source: Adapted from Ref. 73 with permission of *Applied & Environmental Microbiology*.

Nitrate was a potent inhibitor of selenate reduction in anaerobic estuarine sediment slurries (Figure 9) and in anaerobic soil incubations [71].

The commonality of the DSeR reaction in nature poses a philosophical question: Why do bacteria readily express this activity in environments that have no significant levels of selenium? One hypothesis is that the enzyme(s) involved in DSeR have a duality of function and may be capable of reducing a diversity of oxyanions [73]. Preliminary observations with the selenium-respiring isolate SES-3 indicate an ability to grow on several electron acceptors other than selenate (e.g., nitrate, trimethylamine oxide, manganese dioxide, iron oxides), suggesting a niche as a respiratory opportunist.

Results with diverse natural populations demonstrated that removal of nitrogen oxyanions is required before selenate reduction can proceed. This was confirmed with enrichment cultures of selenate respirers (Figure 13), which completely removed nitrate prior to carrying out reduction of selenate [74]. This common observation is of practical importance because nitrate and nitrite are abundant in agricultural runoff, being derived from the application of nitrogenous fertilizers. Hence, bacterial removal processes have stressed the need for a pretreatment step to remove nitrogen oxyanions [75,76]. The basis for the inhibition of DSeR by nitrogen oxyanions lies in their preferential use as electron acceptors for nitrate

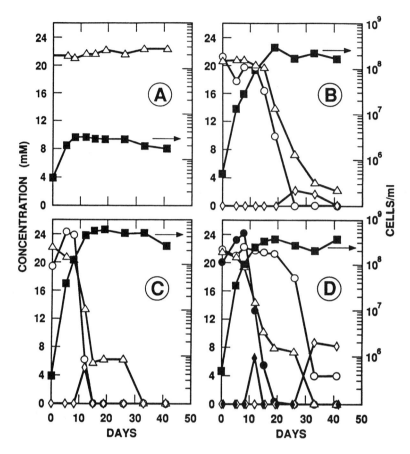

Figure 13 Growth of an anaerobic enrichment culture in mineral medium with and without electron acceptors (A) without electron acceptor, (■) cells, (Δ) acetate; (B) (○) selenate, (◇) selenite, (Δ) acetate, (■) cells; (C) (○) nitrate, (◇) nitrite, (■) cells, (Δ) acetate; (D) (●) nitrate, (◆) nitrite, (○) selenate, (◇) selenite, (■) cells, (Δ) acetate. (From Ref. 74, with permission of *Applied & Environmental Microbiology*.)

respiration. Because isolates of selenate-respiring bacteria are also capable of growth on nitrate in preference to selenate, it appears that this inhibition occurs within the selenate-respiring bacteria themselves rather than being expressed as a competition with a distinct nitrate-respiring population [61,74]. There is a growing body of evidence, much of it circumstantial, that supports the theory that selenate reduction is carried out by one or more of the enzymes of the dissimilatory pathway for nitrate. This evidence can be summarized as follows:

1. DSeR in sediments is inhibited by nitrate [61,71,73].
2. Denitrification and DSeR occur in proximity to each other, and potential denitrification and DSeR have identical depth distributions in sediments [67].
3. In a broad survey of a variety of sediments, potential DSeR correlated significantly with potential denitrification [73].
4. DSeR in some sediments is inhibited by tungstate but not by molybdate, which suggests the involvement of molybdenum-containing enzymes such as nitrate reductase [61,73].
5. Enrichment and pure cultures of selenate respirers will use nitrate in preference to selenate [61,74].

Thus, the possibility exists that the dissimilatory enzymes of nitrate or nitrite reduction to ammonia or of denitrification may also be capable of performing selenate reduction. This must be tempered by the fact that the DSeR activity displayed by two natural populations exhibited significant differences with regard to their susceptibility to various inhibitory oxyanions (Table 1), which may indicate that different metabolic pathways for selenate reduction probably exist in nature. Preliminary investigations with pure cultures of selenate-reducing bacteria suggest that this is indeed the case (see Section V). Further work with novel bacterial isolates is therefore justified and should be encouraged.

V. CHARACTERISTICS OF SELENATE-REDUCING BACTERIA

As yet, there is only a very limited amount of published material concerning the microbiology and biochemistry of selenate-respiring (or reducing) bacteria, primarily because this work is currently under active study in only a few laboratories. The precipitation of Se^0 from selenate was reported in cultures recovered from the Kesterson National Wildlife Refuge [77], although the mechanism(s) for this were not pursued. Oremland et al. [61] first reported the isolation of an anaerobic gram-negative coccus (strain SES-1) that grew in mineral medium with acetate as the electron donor and selenate as the electron acceptor. The extent of acetate oxidation was dependent upon the availability of selenate (Figure 14), and a stoichiometric balance was achieved between selenate reduced, Se^0 recovered, and acetate oxidized. The coccus carried out dissimilatory reduction according to the reaction

$$4\ CH_3COO^- + 3\ SeO_4^{2-} \rightarrow 3\ Se^0 + 8\ CO_2 + 4\ H_2O + 4\ H^+$$

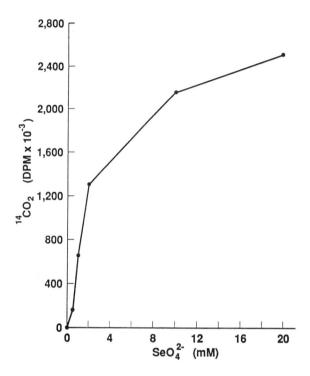

Figure 14 Oxidation of [^{14}C]acetate at different concentrations of selenate by the estuarine selenate respiring bacterium SES-1. (From Ref. 61, with permission of *Applied & Environmental Microbiology*.)

This respiratory mode of growth is highly exergonic (G^o = −326 kJ/mol acetate). Growth was also achieved with nitrate or trimethylamine oxide (TMAO) as electron acceptor, but not oxygen. Cell population densities peaked at about 9×10^8 cells/mL, and growth was more rapid with nitrate or TMAO than with selenate. Growth on nitrate appeared to be by denitrification since cultures produced 25-fold as much N_2O with acetylene as without it, which suggested the presence of N_2O reductase. Strain SES-1 was subsequently lost. Macy et al. [78] reported the isolation of a pseudomonad that grew by oxidation of acetate with reduction of selenate to selenite. Elemental selenium was precipitated by the reduction of selenite carried out by a second organism, a gram-positive rod. However, this is a commonly reported phenomenon carried out by a variety of microorganisms (see Section II). From a geochemical perspective, it is not particularly important whether or not more than one organism is involved in the sequential reduction of selenate. The thermodynamics of selenate reduction to selenite is highly favorable [61,78] and contrasts with the endergonic reaction associated with the reduction of sulfate to sulfite [74]. Rech and Macy [79] reported distinct terminal reductases for nitrate and selenate respiration for their isolate, which they no longer termed a pseudomonad but named *Thauera selenatis*.

Recently, Steinberg et al. [74] screened five common laboratory cultures of pseudomonads (e.g., *Pseudomonas stutzeri*) and *Halobacterium denitrificans* for the ability to grow anaerobically using selenate as an electron acceptor. None of the cultures grew on or achieved chemical reduction of selenate, although they all grew with nitrate, even in the presence of selenate. However, Lortie et al. [80] reported isolation of a wild-type strain of *P. stutzeri* that was capable of vigorous reduction of selenate to Se^0 during aerobic growth. Reduction was inhibited by chromate and tungstate but not by nitrate, nitrite, or sulfate. Similar results with an isolate identified as *P. stutzeri* were reported in a meeting abstract by Barnes et al. [81], although culture collection strains of the organism did not display this ability [J. M. Barnes, personal communication]. All of the above observations suggest that selenate reduction in pseudomonads is associated only with novel isolates that have been taken from Se-contaminated environments and grown with selenate in the medium. Hence, stock cultures that have never been subjected to this selective pressure do not display the ability to reduce selenate. The precise phylogenic grouping of these isolates should be achieved with 16s rRNA base sequencing before unequivocally identifying them as *P. stutzeri*. The reason for selenate reduction by these isolates was not apparent, because they were incubated aerobically and did not use selenate for respiration. Tomei et al. [82] reported that an adapted selenium-resistant strain of *Wolinella succinogenes* could reduce selenate (and selenite) to Se^0 during the stationary phase after anaerobic growth, but it could not grow using selenium oxyanions for respiration. It appears that the above reports of selenate reduction were associated with detoxification mechanisms rather than dissimilatory reduction.

An acetate-utilizing enrichment culture was isolated from a freshwater site in Nevada (Massie Slough) that demonstrated anaerobic growth on selenate, or on nitrate, with consumption of acetate linked to the reduction of the oxyanions (Figure 13). The complete reduction of nitrogen oxyanions preceded the reduction of selenate (Figure 13D). Growth was slower on selenate than on nitrate and could not be explained on the basis of reduction potentials (E_o), which were nearly equivalent (ca. 400 mV) for the two oxyanions. The possibility exists that when respiring selenate, the organism must achieve a balance between energy recovery and the toxicity of the reduction products (e.g., selenite).

A gram-negative anaerobic vibrio was isolated from the enrichment culture that grew by oxidation of lactate to acetate with concomitant reduction of selenate to selenite (Figure 15) and elemental selenium (not shown). The pure culture, designated strain SES-3, is also capable of growth on nitrate, TMAO, manganese dioxide, and iron hydroxides, but not with sulfate or oxygen. Growth on nitrate is achieved by dissimilatory reduction to ammonia rather than by denitrification [R. S. Oremland, J. Switzer Blum, L. G. Miller, and C. W. Culbertson, unpublished data]. Since the organism is capable of achieving growth by the reduction of several types of oxyanions or metal oxides, it occupies an "opportunist" type of niche and is similar to the iron- and manganese-reducing GS-15 [83]. However, GS-15 is not capable of selenate reduction [D. Lovley, personal communication]. Washed cell suspensions of selenate-grown SES-3 reduced selenate but not nitrate, and nitrate-grown cells reduced nitrate but not selenate [J. Switzer Blum, C. Culbertson, and R. S. Oremland, unpublished data]. These observations suggest that separate enzymes are involved in selenate and nitrate respiration, which is in general agreement with the work reported for *T. selenatis* [79].

VI. OTHER REDUCTIVE REACTIONS

Reductive reactions by which selenide is produced from elemental selenium or from oxidized species (selenite and selenate) are theoretically possible but have as yet been poorly characterized. Under strongly reducing conditions (ca. −200 mV) and circumneutral pH, selenides like FeSe are thermodynamically stable [84], although the presence of pyrite favors production of Se^0 from FeSe or $FeSe_2$ [85]. Extracts of *Micrococcus lactilyticus* produced H_2Se from Se^0 formed by reduction of selenite [86], and sulfate reducers can form H_2Se from traces of selenate [60]. However, this latter reaction appears to be of minimal importance in nature (Figure 8). Doran and Alexander [87] reported the formation of H_2Se from anaerobic soil amended with Se^0. Addition of thiols or inorganic sulfides to soils with Se^0 may result in its reduction to soluble forms of selenide [88]. Clearly, further investigations on the chemical and bacterial reductive reactions of Se^0 are warranted.

It appears likely that Se^0, although far less reactive than the speciated forms of this element [e.g., Se(VI), Se(IV), Se(-II)], is still bioavailable for uptake, assimilation, and biotransformation by cells. For example, Doran

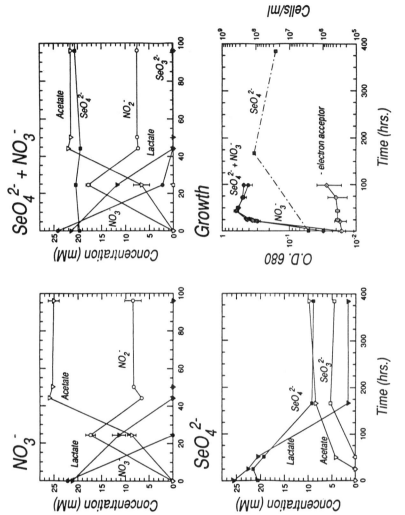

Figure 15 Growth of freshwater isolate SES-3 on lactate and various electron acceptors.

and Alexander [87] reported that *Corynebacterium* sp. could form DMSe from Se^0, and Bacon and Ingledew [89] observed that *Thiobacillus ferrooxidans* can form H_2Se from Se^0 under anaerobic conditions. Recently, Luoma et al. [90] investigated the uptake and incorporation of both organoselenium and Se^0 by the suspension-feeding bivalve *Macoma balthica*. The organoselenium was in the form of ^{75}Se-labeled diatoms, while $^{75}Se^0$ was generated from anaerobic sediments amended with [^{75}Se]selenate. After ingestion of the particles, *M. balthica* retained about 90% and about 17% of the ingested organoselenium and Se^0, respectively, compared with ^{51}Cr-labeled glass beads (Figure 16). The ingested $^{75}Se^0$ was detected in various soft tissues of the animal, indicating translocation from the gastrointestinal tract and incorporation into protein.

The above results have important implications for the management and treatment of seleniferous wastewaters. Even under reducing conditions, Se^0 is clearly reactive and bioavailable. Hence inexpensive and simple treatment schemes such as "wet flux" strategies, in which selenium is allowed to accumulate in the reducing sediments of evaporation ponds

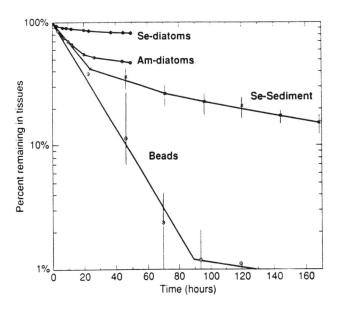

Figure 16 Bioavailability (percent retention) of various forms of selenium ingested by *M. balthica*, including $^{75}Se^0$ associated with sediment. (From Ref. 86, with permission of *Environmental Science & Technology*.)

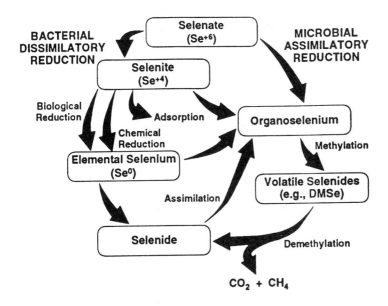

Figure 17 The reductive portion of the biogeochemical cycle of selenium.

or marshes primarily as Se^0, are inherently flawed. Under such strategies, the element is still available to the biota, and selenium contamination of the surface waters will persist due to a continual slow oxidation of the large pool of reduced forms of the element. To overcome this, treatment processes that are segregated from the biota and can immobilize selenium as Se^0 [75] or as adsorbed selenite [76] are desirable, although they are complex, not fully proven, and expensive. The environmental issue of selenium remediation can then be presented in simplified economic terms: Is the cost of such proposed treatment and the water consumed by irrigation of the crops justified by the value of the crops produced? The political ramifications of this problem were recently reported in the popular press [91].

VII. CONCLUSIONS

In the decade since Doran's review on selenium cycling [16] appeared, there have been many developments in the field. These include a reevaluation of the dynamics of selenium cycling in the oceans [18], numerous environmental surveys on selenium chemistry and contamination of biota, environmental assessments of selenium methylation and demethylation [40,55], the discovery of dissimilatory selenate reduction and its biogeochemical significance [61], and the potential for Se^0 to be absorbed by biota [86]. It now appears that with regard to the behavior of selenium in anoxic environments, a reductive cycle is operative (Figure 17), and hence this aspect of the hypothesis first outlined by Shrift [17] 30 years ago has been confirmed. Considerably more work needs to be done on the oxidative part of the cycle, in addition to eliminating significant gaps in knowledge about the reductive side.

ACKNOWLEDGMENTS

I am grateful to B. F. Taylor and D. Lovley for reviewing this manuscript. Our research was funded by the Irrigation Drainage Program of the U.S. Department of the Interior.

REFERENCES

1. T. C. Stadtman, *Annu. Rev. Biochem.* 59:111–127 (1990).
2. J. G. Morris, W. S. Cripe, H. L. Chapman, Jr., D. F. Walker, J. B. Armstrong, J. D. Alexander, Jr., R. Miranda, A. Sanchez, Jr., B. Sanchez, J. R. Blair-West, and D. A. Denton, *Science* 223:491–493 (1984).
3. J. E. Oldfield, *Biol. Trace Element Res.* 20:23–29 (1989).
4. K. Lindstrom, *Hydrobiologia* 101:35–48 (1983).
5. N. M. Price, P. A. Thompson, and P. J. Harrison, *J. Phycol.* 23:1–9 (1987).
6. P. J. Harrison, P. W. Yu, P. A. Thompson, N. M. Price, and D. J. Phillips, *Mar. Ecol. Prog. Ser.* 47:89–96 (1988).
7. J. B. Jones and T. C. Stadtman, *J. Bacteriol* 130:1404–1406 (1977).
8. C. Lindblow-Kull, A. Shrift, and R. L. Gherna, *Appl. Environ. Microbiol.* 44:737–743 (1982).
9. W. O. Robinson, *J. Assoc. Offic. Agric. Chem.* 16:423–442 (1933).
10. K. W. Franke, *J. Nutrit.* 8:597–608 (1934).
11. I. G. Fels and V. H. Cheldelin, *J. Biol. Chem.* 176:819–828 (1948).
12. A. Shrift, *Am. J. Bot.* 41:223–230 (1954).
13. A. Shrift and E. Kelly, *Nature (Lond.)* 195:732–733 (1962).
14. H. D. Kumar, *Plant Cell Physiol.* 5:465–477 (1964).
15. R. G. L. McCready, J. N. Campbell, and J. I. Payne, *Can. M. Microbiol.* 12:703–714 (1966).
16. J. W. Doran, *Adv. Microb. Ecol.* 6:17–32 (1982).
17. A. Shrift, *Nature (Lond.)* 201:1304–1305 (1964).
18. G. A. Cutter and K. W. Bruland, *Limnol. Oceanogr.* 29:1179–1192 (1984).
19. J. Scala and H. H. Williams, *Arch. Biochem. Biophys.* 101:319–324 (1963).
20. A. E. Wheeler, R. A. Zingaro, and K. Irgolic, *J. Exp. Mar. Biol. Ecol.* 57:181–194 (1982).

21. T. A. Brown and A. Shrift, *Can. J. Microbiol.* 28:307–310 (1980).
22. J. F. Hudman and A. R. Glenn, *Arch. Microbiol.* 140:252–256 (1984).
23. R. D. Bryant and E. J. Laishley, *Can. J. Microbiol.* 34:700–703 (1988).
24. J. J. Wrench, *Mar. Biol.* 49:231–236 (1978).
25. J. A. Karle and A. Shrift, *Biol. Trace Element Res.* 11:27–35 (1986).
26. I. G. Fels and V. H. Cheldelin, *Arch. Biochem.* 22:323–324 (1949).
27. A. Foda, J. H. Vandermeulin, and J. J. Wrench, *Can J. Fish. Aquat. Sci.* 40:215–220 (1938).
28. J. F. Hudman and A. R. Glenn, *FEMS Microbiol. Lett.* 27:215–220 (1985).
29. M. Zolokar, *Arch. Biochem.* 44:330–337 (1953).
30. G. Falcone and W. J. Nickerson, *J. Bacteriol.* 85:763–771 (1963).
31. M. O. Moss, F. Badii, and G. Gibbs, *Trans. Br. Mycol. Soc.* 4:578–580 (1987).
32. M. Sielicki and J. C. Burnham, *J. Phycol.* 9:509–514 (1973).
33. K. Rashid and H. R. Krouse, *Can. J. Chem.* 63:3195–3199 (1985).
34. H. E. Ganther, *Biochemistry* 7:2898–2905 (1968).
35. H. E. Ganther, *Biochemistry* 10:4089–4098 (1971).
36. F. Challenger and H. E. North, *J. Chem. Soc.* 1934:68–71.
37. R. W. Fleming and M. Alexander, *Appl. Microbiol.* 24:424–429 (1972).
38. L. Barkes and R. W. Fleming, *Bull. Environ. Contam. Toxicol.* 12:308–311 (1974).
39. J. W. Doran and M. Alexander, *Appl. Environ. Microbiol.* 33:31–37 (1977).
40. T. D. Cooke and K. W. Bruland, *Environ. Sci. Technol.* 21:1214–1219 (1987).
41. D. C. Reamer and W. H. Zoller, *Science* 208:500–502 (1980).
42. A. J. Francis, M. Duxbury, and M. Alexander, *Appl. Microbiol.* 28:248–250 (1974).
43. U. Karlson and W. T. Frankenberger, Jr., *Soil Sci. Soc. Am. J.* 52:1640–1644 (1988).
44. U. Karlson and W. T. Frankenberger, Jr., *Soil Sci. Soc. Am. J.* 53:749–753 (1989).
45. E. T. Thompson-Eagle and W. T. Frankenberger, Jr., *Water Res.* 25:231–240 (1991).
46. G. A. Cutter, *Science* 217:829–831 (1982).
47. J. E. Lovelock, R. J. Maggs, and R. A. Rasmussen, *Nature (Lond.)* 237:452–453 (1972).
48. M. O. Andreae and H. Raemdonck, *Science* 221:744–747 (1983).
49. A. Vairavamurthy, M. O. Andreae, and R. L. Iverson, *Limnol. Oceanogr.* 30:59–70 (1985).
50. T. Kanagawa and D. P. Kelly, *FEMS Microbiol. Lett* 34:13–19 (1986).
51. G. M. Suylen, G. C. Stefess, and J. G. Kuenen, *Arch. Microbiol.* 146:192–198 (1986).
52. R. P. Kiene, R. S. Oremland, A. Catena, L. G. Miller, and D. G. Capone, *Appl. Environ. Microbiol.* 52:1037–1045 (1986).
53. P. T. Visscher and H. van Gemerden, *FEMS Microbiol. Lett.* 81:247–250 (1991).
54. R. S. Oremland, R. P. Kiene, I Mathrani, M. J. Whiticar, and D. R. Boone, *Appl. Environ. Microbiol.* 55:994–1002 (1989).
55. R. S. Oremland and J. P. Zehr, *Appl. Environ. Microbiol.* 52:1031–1036 (1986).
56. B. C. McBride and R. S. Wolfe, *Biochemistry* 10:4312–4317 (1971).
57. R. P. Kiene and T. S. Bates, *Nature (Lond.)* 345:702–705 (1990).
58. L. C. Alderman and J. J. Bergin, *Arch. Environ. Health* 41:354–358 (1986).

59. R. S. Oremland, C. W. Culbertson, and M. R. Winfrey, *Appl. Environ. Microbiol.* 57:130–137 (1991).
60. J. P. Zehr and R. S. Oremland, *Appl. Environ. Microbiol.* 53:1365–1369 (1987).
61. R. S. Oremland, J. T. Hollibaugh, A. S. Maest, T. S. Presser, L. G. Miller, and C. W. Culbertson, *Appl. Environ. Microbiol.* 55:2333–2343 (1989).
62. J. R. Postgate, *Nature (Lond.)* 164:670–671 (1949).
63. J. R. Postgate, *J. Gen. Microbiol.* 6:128–142 (1952).
64. H. D. Peck, Jr., *Proc. Natl. Acad. Sci. U.S.A.* 45:701–708 (1959).
65. B. F. Taylor and R. S. Oremland, *Curr. Microbiol.* 3:101–103 (1979).
66. L. G. Wilson and R. S. Bandurski, *J. Biol. Chem.* 233:975–981 (1958).
67. R. S. Oremland, N. A. Steinberg, A. S. Maest, L. G. Miller, and J. T. Hollibaugh, *Environ. Sci. Technol.* 24:1157–1164 (1990).
68. R. S. Oremland, N. A. Steinberg, T. S. Presser, and L. G. Miller, *Appl. Environ. Microbiol.* 57:615–617 (1991).
69. N. M. Dubrovsky, J. M. Niel, R. Fujii, R. S. Oremland, and J. T. Hollibaugh, U.S. Geol. Surv. Open File Rep. 90-138, Sacramento, Calif. 1990.
70. R. H. B. Long, S. M. Benson, T. K. Kokunaga, and A. Yee, *J. Environ. Qual.* 19:302–311 (1990).
71. G. Sposito, A. Yang, R. H. Neal, and A. Mackzum, *Soil Sci. Soc. Am. J.* 55:1597–1602 (1991).
72. P. H. Masscheleyn, R. D. Delaune, and W. H. Patrick, Jr., *Environ. Sci. Technol.* 24:91–96 (1989).
73. N. A. Steinberg and R. S. Oremland, *Appl. Environ. Microbiol.* 56:3550–3557 (1990).
74. N. A. Steinberg, J. Switzer Blum, L. Hochstein, and R. S. Oremland, *Appl. Environ. Microbiol.* 58:426–428 (1992).
75. R. S. Oremland, Selenate removal from wastewater, U.S. Patent 5,009,786 (1991).
76. M. B. Gerhard, F. B. Green, R. D. Newman, T. J. Lundquist, R. B. Tresan, and W. J. Oswald, *Res. J. Water Pollut. Control Fed.* 63:799–805 (1991).
77. D. T. Maiers, P. L. Wichlacz, D. L. Thompson, and D. F. Bruhn, *Appl. Environ. Microbiol.* 54:2591–2593 (1988).
78. J. M. Macy, T. A. Michel, and D. G. Kirsch, *FEMS Microbiol. Lett.* 61:195–198 (1989).
79. S. A. Rech and J. M. Macy, *J. Bacteriol.* 174:7316–7320 (1992).
80. L. Lortie, W. D. Gould, S. Rajan, R. G. L. McCready, and K.-J. Cheng, *Appl. Environ. Microbiol.* 58:4042–4044 (1992).
81. J. M. Barnes, J. K. Polman, and J. H. McCune, *Abstr. Ann. Meeting Am. Soc. Microbiol.*, 1992, p. 384.
82. F. A. Tomei, L. L. Barton, C. L. Lemanski, and T. G. Zocco, *Can. J. Microbiol.* 38:1328–1333 (1992).
83. D. R. Lovley and E. J. P. Phillips, *Appl. Environ. Microbiol.* 54:1472–1480 (1988).
84. J. H. Howard, *Geochim. Cosmochim. Acta* 41:1665–1678 (1977).
85. P. H. Masscheleyn, R. D. Delaune, and W. H. Patrick, Jr., *J. Environ. Sci. Health* A26:555–573 (1991).
86. C. A. Woolfolk and H. R. Whiteley, *J. Bacteriol.* 84:647–658 (1962).

87. J. W. Doran and M. Alexander, *Soil Sci. Soc. Am. J.* 40:687–690 (1977).
88. O. Weres, A. R. Jaouni, and L. Tsao, *Appl. Geochem.* 4:543–563 (1989).
89. M. Bacon and W. J. Ingledew, *FEMS Microbiol. Lett.* 58:189–194 (1989).
90. S. N. Luoma, C. Johns, N. S. Fischer, N. A. Steinberg, R. S. Oremland, and J. R. Reinfelder, *Environ. Sci. Technol.* 26:485–491 (1992).
91. R. Boyle, *Sports Illustrated*, Mar. 22, 1993, pp. 62–69.

17
Biochemistry of Selenium Metabolism by *Thauera selenatis* gen. nov. sp. nov. and Use of the Organism for Bioremediation of Selenium Oxyanions in San Joaquin Valley Drainage Water

Joan M. Macy

University of California
Davis, California

I. INTRODUCTION

The goal of the work described here was to find an organism that might be used to detoxify the selenium found in saline drainage water collected by subsurface drains in the western side of the San Joaquin Valley, California. The transport of this water via the San Luis Drain to the Kesterson Reservoir from 1978 to 1983 led to the selenium contamination problem discovered there in 1983 [1–3]. The drainage water was found to contain about 300 µg/L selenium-Se (primarily as selenate) and 50 mg/L nitrate-N [3].

It was known that elemental selenium could be removed from water by filtration [4]. Therefore, if an organism could be found that is able to reduce selenate to elemental selenium, uninhibited by the presence of nitrate, such an organism could be used in a biological reactor system for bioremediation of toxic selenium oxyanions in agricultural drainage water.

Such an organism was found, *Thauera selenatis* gen. nov. sp. nov. [5]. This chapter describes the organism's phylogenetic and physiological characteristics, the mechanisms of selenate and selenite reduction to elemental selenium, and the effectiveness of selenium oxyanion bioremediation by this organism in a biological reactor system treating drainage water from the Westlands Water District of the San Joaquin Valley.

II. ISOLATION AND CHARACTERIZATION OF A NEW SELENATE-RESPIRING BACTERIUM

Because the chemistry of selenate is similar to that of sulfate, and because sulfate can act as the terminal electron acceptor for a specialized group of strictly anaerobic bacteria—the sulfate-reducing bacteria [6]—it seemed possible that there were also naturally occurring organisms that could respire anaerobically using selenate as the terminal electron acceptor. Like sulfur, selenium can be found in four different oxidation states: selenate [Se(VI); SeO_4^{2-}], selenite [Se(IV); SeO_3^{2-}], elemental selenium [Se(0); Se^0], and selenide [Se(-II); Se^{2-}] [7]. In comparison to sulfate [Ss(VI)], selenate is theoretically a better electron acceptor (Table 1); it is even slightly better than nitrate, which is a well-known terminal electron acceptor for respiration in the absence of oxygen [6].

To isolate such a selenate-respiring bacterium, anaerobic enrichments inoculated with selenium-contaminated sediment were made that contained selenate and the respiratory substrate acetate [8]. Positive enrichments were observed (i.e., the formation of elemental selenium, seen as a red color) whenever selenium-contaminated material was used as an inoculum, whereas negative enrichments were observed when the inoculum was from noncontaminated areas (e.g., Sacramento River sediment). From one of these enrichments a motile rod was isolated that could grow in an anaerobic defined minimal salts medium with acetate as the

Table 1 Oxidation–Reduction Potentials of Selected Redox Pairs

Redox couple	Eo' (V)
SeO_4^{2-}/SeO_3^{2-}	+0.44
NO_3^-/NO_2^-	+0.42
SeO_3^{2-}/Se^0	+0.21
SO_4^{2-}/H_2S	−0.22
S^0/H_2S	−0.27
SO_4^{2-}/SO_3^{2-}	−0.52
Se^0/H_2Se	−0.73

Sources: Refs. 7 and 13.

electron donor and carbon source and selenate as the terminal electron acceptor. The organism was designated AX [8].

In addition to selenate, nitrate or nitrite can also be used as a terminal electron acceptor in the absence of oxygen; respiration with oxygen also occurs [8–10]. The presence of nitrate does not inhibit selenate respiration, suggesting that the nitrate and selenate respiratory systems of this organism are different [9] (see Section III).

That AX respires using selenate as the terminal electron acceptor and acetate as the electron donor was shown by demonstrating that (a) the respiratory substrate acetate is oxidized to CO_2 during selenate reduction [8] and (b) energy is conserved when selenate is used as the terminal electron acceptor [11].

Acetate oxidation to CO_2 was demonstrated by using resting cell suspensions of the organism grown with selenate plus acetate. Growth, selenate reduction, and selenite formation are shown in Figure 1 [11]. In cell suspensions made from such cultures, uniformly labeled [^{14}C]acetate was oxidized to CO_2 with concomitant reduction of selenate to selenite [8]. The stoichiometry of the reaction is shown in equation (1) ($\Delta G^{0\prime} = -575$ kJ/mol acetate or 144 kJ/mol selenate; calculated from free energies of formation) [12,13]:

$$CH_3COO^- + 4\, SeO_4^{2-} \rightarrow 4\, SeO_3^{2-} + H^+ + 2\, HCO_3^- \qquad (1)$$

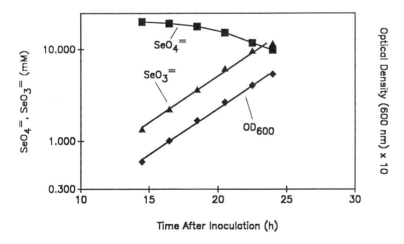

Figure 1 Growth of AX with selenate (20 mM) and acetate (20 mM) in minimal medium [8]. The inoculum was 5 mL of an overnight culture grown in the same medium.

Energy is generated during growth on acetate with selenate as the terminal electron acceptor [equation (1)], as the cell yield (Y_M) was found to be 57 g (dry weight) cells/mol acetate [11]. This equates to a theoretical ATP yield of 5.7 ATP per acetate oxidized [11]. Assuming that 50 kJ of standard free energy is required to synthesize 1 ATP [14] and that energy conservation is 50% efficient, the theoretical number of ATPs that can be made during growth with selenate is 5.8 ATP/acetate [11]. This value compares well with the value estimated from the Y_M value [11].

Most, if not all, of the energy conserved during growth with selenate and acetate (i.e., 5.7 ATP/acetate) must have come from selenate respiration. Oxidation of acetate to CO_2, presumably via the TCA cycle (see Section III.D), could have yielded a maximum of only 1 ATP [14,15], and this only if the organism AX possesses an unusual TCA cycle involving ATP citrate lyase instead of citrate synthase [15]. With a "normal" TCA cycle, no net ATP could be produced.

Growth with nitrate as the terminal electron acceptor also occurs, according to the stoichiometry shown in equation (2); nitrous oxide is the product, not nitrogen gas [5]:

$$CH_3COO^- + H^+ + 2 NO_3^- \rightarrow N_2O + 2 HCO_3^- + H_2O \qquad (2)$$

During growth with nitrate, the nitrate is first reduced to nitrite ($\Delta G^{0'}$ = -547.6 kJ/mol acetate [12,13]) [10]:

$$CH_3COO^- + 4 NO_3^- \rightarrow 4 NO_2^- + H^+ + 2 HCO_3^- \qquad (3)$$

The nitrite is further reduced to nitrous oxide [5,10]:

$$CH_3COO^- + 4 NO_2^- + 3 H^+ \rightarrow 2 N_2O + 2 HCO_3^- + 2 H_2O \qquad (4)$$

Based upon its genotypic and phenotypic characteristics, this selenate-respiring organism (AX) was found to represent a new genus and species [5]. The organism is rod-shaped and has a single polar flagellum (Figure 2). Comparison of the 16s rRNA sequence of AX with other organisms indicated that it has the highest similarity to the β subclass (86.8% similarity) and not the γ subclass (80.2% similarity) of the *Proteobacteria*; the γ subclass contains the authentic genus *Pseudomonas* [5]. The presence of the specific polyamine 2-hydroxyputrescine and the presence of a ubiquinone with eight isoprenoid units in the side chain (Q-8) also placed AX in the β subclass of the *Proteobacteria* [5]. Within the β subclass, AX is related most closely to *Iodobacterium fluvatile*.

But AX cannot be placed in the genus *Iodobacterium* because AX lacks common phenotypic characteristics with this genus and the phylogenetic distance between AX and *Iodobacterium* is not large enough (i.e., less than

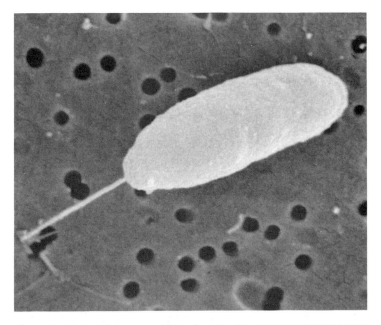
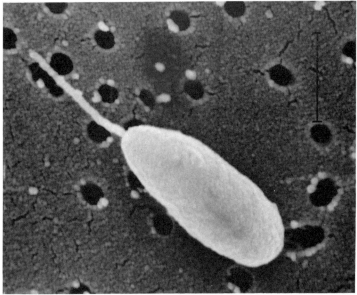

Figure 2 Scanning electron micrographs of *Thauera selenatis* grown anaerobically in minimal medium with nitrate (10 mM) and acetate (10 mM). Bar represents 1 μm.

90% similarity) [5]. Therefore, because of the above and because of its unique mode of anaerobic respiration with selenate (not inhibited by nitrate), AX has been placed in the new genus and species *Thauera selenatis* [5].

Two other pure cultures of selenate-respiring bacteria have been isolated [16,17]. The mechanism of selenate respiration in these two organisms appears to be quite different from that of the organism described above. Unlike *Thauera selenatis*, they both prefer using nitrate as the terminal electron acceptor, and when they are grown with both nitrate and selenate, respiration with selenate is inhibited. The bases and significance of these differences are discussed in the next section.

III. BIOCHEMISTRY OF SELENATE REDUCTION

A. Involvement of Nitrate in Selenate Reduction

Since both selenate and nitrate can be used as terminal electron acceptors by *T. selenatis*, the question arose as to whether a single terminal reductase (e.g., nitrate reductase) catalyzes the reduction of both electron acceptors. Based upon the findings discussed below, it was found that this organism possesses a different terminal reductase for each of these electron acceptors—a selenate reductase (SR) that catalyzes the reduction of selenate to selenite and a nitrate reductase (NR) that catalyzes the reduction of nitrate to nitrite [9]. The findings concerning these terminal reductases are discussed below.

1. The pH optima of the selenate and nitrate reductases are different; the pH optimum for the SR is 6.0, while that of the NR is 7.0 [9]. If the reduction of selenate and nitrate were carried out by the same enzyme, the pH optimum for both reduction reactions would be the same.
2. If a single reductase (e.g., NR) catalyzes the reduction of both, it would be expected that the presence of nitrate would inhibit selenate reduction. In resting cell suspensions of *T. selenatis* grown with selenate, the presence of nitrate does not inhibit selenate reduction (Figure 3) [9]. Nor does the presence of nitrate inhibit the reduction of selenate in resting cell suspensions of cells grown with selenate plus nitrate; instead, both selenate and nitrate are reduced concomitantly (Figure 4) [9].
3. Additionally, if the reduction of both selenate and nitrate were catalyzed by a single reductase, the activity of both reductase reactions would be the same at any given stage of growth, regardless of whether nitrate or selenate were present during growth. This is not the case for *T. selenatis*, as SR activity is at its highest and NR at its lowest when the organism is grown with selenate as the terminal electron acceptor (Table 2).

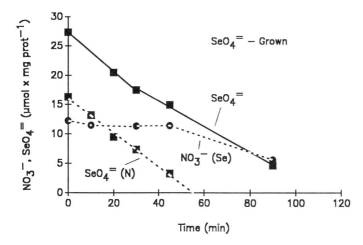

Figure 3 Rates of selenate and nitrate reduction by resting cell suspensions of *T. selenatis* grown with selenate as the terminal electron acceptor. The dashed line indicates that both nitrate and selenate were present during the incubation.

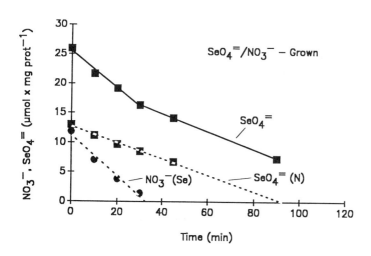

Figure 4 Rates of selenate and nitrate reduction by resting cell suspensions of *T. selenatis* grown with both selenate and nitrate as the terminal electron acceptors. The dashed line indicates that both nitrate and selenate were present during the incubation.

Accordingly, NR activity is at its highest and SR activity at its lowest when nitrate is the electron acceptor present during growth [9].

4. The isolation of mutants lacking NR activity (i.e., unable to grow with nitrate as the terminal electron acceptor) and the finding that these mutants grow normally with selenate and form active SR are further proof that *T. selenatis* synthesizes both a selenate reductase and a nitrate reductase (Table 2) [9].

5. Finally, the locations of the SR and NR activities in the cell are different. NR activity is present in the cytoplasmic membrane, as are most other dissimilatory nitrate reductases [9], whereas the SR activity is located in the periplasmic space of *T. selenatis*, loosely associated with the cytoplasmic membrane [9]. As will be discussed below (see Section III.C), the nitrite reductase (NIR) of *Thauera selenatis* is also located in the periplasmic space, but this enzyme is not responsible for selenate reduction. The NIR does, however, catalyze the reduction of selenite to elemental selenium [10].

In comparison with *T. selenatis*, the other two pure cultures of bacteria, mentioned above, that are able to grow using selenate as the terminal electron acceptor prefer using nitrate as the terminal electron acceptor. In the presence of nitrate, the reduction of selenate is inhibited [16,17]. It seems, therefore, that both of these organisms synthesize a single reductase, NR, which serves as the terminal reductase for respiration with both nitrate and selenate. When both nitrate and selenate are present, only nitrate can be used as the terminal electron acceptor; selenate can serve as an electron acceptor only when nitrate is absent. It would be interesting to compare the SR and NR activities and the rates of nitrate and selenate

Table 2 Activity of SR and NR in Cells of *T. selenatis* Grown with Selenate, Nitrate, or Selenate plus Nitrate as Electron Acceptors[a]

Growth electron acceptor	Nitrate reductase			Selenate reductase		
	Wild type	Mutant A	Mutant B	Wild type	Mutant A	Mutant B
SeO_4^{2-}	0.24 (0.12)	<0.01	<0.01	3.84 (0.33)	7.20 (0.33)	3.66 (0.13)
NO_3^-	3.59 (0.15)	NG^b	NG^b	0.76 (0.05)	NG^b	NG^b
SeO_4^{2-} + NO_3^-	3.77 (0.21)	<0.01	<0.01	1.60 (0.27)	3.04 (0.19)	5.21 (0.53)

[a]Micromoles per minute per milligram of protein. Each number represents an average of at least three analyses, and the number in parentheses is the standard deviation.
[b]NG = organisms cannot grow using nitrate as the terminal electron acceptor.

reduction by these organisms when they are grown with the respective electron acceptor.

B. Characterization of the Selenate Reductase

The SR of *T. selenatis*, released from the periplasmic space by treatment of the cells with lysozyme-EDTA, has been partially purified [Rech and Macy, manuscript in preparation]. Selenate was the preferred electron acceptor for the SR, and neither nitrate, sulfate chlorate nor nitrite could be used as an electron acceptor [Rech and Macy, manuscript in preparation].

The periplasmic cytochrome C_{551} copurifies with the SR [Rech and Macy, manuscript in preparation], suggesting that this cytochrome is involved in electron transport, as discussed in Section III.D.

C. Involvement of Nitrite in Selenite Reduction

As described above, nitrate does not inhibit selenate reduction in *T. selenatis*. In addition, the presence of nitrate in selenate-respiring cultures is necessary if elemental selenium is to be formed from selenate. For example, when *T. selenatis* is grown with selenate as the sole electron acceptor, very little of the selenite formed during respiration is further reduced to elemental selenium. When grown with both selenate and nitrate, however, much of the selenite formed during selenate respiration is further reduced to elemental selenium.

It was observed (Figure 5) that cultures grown with nitrate did not begin reducing nitrite until after all the nitrate had first been reduced [10]. In cultures grown with nitrate plus selenate, these electron acceptors were reduced concomitantly, and selenite reduction began when nitrite reduction commenced [10]. Activity of the NIR was not detected in cultures until after the 18th hour of growth (i.e., late log phase of growth [10]). These observations suggested that there may be some relationship between nitrite and selenite reduction. For instance, perhaps the nitrite reductase (NIR) catalyzes the reduction of both nitrite and selenite. To investigate this possibility, the NIR was characterized, and then mutants were isolated that either lacked active NIR or possessed greater NIR activity than the wild type. The ability of these mutants and the wild type to reduce selenite and nitrite were compared. From these studies, described below, it was concluded that either the NIR of *T. selenatis* or some component of the NIR respiratory system can also reduce selenite [10].

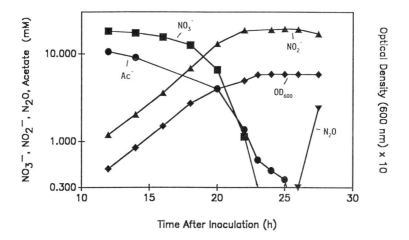

Figure 5 Anaerobic growth of *T. selenatis* with nitrate (20 mM) and acetate (10 mM) in minimal medium.

Like the SR, and along with cytochrome C_{551}, the NIR is located in the periplasmic space, but it is more tightly bound to the cytoplasmic membrane than the SR [10]. The pH optimum of the NIR is 7.5 [10]; pH optima of the NR and SR are 7.0 and 6.0, respectively [9]. NIR is a copper-containing enzyme [10], as are the NIRs of certain other denitrifying bacteria, because it is inhibited by the copper-binding inhibitor diethyldithiocarbamate (DDC) [18]. Those denitrifying bacteria not having a copper-type NIR have a cd_1-type NIR [18].

Based upon the recent findings of Zumpft and Kroneck [19] the NIR probably catalyzes the following reaction:

$$4 H^+ + 4 NO_2^- + 4 H \rightarrow 4 NO + 4 H_2O \tag{5}$$

A nitric oxide reductase would then reduce the NO to N_2O [19]:

$$4 NO + 4 H \longrightarrow 2 N_2O + 2 H_2O \tag{6}$$

Reduction of selenite to elemental selenium would be as follows:

$$2 H^+ + SeO_3^{2-} + 4 H \rightarrow Se^0 + 3 H_2O \tag{7}$$

If the NIR catalyzed both reactions (5) and (7), mutants lacking NIR activity would no longer be able to reduce selenite to elemental selenium. This seems to be the case. Mutants having no NIR activity (Table 3) are unable to reduce nitrite and are also incapable of reducing selenite to

Table 3 Activities of Selenate Reductase (SR), Nitrate Reductase (NR), and Nitrite Reductase (NIR) in Wild Type and Mutants of *T. selenatis* Grown with Nitrate Plus Selenate[a]

Enzyme	Wild type	NIR⁻ mutants			NIR overproducers	
		D6	D8	E3	OP1	OP2
NR	855	1400	2430	1730	530	765
NIR	159	<0.01	<0.01	<0.01	205	253
SR	177	24	47	76	120	132

[a]Nanomoles per minute per milligram of protein. Organism grown with 20 mM nitrate plus 20 mM selenate. Each number represents an average of two analyses that did not differ by more than 3%.

elemental selenium [10]. These mutants grow normally with selenate or nitrate as electron acceptor (i.e., they can form both selenite and nitrite from selenate and nitrate, respectively [10]). Similarly, mutants with increased NIR activity (Table 3) not only reduce nitrite more rapidly than the wild type but also reduce selenite and form elemental selenium more rapidly [10]. It is important to note that selenite reduction by NIR is optimal only if 14 mM nitrite is present also. Figure 6 shows growth of NIR⁻ mutants in medium containing selenate and nitrate. These mutants (D6, E3, and D8) formed no elemental selenium after 48 h of growth, whereas the wild type (WT) formed large amounts (Figure 6). Mutants having increased levels of NIR formed elemental selenium more rapidly than the wild type [10].

It would therefore appear that the NIR (or a component of the NIR respiratory system) of *T. selenatis* is involved in catalyzing the reduction of selenite to elemental selenium while also reducing nitrite. A number of other denitrifying bacteria are also able to reduce selenite under denitrifying conditions, although not as effectively as *T. selenatis*, and none of the organisms was able to respire using selenate [10].

The role of nitrate metabolism (i.e., denitrification) in selenate reduction to elemental selenium in *T. selenatis* is therefore quite important. Selenite reduction is dependent upon the presence of nitrite, formed from nitrate reduction, and an active NIR. This is not the case for the two selenate-respiring bacteria described above [16,17]. For these organisms, nitrate must be absent before selenate can be reduced [16,17].

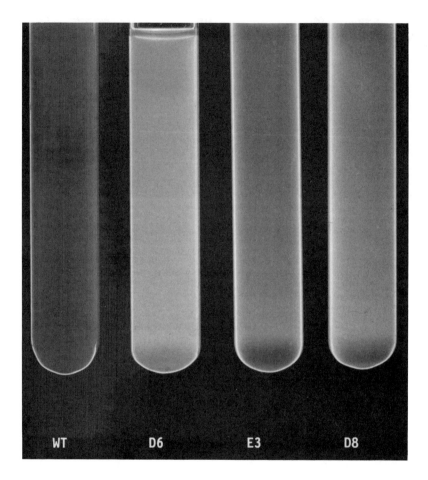

Figure 6 Growth of wild-type (WT), NIR⁻ mutants (D6, E3, D8) in minimal medium plus nitrate (20 mM), selenate (20 mM), and acetate (20 mM). Incubation time: 48 h. (See color plate.)

D. The Selenate Respiratory System

In preliminary experiments with membrane preparations containing active NADH and succinate dehydrogenase activities, as well as associated SR activity and cytochrome C_{551}, both NADH and succinate reduced cytochrome C_{551}; the cytochrome was reoxidized by the addition of selenate (data not shown). It therefore appears that acetate is oxidized via the TCA cycle and that the NADH and succinate formed during this oxidation are used as electron donors to reduce selenate. The electrons are transferred via an electron transport system that is either a part of (e.g., the NADH dehydrogenase), or loosely bound (e.g., SR and C_{551}) to the cytoplasmic membrane. In the presence of nitrate, when denitrification is occurring, both a NR and NIR respiratory system must also be present; the NIR (or a component of the system) would then reduce the selenite formed via selenate respiration to elemental selenium.

A hypothetical view of the selenate respiratory system, with NADH as the electron donor, is presented in Figure 7. The further reduction of selenite via the NIR is also shown.

IV. BIOLOGICAL REACTOR STUDIES— BIOREMEDIATION OF SELENIUM OXYANIONS IN SAN JOAQUIN VALLEY DRAINAGE WATER

T. selenatis is well suited for bioremediation of selenium oxyanions in San Joaquin Valley drainage water—wastewater that contains both selenate (300–500 µgSe/L) and nitrate (45–65 mgN/L) [1]. This is so because in this organism selenate and nitrate metabolism complement one another; as only when both electron acceptors are present can selenate be completely reduced to elemental selenium. As described above, when grown with both selenate and nitrate, these electron acceptors are reduced at the same time, each by a different terminal reductase (SR and NR, respectively). The selenite and nitrite formed during these reduction reactions are further reduced by the NIR (or a component of the NIR respiratory system); the selenite is reduced to elemental selenium, while the product of nitrite reduction by NIR is not known, although the final product of denitrification is N_2O [5]. If the nitrate were not present in drainage water, *T. selenatis* would only reduce selenate to selenite; very little selenite would be reduced further to elemental selenium. Therefore, for complete reduction of selenate in drainage water, the presence of nitrate is a necessity.

The effectiveness of *T. selenatis* in detoxifying selenium oxyanions in San Joaquin Valley drainage water was tested in a biological reactor system inoculated with this organism. The experiments are described below. The drainage water was obtained from the Westlands Water District of the San Joaquin Valley of California.

A. Growth of *T. selenatis* in Drainage Water

In order to be certain that *T. selenatis* could be maintained in a biological reactor system, it was necessary to determine whether this organism could

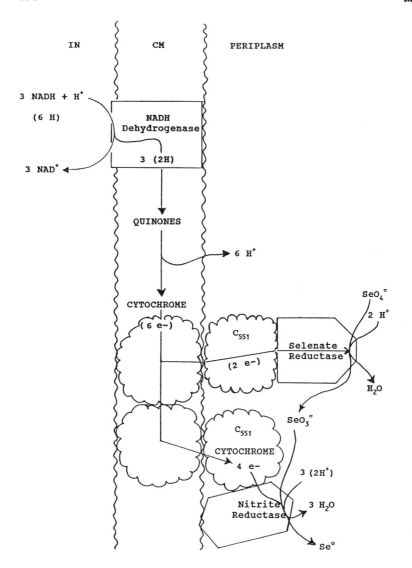

Figure 7 Hypothetical view of selenate reduction to elemental selenium involving periplasmic selenate reductase, cytochrome C_{551}, and nitrite reductase.

grow in drainage water. This was tested, and it was found that in order for *T. selenatis* to grow in anaerobic drainage water to which acetate (10 mM), selenate (5 mM), and nitrate (5 mM) are added, it is also necessary to add ammonium chloride to the water; 0.56 mM ammonium chloride is required for optimum growth (i.e., maximum optical density at 600 nm) [20]. Oddly, the same requirement is not observed when the organism is grown with the same electron donor and electron acceptors in minimal medium [20]. Therefore, while nitrate assimilation functions normally in minimal medium, something in the drainage water inhibits this assimilation, so that ammonia must be added to permit growth [20].

B. Reactor Design

The biological reactor system being used is very similar to that described by Squires et al. [4] and was designed by R. Groves of EPOC Water Inc., Fresno, California (Figure 8). It consists of a sludge-blanket reactor (SBR; 1 L containing 400 g of sand) and a fluidized-bed reactor (FBR; 1 L containing 300 g of sand). For the experiments described here, water passing through each of these reactors is recycled (Figure 8). The drainage water is pumped from either 10-, 20-, or 40-L carboys, where it is kept under an atmosphere of 100% nitrogen gas. The pH of the drainage water was adjusted to 5.9, 6.9, or 7.9 (optimum pH for growth with selenate, 7.0; for growth with nitrate, 8.0). Acetate and ammonium chloride were fed into the drainage water before it entered the SBR; the volume of feed was 100 fold less than the volume of drainage water. For initial experiments, the concentrations of acetate and ammonium chloride flowing into the SBR were 11 and 2.9 mM, respectively. After 275 days the acetate concentration was lowered to 3 mM; at 286 days the level of ammonia was decreased to 0.56 mM. At 395 days the acetate level was again lowered to 2 mM. Flow rates through the reactor were either 6.5 or 13 mL/min. Residence time in the reactor when the flow rate was 6.5 mL/min was 140 min [20].

C. Bioremediation of Selenate and Selenite in Drainage Water

It is not possible to present here all the results of the biological reactor experiment, which lasted 507 days. Therefore, some of the results will be summarized.

During the first 120 days of operation, the pH of the inflowing water was adjusted to 6.9, the flow rate was 6.5 mL/min, and the final concentrations of acetate and ammonia in the drainage water flowing into the SBR were 11 and 2.9 mM, respectively. After about 2 months of adaptation, the

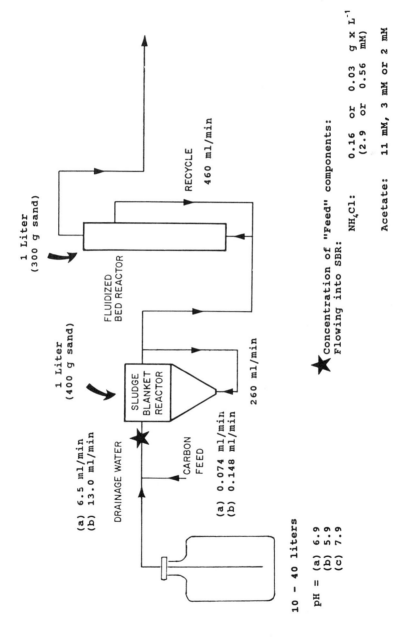

Figure 8 Biological reactor system for reduction of selenium oxyanions in agricultural drainage water.

selenate in the drainage water had been reduced such that the level of selenium oxyanions remaining in the outflow water, from day 62 to day 120, was on the average less than 10 µgSe/L [20].

From day 120 to day 260, the effects of different inflowing water pH values and different flow rates on selenate reduction were tested. Increasing the flow rate to 13 mL/min, with the pH of inflowing water remaining at 6.9, had no significant effect on the level of selenium oxyanions remaining in the outflow water. Reducing the pH of inflowing water to 5.9, at flow rates of 6.5 or 13 mL/min, resulted in only slight increases in the levels of selenium oxyanions remaining in the outflow water, but levels were always, on the average, less than 15 µgSe/L [20].

Increasing the pH of inflowing water to 7.9 had a detrimental effect on the efficient functioning of the reactor system. At the slower flow rate of 6.5 mL/min, a precipitate (which forms gradually in tubing at the lower pHs, requiring periodic changing) began to build up rapidly in the SBR; this buildup was accelerated when the flow rate was increased to 13 mL/min. The effectiveness of selenium oxyanion reduction in the SBR was decreased as the precipitate built up, although the level flowing out of the reactor was, on the average, less than 20 µgSe/L. Once the SBR became full, the top came off and the reactor could no longer function. At this point, the pH was reduced to 5.9 (on day 225), the flow rate was reduced to 6.5 mL/min, the precipitate in the SBR was broken up, and the reactor was given time to recover. The lower pH of the inflow water was chosen, as it was hoped that the slightly more acidic inflowing water might permit more rapid dissolution of the precipitate in the SBR. The amount of material in the SBR began to decrease and reached a normal level (i.e., similar to the level before pH increase to 7.9) at 265 days. At this point, the level of selenium oxyanions flowing out of the reactor was again less than 10 µgSe/L (Figure 9) [20].

At day 276, the concentration of acetate in the water flowing into the SBR was decreased from 11 to 3 mM (Figure 9). This was done because the levels of nitrate and selenate in the drainage water never exceeded 5 mM and 0.006 mM, respectively; therefore, the 11 mM acetate used during the first 276 days of the reactor experiments exceeded the requirements for complete reduction of this concentration of nitrate [see equation (2)] [20].

At day 286, the level of ammonium chloride in the water flowing into the reactor was decreased to the optimum level required for growth of *T. selenatis* in drainage water (i.e., from 2.9 to 0.56 mM). The 2.9 mM used during the first 285 days represented half the concentration present in minimal medium [8]. The changes in acetate and ammonia concentrations had a slight effect on the level of selenium oxyanions removed from the

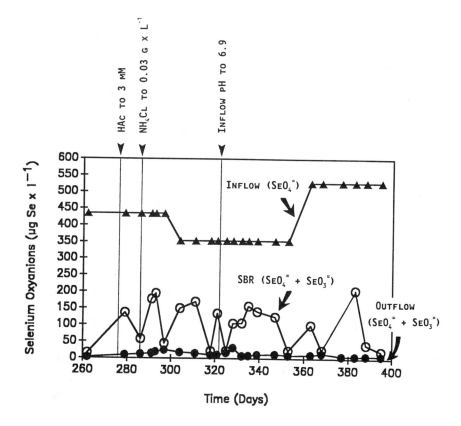

Figure 9 Concentration of selenium oxyanions in drainage water flowing into and out of the SBR and the biological reactor system. The pH of the inflowing water was adjusted to either 5.9 or 6.9; the flow rate through the reactor was 6.5 mL/min. The concentrations of acetate and ammonia are indicated.

drainage water in the reactor system; on day 296 the level reached 24 µg Se/L (Figure 9). The concentration of selenium oxyanions remaining in the outflowing water dropped to levels of about 10 µgSe/L again by day 318 (Figure 9) [20].

Because the reactor system functioned slightly better when the pH of the inflowing water was at 6.9, on day 322 the pH of the inflowing water was changed from 5.9 to 6.9 (Figure 9); adjustment to this pH also required addition of less acid and was near the optimum pH for growth with selenate. This change also had an effect on selenium oxyanion reduction in the reactor system (Figure 9); the level in the water flowing out of the reactor system had increased to 31 µgSe/L by day 325. Within 10 days after the change (i.e., day 332), however, the reactor system was back to normal (i.e., 10 µgSe/L or less) and remained so until the concentration of acetate was changed again on day 395. During the period between days 322 and 395 (i.e., 62 days), the average concentration of selenium oxyanions remaining in the water flowing out of the biological reactor system was 9.8 µgSe/L. During this time, the concentration of selenium oxyanions flowing out of the SBR fluctuated, but there seemed to be no distinguishable pattern. Any selenium oxyanions not reduced in the SBR were, however, effectively reduced in the FBR so that the level flowing out of the reactor represented a 98% reduction in the selenate present in the drainage water [20].

On day 395, the concentration of acetate fed into the reactor was reduced to 2 mM; of the 3 mM that had been fed between days 276 and 395, approximately 1 mM had not been used. This change had a profound effect on the level of selenium oxyanions reduced in the reactor system; by day 402 the level had increased to almost 300, and 59 days of gradual decrease was required before a constant level was again achieved on day 461 (Figure 10). From day 461 to the end of the experiment (day 507), the level of selenium oxyanions remaining in the water after treatment in the reactor system averaged 5 µgSe/L, except for a sample taken on day 489, two days after material was removed from the SBR and FBR for microbiological analysis (the system was probably disturbed by the introduction of oxygen) (Figure 10).

Therefore, it can be concluded that the optimum conditions for reduction of selenium oxides in the biological reactor system described include the following: (a) having the pH of inflowing water near that of the optimum pH for growth of *T. selenatis* with selenate (i.e., 7.0); (b) using 2 mM acetate as the carbon and energy source, as with this respiratory substrate the organism is selectively maintained in the reactor system (using a nonrespiratory substrate such as glucose or molasses would permit the growth of many different microorganisms, and *T. selenatis* would quickly be excluded as it grows poorly on sugars and not at all on methanol [5]); and (c) including a small amount of ammonia in the water, as something in drainage water inhibits nitrate assimilation.

During the last experimental period, when the system was constant (i.e., days 461–507), it was found that not all of the selenium flowing into the reactor as selenate could be recovered in the outflow as elemental selenium. The unrecovered selenium had been deposited as part of the precipitate that forms in the reactor tubing with time (see above). A

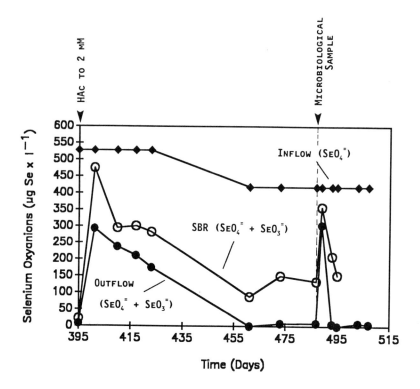

Figure 10 Concentration of selenium oxyanions in drainage water flowing out of the biological reactor system. The pH of the inflowing water was adjusted to 6.9; the flow rate through the reactor was 6.5 mL/min. The concentration of acetate was changed to 2 mM on day 395. A sample for microbiological analysis was taken on day 487.

portion of the precipitate was carbonate. Of the remainder, calcium accounted for 16%, and elemental selenium, 79%. The precipitate therefore appears to be composed primarily of calcium carbonate and elemental selenium [20]; the selenium not recovered as elemental selenium in the outflow had been trapped in the calcium carbonate precipitate formed in the reactor tubing. The drainage water contains high levels of calcium [1,4].

Also during this last experimental period when conditions were optimal, the numbers of selenate-respiring bacteria found in the SBR and FBR were 9×10^5 and 6×10^6, respectively, per gram of sand [20]. The only selenate-respiring bacterium present was *T. selenatis* [20]. The numbers of nitrate-respiring bacteria found in the SBR and FBR were 3×10^8 and 3×10^9, respectively, per gram of sand [20]. Only nanomole amounts of sulfide were detected. Therefore, it appears that in the biological reactor system, *T. selenatis* reduced selenate to selenite. Selenite reduction to elemental selenium was probably catalyzed by the nitrite reductases of both *T. selenatis* and the more numerous denitrifying population [10,20]. The same denitrifying population effectively reduced the nitrate present in the drainage water.

D. Bioremediation of Nitrate

The level of nitrate in the drainage water used in the experiments described above ranged between 35 and 65 mgN/L. Treatment of drainage water in the biological reactor system from day 1 to day 286 resulted in the reduction of nitrate to less than 2 mgN/L. This represented the removal of 97% of the nitrate from the water. After day 286, when the ammonia concentration in the water was lowered to 0.56 mM, the nitrate levels were reduced to less than 1 mgN/L (or less than 0.09 mM) and remained at that level until the end of the experiment. This represented a reduction in nitrate of 98% and was achieved by the population of denitrifying bacteria living in the reactor system with *T. selenatis*. The nitrate was presumably reduced to N_2; about 1 µM nitrite was detected in the water flowing out of the reactor [20].

V. CONCLUSIONS

A new organism, one that respires anaerobically using selenate as the terminal electron acceptor without inhibition by nitrate, has been isolated and characterized. Extensive biochemical and physiological studies have revealed the mechanism of selenate and selenite reduction and the involvement of nitrate metabolism in this reduction. With these insights, the organism has been tested for its ability to reduce selenium oxyanions (together with a denitrifying population that assisted in selenite reduction) in a biological reactor system for treatment of selenate-containing San Joaquin Valley drainage waters.

In the reactor studies of Squires et al. [4], where methanol or molasses was the carbon source (substrates not utilized by *T. selenatis*), selenate could be reduced only after all the nitrate had first been reduced, and then only to levels between 10 and 30 µgSe/L. The organisms present in the Squires et al. [4] reactor were probably denitrifying bacteria similar to the organisms described by Barnes et al. [16] and Steinberg et al. [17] that reduced selenate only when extra electrons were available because nitrate was no longer present. Such reduction, most likely using a nitrate reductase, cannot be as rapid or as effective as the approach used by *T. selenatis*. This organism carries out this important first step in selenium bioremediation using a selenate reductase, a specific reductase that catalyzes the reduction rapidly, leaving only low levels of selenate unreduced. The different approaches used by *T. selenatis* and denitrifiers are summarized in Figure 11.

In conclusion, because of its ability to reduce selenate with a specific selenate reductase, *T. selenatis* gen. nov. sp. nov. is well suited for use in a

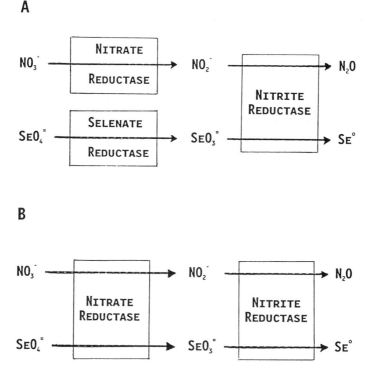

Figure 11 (A) Schematic diagram of enzymes involved in selenate, selenite, nitrate, and nitrite reduction by *T. selenatis*. (B) Schematic diagram of enzymes that appear to be involved in selenate, selenite, nitrate, and nitrite reduction by those denitrifying bacteria able to reduce selenate and selenite.

biological reactor system for the reduction of selenium oxyanions in agricultural drainage waters from the San Joaquin Valley. Under optimum conditions in a biological reactor system (see above), *T. selenatis* reduces selenate and (presumably with the assistance of a selenite- reducing denitrifying population [10]) selenite in drainage water from 350–450 µgSe/L to an average of 5 µgSe/L. In the same system, nitrate levels in the water are reduced from 260–380 mgN to less than 1 mgN/L, presumably by the denitrifying population. All that is required to keep *T. selenatis* in the reactor system is that it be able to grow in the water (i.e., correct pH and ammonia provided) and that acetate and selenate be present to apply a selective pressure. With this in mind, it should be a simple matter to use the organism in reactor systems of any scale.

This newly isolated and characterized organism may help provide a solution to the selenium problem in the San Joaquin Valley.

ACKNOWLEDGMENTS

Part of this work was supported by the Department of Water Resources, State of California (contracts DWR B-56542, B-57045, and B-57845) and by a grant from the University of California Salinity Task Force.

REFERENCES

1. R. L. Brown and L. A. Beck, Subsurface agricultural drainage in California's San Joaquin Valley, in *Biotreatment of Agricultural Wastewater* (M. E. Huntley, Ed.), CRC Press, Boca Raton, Fla., 1989, pp. 1–13.
2. M. K. Saike and T. P. Lowe, *Arch. Environ. Contam. Toxicol.* 16:657–670 (1987).
3. O. Weres, A.-R. Jaouni, and L. Tsao, *Appl. Geochem.* 4:543–563 (1989).
4. R. C. Squires, G. R. Groves, and W. R. Johnston, *J. Irrigat. Drainage Eng.* 115:48–57 (1989).
5. J. M. Macy, S. Rech, G. Auling, M. Dorsch, E. Stackebrandt, and L. Sly, *Int. J. Syst. Bacteriol.* 43:135–142 (1993).
6. G. Gottschalk, *Bacterial Metabolism*, Springer-Verlag, New York, 1986.
7. J. W. Doran, *Adv. Microb. Ecol.* 6:1–32 (1982).
8. J. M. Macy, T. A. Michel, and D. G. Kirsch, *FEMS Microbiol. Lett.* 61:195–198 (1989).
9. S. Rech and J. M. Macy, *J. Bacteriol.* 174:7316–7320 (1992).
10. H. DeMoll-Decker and J. M. Macy, *Arch. Microbiol.* 160:241–247 (1993).
11. J. M. Macy and S. Lawson, *Arch. Microbiol.* 160:295–298 (1993).
12. W. M. Latimer, *The Oxidation States of the Elements and Their Potentials in Aqueous Solution*, Prentice-Hall, Englewood Cliffs, N.J., 1938.
13. R. K. Thauer, K. Jungermann, and K. Decker, *Bacteriol. Rev.* 41:100–180 (1977).

14. R. K. Thauer and J. G. Morris, Metabolism of chemotrophic anaerobes. Old views and new aspects, in *The Microbe*, Part II: *Prokaryotes and Eukaryotes* (D. P. Kelly and N. G. Carr, Eds.), Cambridge Univ. Press, Cambridge, U.K., 1984, pp. 123–168.
15. R. K. Thauer, D. Moeller-Zinkhan, and A. M. Spormann, *Annu. Rev. Microbiol.* 43:43–67 (1989).
16. J. M. Barnes, J. K. Polman, and J. H. McCune, Bioremediation of selenate and selenite by *Pseudomonas stutzeri*, JB1, Abstr. Q-294, Abstr. Am. Soc. Microbiol., Annu. Meeting, New Orleans, 1992.
17. N. A. Steinberg, J. S. Blum, L. Hochstein, and R. S. Oremland, *Appl. Environ. Microbiol.* 58:426–428 (1992).
18. M. S. Coyne and J. A. Tiedje, Distribution and diversity of dissimilatory NO_2-reductases in denitrifying bacteria, in *FEMS Symposium 56: Denitrification in Soil and Sediment* (N. P. Revsbech and J. Sorensen, Eds.), Plenum, New York, 1990, pp. 21-35.
19. W. Zumpft and P. M. H. Kroneck, Metabolism of nitrous oxide, in *FEMS Symposium 56: Denitrification in Soil and Sediment* (N. P. Revsbech and J. Sorensen, Eds.), Plenum, New York, 1990, pp. 37–55.
20. J. M. Macy, S. Lawson, and H. De Moll-Decker, *Appl. Microbiol. Biotechnol.*, in press (1993).

Index

Adsorption, selenium (*see also*
  Selenate; Selenite):
  onto metal oxides, 53–54
  onto soil minerals, 18–19, 52–54
  onto soil organic matter, 19
  by soils in China, 54–56
Agriculture (*see also* Agroforestry):
  bioredmediation of selenium
    oxyanions, 421–443
  irrigation (*see also* San Joaquin
    Valley), 158–161, 238–240, 247,
    249, 389
  livestock, 8–9, 15, 17–18, 21–23,
    238
  plant uptake of selenium in, 14–15,
    20, 24, 32–37, 39–40, 59, 72
  selenium deficiency in, 14–18, 21,
    23, 24, 30, 47–51, 55–58, 238
  selenium distribution in, 35
  selenium in drainage water from,
    31
  selenium in fertilizer inputs to,
    15–17, 41, 43
  selenium in sludge inputs to, 17–18

Agroforestry:
  concentration of selenium and salt
    in, 244, 247
  use of halophytes, 237–242, 244–249
  and irrigated agriculture, 238–240,
    247, 249
  management of selenium and salt,
    237–240, 244, 247–249, 329
  principles, 237–240, 247–249, 329
  reuse of drainage water, 238–241,
    244–245, 247, 249, 329
  use of salt-tolerant trees, 237–241,
    243–249
  and wildlife, 245, 249
Alfalfa, 33–34, 36–39, 246, 336,
  344–345
Algae (*see* Kesterson Reservoir,
  microbial community)
Algal-bacterial selenium removal
  system:
  field study
    methane fermentation, 274–275
    nitrate removal by, 265–269, 275
    selenium removal by, 268–275

445

[Algal-bacterial selenium removal system, *continued*]
  microbes involved in, 251–261, 264–267, 274–275
  pilot plant
    high-rate algae production ponds, 253–255, 266–267, 275
    methane fermentation by, 255
    nitrate reduction by, 255
    schematic diagram, 254
    soluble selenium removal by, 255
  principles of, 251–253, 275–276
  research needs, 276
  selenium reduction mechanisms
    assimilatory, 261–262
    dissimilatory, 261–263, 275, 370
    inhibition by atmospheric oxygen, 257
    inhibition by nitrate, 258–261, 275
    inhibition by sulfate, 261–262, 275
    microbes involved, 256
    redox couples, 262, 264
    speciation, 252, 256, 260, 263, 269–270, 272–273, 275

Bacteria:
  denitrifying, 431
  selenate-respiring (*see Thauera selenatis*)
  selenium-respiring
    characteristics of, 409–412
    inhibition by nitrates, 407–409
  sulfate-reducing, 397–402
Bioremediation (*see also* Kesterson Reservoir, bioremediation; Volatilization, selenium):
  of nitrates, 441
  of selenium oxyanions (*see Thauera selenatis*)
Birds, aquatic, 69, 71–72, 74, 78, 80–81, 94–95
  bioaccumulation of selenium (*see also* Selenium, contamination), 80, 100, 102–103

[Birds, aquatic, *continued*]
  effect of selenium on reproduction of, 69, 73, 80, 100–104, 107, 111–112, 139, 153, 251, 280
  habitats at Kesterson Reservoir, 93–95, 100–104
  monitoring at Kesterson Reservoir, 106, 108–109, 113

California Coast Ranges (*see also* San Joaquin Valley):
  chemistry of streams, seeps of drainage basins, 146–152
  evolution of, 142–143
  geologic formation, 142–143
  geologic map of, 140, 142
  geologic origins of Se contamination at, 141
  geologic sources of Se in concentration in bedrock, 144–146
  lithology of seleniferous shales, 145
  mobility of selenium from, 139–142, 146, 148, 150–151, 153
  selenium concentration at, 139–141, 144, 148–151
California Department of Fish and Game, 73
  Wildlife Habitat Relationships (WHR) database, 73, 75, 78–79
California Department of Food and Agriculture, 247
Chara (*see* Kesterson Reservoir, microbial community)

Dimethyl diselenide (*see also* Volatilization, selenium):
  in atmospheric vapor, 12, 384
  biochemical pathway, 398
  in natural waters, 357
  respiratory product, 37, 39
  volatile methylated species, 187, 369, 398
  volatility of, 3, 233, 311, 344, 371
    by microorganisms, 37, 311, 357–358, 372, 374, 393

INDEX

[Dimethyl diselenide, *continued*]
[volatility of, *continued*]
  by plants, 3, 37
  by Se-accumulator plants, 344, 360–363
Dimethylselenide (*see also* Volatilization, selenium):
  in atmospheric vapor, 12, 384–385
  biochemical pathway, 398
  characteristic odor of, 34, 37
  demethylation of, 395–397
  effect of aeration on, 379, 394
  effect of protein amendments on, 374
  effect of temperature on, 375
  and garlic breath, 7
  respiratory product, 37, 39
  toxicity of, 373–375, 393
  volatile methylated species, 187, 369, 398, 414–415
  volatility of, 3, 233, 311, 344, 371
    by microorganisms, 37, 56, 311, 357–358, 372–375, 393–396
    by nonaccumulator plants, 344
    by plants, 3, 360–363
Dimethylselenone (*see also* Volatilization, selenium):
  in atmospheric vapor, 12, 384
  volatility of, 3, 344, 371
    by microorganisms, 357–358, 374
Dimethylselenonium, 187
Dimethylsulfide, 384–385, 394, 396
Diseases, selenium-responsive, 41

Environmental Trace Substances Research Center, 84

Geoecosystem, selenium:
  biogeochemical cycle, reductive portion, 415
  in China, 47–65
    adequate Se belt, 50
    low Se belt, 47, 50–51, 60–61, 63–65
  cycling of, 1, 4–5, 9–24, 31, 51–52, 55–59, 228–234

Glutathione peroxidase, 6–7, 9, 30, 37–38, 41, 43, 345, 361
Groundwater systems (*see also* San Joaquin Valley, groundwater system; Kesterson Reservoir, groundwater system):
  microbial activity in, 215–219
  selenium geochemistry in, 185
  selenium mobility in, 186, 194–213
  selenium uptake from, 213–218

Hydride generation atomic absorption spectrophotometry, 142

Kaschin-Beck disease, 30, 47–51, 56–58
  disease belt, 47, 50–51
  geographical distribution of, in China, 47, 49, 55–58, 63
  identification, 42
  prevention, 42
  treatment, 42
Keshan disease, 5, 30, 47–51, 56–58
  disease belt, 47, 50–51
  geographical distribution of, in China, 47–48, 55–58, 63
  identification, 42
  prevention, 42
  treatment, 42
Kesterson National Wildlife Refuge, 8, 71, 95, 103, 112, 123
Kesterson Reservoir, 8, 24, 69–114
  Biological Monitoring Program, 72, 78, 85, 90, 108–109
  bioremediation, 233, 311–312, 339, 394
    by dissimilatory reduction of selenate, 403
    by microbial volatilization, 382–384
    by selenate-respiring bacteria (*see Thauera selenatis*)
    using tall fescue grass, 287, 290
    using vegetation management strategies, 290, 320–321, 328–329, 331–340, 363–365

[Kesterson Reservoir, *continued*]
cleanup, 223–224, 233–234, 312, 322
colonization of Se-contaminated soils
  effects of saline irrigation water on, 302–305, 329
  by salt grass, 305–313, 316–317, 321–322
  succession of grassland communities, 312–320
ecological risk assessment, 69–114
  conclusions and limitations, 104–106
  recommendations, based on, 106–109
  summary of, 109–114
groundwater system, 188
  chemistry of, 202
  microbial processes in, 215–219
  relationship between nitrate and selenate, 217–219
  selenium distribution in, 202–207
habitat types, 76–80, 82, 84–85, 87, 110
  aquatic habitats, 93–95
  terrestrial habitats, 85–93
history, 69, 75–76, 109, 123, 201, 224, 280
hydrogeology, 201–202
kinetics of selenium flux
  correlation with global cycle, 232
  depuration rates, 223–234
  by microbial community, 225–234
  by mosquitofish, 224–234
  seasonal cycling, 228–234
  uptake rates, 224, 226–230, 233–234
location, 69–70, 75, 123, 200, 203
management of rainwater pools, 106–108, 111–113
microbial community, 225–234, 284
mosquitofish, 224–234
mushrooms and selenium risk at, 99, 104–106, 108–112

[Kesterson Reservoir, *continued*]
phosphate-extractable selenium, 123, 125, 129, 133
reoxidation and movement of Se at, 130–135
restoration, 280, 298
selenate concentration, 123, 125, 191–193, 212, 217–218, 251, 305–308, 331
selenite concentration, 123, 125, 191–193, 305–308, 331
selenium concentrations, 73, 104–107, 109, 123, 126
  in aquatic habitats, 81–83, 93–95, 100–104
  correlation with redox chemistry, 210–213
  expected in the future, 95–99, 104, 110
  in groundwater, 203–207
  relationship to boron, 204–206, 209–210, 212
  temporal variations in, 205, 208–210, 233
  in terrestrial habitats, 85, 87, 88–93
  in vegetation, 313–319
selenium contamination, 139, 143, 151, 223, 251, 421
selenium exposure pathways, 85–95, 99, 104, 106, 110–111
selenium speciation, 187, 189–193, 217, 233
soil map, 123
soil profile, 122–123, 125–134
succession at, 78, 80–81, 105, 109
vegetation types, 80–81, 85, 87–92, 110, 302–320
volatilization from, 233, 311–312, 339, 377, 379, 382–385, 394
water-extractable selenium at, 123, 125–135, 313–314

Lahontan Valley, 119–120
phosphate-extractable selenium, 123–124, 126–127

[Lahontan Valley, *continued*]
  selenate concentration, 125
  selenite concentration, 125
  selenite/selenate ratio, 124, 126
  soil map, 123
  soil profile, 122–124, 126–127
  water-extractable selenium,
    123–124, 126

Methylselenocysteine, 334
Migratory Bird Treaty Act, 103, 112
Mosquitofish (*see* Kesterson Reservoir, mosquitofish)
Mushrooms (*see also* Kesterson Reservoir, mushrooms and selenium risk at), 99

National Wildlife Refuge System, 76

Phosphates, 17, 19, 20, 63–64
  extraction of selenium, 120–122
Plants, salt tolerant (*see also*
  Agroforestry, use of salt-tolerant trees), 285–286,
  288–290
Plants, selenium tolerant (*see also*
  Agroforestry; Kesterson
  Reservoir, bioremediation):
  accumulators (*see* primary
    accumulators)
  agricultural crops, 281
  and boron tolerance, 331–332,
    335–336, 339
  for decontamination, 329
  forages, 280, 282–286, 291–302
  genetic variation in tall fescues,
    287–290
  nonaccumulators, 281, 322, 334,
    339, 344–345
  for prevention of contamination,
    328
  primary accumulators, 281,
    328–329, 332–337, 339,
    344–345, 360–361
  for remediation of contamination,
    328–329

[Plants, selenium tolerant, *continued*]
  saline irrigation water
    effects on forages, 298–302
    effects on naturally established
      species, 302–305
  secondary accumulators, 281, 334
  selenium speciation in, 283–286
  sulfate effects on, 290–298, 301, 304,
    335–337, 339
  turfgrasses, 280, 282–286,
    291–302

Reclamation (*see* U.S. Bureau of
  Reclamation)
Rocks:
  parent, 10, 15
  selenium concentration of, 10, 64
  pyrite-containing, 279, 369

Safe Drinking Water Act, 31
San Joaquin Valley (*see also* California Coast Ranges;
  Kesterson Reservoir), 8, 75,
  101, 122–123
  agricultural drainage systems, 31,
    161, 168, 171–174, 176, 179,
    181, 189, 251
  bioremediation of selenium
    oxyanions in (*see also* Thauera
    selenatis), 421–443
  geohydrology of, 158–161, 171, 175,
    194–196
  groundwater system
    chemical composition, 188
    flow, 157–161, 173–174, 176, 178,
      180–181
    redox potential, 198–200
    selenate concentration (*see also*
      Selenate), 198
    selenite concentration (*see also*
      Selenite), 198
    selenium concentration in, 157,
      162–171, 174, 176–177, 179,
      181, 194, 196, 198, 200
    Se distribution and mobility in,
      157, 194–200

[San Joaquin Valley, continued]
  [Se distribution and mobility in, continued]
    local (farm-field) scale, 157–158, 161, 167–174
    redox controls on, 185–194
    regional scale, 157–158, 161–167
    subregional scale, 157, 159, 161, 174–180
  irrigation, 157–164, 166, 170–171, 173, 176, 179–180
  location, 157–158
  management of salts (see Plants, salt tolerant)
  management of selenium (see also Plants, selenium tolerant; Kesterson Reservoir, bioremediation)
    by dissimilatory reduction of selenate, 403
  nitrates in drainage water, 251, 421, 433
  selenate concentration, 433
  selenium concentration, 421
  Se contamination of, 8, 31, 139, 141, 152, 185, 251
  Se speciation in surface drainage water, 186–187, 189, 251, 331
  Se transport to, 139–140, 142, 151, 153
  soil salinity, 160–166, 168, 170–171, 181, 280
San Luis Drain, 69, 75–76, 84, 95, 327, 377, 421
Selenate:
  adsorption, 3, 18–19, 120–122, 133–134, 190–191
  assimilatory reduction, 390–391, 398, 415
  at California Coast Ranges, 146–148, 150–152
  chemistry of, 20, 422
  competition with sulfate, 20, 346, 349, 391–392, 397, 399–409, 422
  dissimilatory reduction of, 397, 399–413, 415

[Selenate, continued]
  at Kesterson Reservoir, 93, 109, 123, 125–127, 130–135, 190–193, 331
  at Lahontan Valley, 124, 126–127
  metabolism in plants, 347, 360–362
  mobility of, 3, 93, 119, 133–136, 180, 218, 370
  oxidation state, 3, 186, 370
  phosphate-extracted, 121–122
  ratio to selenite, 63, 123–124, 126, 191–192, 198, 370
  reduction
    biochemistry, 426–429
    enzymes involved, 442
    by microbes, 36
    role of nitrate metabolism, 431, 433
    to selenite, 360–361, 411–412, 423, 426–429
  relationship to nitrates, 217–219, 403–405, 407–409, 411–413, 423–429, 431
  removal by microbes, 216–219
  respiratory system, 432–434
  respiring bacteria (see *Thauera selenatis*)
  in San Joaquin Valley (see also San Joaquin Valley; Algal-bacterial selenium removal system), 187
  source of, 186
  synergism with sulfate, 347
  toxicity of, 390–391
  uptake by plants, 20, 346, 370
  volatilization, 351, 359–361, 371–372, 375
Selenate reductase, 426, 428–429, 431–432, 442
Selenide:
  biochemical pathway, 415
  mobility of, 3
  oxidation state, 3, 186, 369, 422
  oxyanion reduction product, 360–361, 390, 397, 412, 415
  in particulate Se compounds, 152
  plant metabolism of, 360–362

[Selenide, *continued*]
  precipitation with metals, 19, 186
  solubility, 370
  source of, 36
Selenite:
  adsorption, 18–19, 120–122, 127,
    133–134, 191, 370, 415
  assimilatory reduction, 390–391
  at California Coast Ranges, 148,
    152, 167
  at Kesterson Reservoir, 123, 125,
    127, 133–135, 190–193,
    305–308, 331
  at Lahontan Valley, 124, 126–127
  mobility of, 18, 119, 133–136, 167,
    370
  oxidation state, 186, 370
  and phosphate exchange, 120
  phosphate-extracted, 121–122, 127
  precipitation from, 19, 391, 393
  ratio to selenate, 63, 123–124, 126,
    191–192, 198, 370
  reduction of
    by denitrifying bacteria, 431
    nitrite involvement in, 429–432,
      434
    to selenide, 360–361
  in San Joaquin Valley (*see also*
    San Joaquin Valley; Algal-
    bacterial selenium removal
    system), 187
  solubility, 3
  source of, 186
  supplement tablets, 42
  toxicity, 390–391
  volatilization, 351, 359–361,
    371–372, 375
Selenium (Se):
  abundance, 369, 389
  adsorption (*see also* Selenate and
    Selenite), 18–20, 23, 52–53,
    119–120
    onto metal oxides, 53–54
    onto soil minerals, 18–19, 52–54
    onto soil organic matter, 19
    by soils in China, 54–56

[Selenium (Se), *continued*]
  in anoxic environments, 389–416
  anticarcinogenicity, 6, 22
  atmospheric dissipation of, 384
  bioaccumulation in the food chain,
    33–34, 44, 153, 233, 280, 287
  bioavailability
    in feedstuffs, 37–39, 41, 43
    in sodium selenite, 37–38, 42
  bioextraction (*see* Plants, selenium
    tolerant)
  biomethylation (*see also*
    Volatilization, selenium), 12,
    343, 371, 375, 378–379,
    393–398
  bioremediation (*see also* Kesterson
    Reservoir, bioremediation;
    Plants, selenium tolerant;
    Volatilization, selenium), 370,
    382, 403, 415
  chemical properties, 1–3, 19, 23
  chemistry, 185–186, 223, 369
  classification of, 369, 390
  cleanup (*see* Kesterson Reservoir)
  compounds, 2–4, 19
    usages of, 4
  concentration
    in aerosols, 31, 384
    in agricultural drainage waters
      (*see also* Kesterson Reservoir;
      Agroforestry), 31
    in blood, 38, 40–41, 43, 245–246
    in Chinese landscapes, 50–52, 59,
      63–65
    at Kesterson Reservoir (*see*
      Kesterson Reservoir, selenium
      concentrations)
    in natural waters, 31
    in parent rocks, 10, 64, 186
    in plants (*see also* Plants, selenium
      tolerant), 33
    in seleniferous soils, 7, 10, 32–34
    in sewage sludge, 17–18
  contamination, 8
    of agricultural drainage waters, 8,
      24, 31, 69, 139

[Selenium (Se), *continued*]
[contamination, *continued*]
distribution of, 369
geologic origin of, 31, 141
at Kesterson Reservoir (*see also* Kesterson Reservoir), 8, 69–114, 139
in San Joaquin Valley (*see* San Joaquin Valley)
in wildlife populations, 8, 24, 69–114
cycle (*see also* Selenium, global cycling; Kesterson Reservoir, seasonal cycling), 370, 390
biogeochemical, reductive portion, 415
deficiency (*see also* Kaschin-Beck disease and Keshan disease)
location, 245
management, 41, 43–44, 238, 241, 245–246
deficiency symptoms (*see also* Kaschin-Beck disease and Keshan disease), 1, 5–7, 389
in livestock, 8–9, 15–17, 21, 29–30, 40–41
in humans, 30
management of, 41, 43–44, 238, 241, 245–246
metabolic effects of, 6–7, 30
demethylation, 394–398
deposition, 13–15, 21, 24
desorption, 18–19
dietary benefits (*see also* Selenium, nutritional needs), 5–7
dietary supplements (*see also* Plants, selenium tolerant), 7, 9, 41–44, 245–246
digestibility, 37
discovery of, 370
distribution of, 1–3, 7, 33–35, 122–129, 279, 384
elemental, 3–4, 61, 186, 412, 414–416, 421–422, 439–440
excretion, 39
extraction methods, 63–65, 119–120, 122

[Selenium (Se), *continued*]
[extraction methods, *continued*]
phosphate-extractable, 120–127, 129
water-extractable, 122–132, 134
fertilizer inputs, 15–17
fluxes, 10–14, 20–23
foliar application, 17
geochemistry, 185–186
geographical surveys, 1–3, 7, 33
in human health (*see also* Geoecosystem, in China), 7
global cycling (*see also* Geoecosystem, selenium), 1, 4–5, 9–24, 31, 186
anthropogenic interference in, 4–5, 21–24, 56, 328, 389
aquatic and marine pathways, 21–24
biogeochemical, reductive portion, 415
biosphere enrichment factor, 4–5
concentration, 384
correlation to flux at Kesterson Reservoir, 232
historical changes in, 5
industries involved in, 5, 22–23
seasonal cycling at Kesterson Reservoir, 228–234
surface to atmosphere fluxes, 10–14, 23
through soil-plant systems, 14–21, 59
global importance, 1
grazing management of, 15, 24
half-life, 19
industrial uses, 2–4
isotopes, 2, 134, 136, 167
kinetics of uptake and loss, 223–234
leaching, 21, 389
mass balance, 14–18, 20, 23
metabolism, 37, 43
microbial transformation of
oxidation, 371
reduction, 370
volatilization (*see also* Volatilization, selenium), 371–373, 375

INDEX

[Selenium (Se), *continued*]
as micronutrient in plants, 345–346, 389
mining of, 2, 4, 7
mobilization in soil (*see also* Selenate, mobility of), 18–20, 139
new markets for, 237–238
nutritional needs (*see also* Selenium, bioavailability)
  in animals, 8–9, 29, 43–44, 59, 280–281, 389
  in birds, 29
  diagnostics for, 40–41
  in fish, 29
  historical perspective, 29–31
  in man, 29, 42, 43, 59
  management of, 41–44
  in plants, 29, 34, 43, 345–346, 389
  of Se-accumulator plants, 345, 347
oxidation states, 1–2, 61, 63, 120, 127, 185–186, 331, 422
oxyanions (*see also* Selenate; Selenite), 389–390
particle suspension, 12–13
phosphate-extractable (*see* Selenium, extraction methods)
physical properties, 1–3
physicochemical properties, 1–2
plant uptake of (*see also* Plants, selenium tolerant; Agroforestry; Volatilization, selenium), 14–15, 20, 24, 32–37, 39–40, 59, 72, 87–93, 110, 346, 349
precipitates, 19, 120, 133
production, 2
redox processes, 185–219, 370, 390, 422
relationship to lipid metabolism, 6–7
relationship to sulfur (*see* Sulfur; Sulfates)
reproductive performance, effects on, 8–9, 69, 73, 100–103

[Selenium (Se), *continued*]
research needs, 19–24, 44, 65, 72, 113–114, 238, 244, 276, 305, 311, 339–340, 350, 360, 364–365, 384
respiring bacteria (*see Thauera selenatis*)
seasonal cycling (*see also* Kesterson Reservoir), 228–234
solubility, 32, 186, 370
speciation (*see also* Soils, speciation of Se in; San Joaquin Valley, Se speciation; Kesterson Reservoir, Se speciation; Algal-bacterial selenium removal system)
  aqueous, 186–187, 189, 190
  relationship to groundwater redox couples, 191–194
  solid compounds, 186, 189
  solubility controls on, 187, 190
  sorption controls on, 190–191
  in surface drainage waters, 186–187, 189–190
tolerance (*see* Plants, selenium tolerant)
toxicity (*see also* Selenium, contamination)
  acclimation to, 39
  in animals, 1, 8–9, 24, 29–31, 43, 71–73, 141, 279, 281, 389
  anthropogenic effects on, 185
  in birds (*see also* Birds, aquatic; Kesterson Reservoir), 31, 43, 71–73, 139
  in humans, 7–8, 30
  measurement of, 7
  of methylated species, 373
  oxidation state effect on, 43
  in plants, 282, 347
  protection from, 38, 371
  relationship to sulfur metabolism, 347–348
  to Se-accumulator plants, 347
  symptoms of, 7–8, 30–31, 42–43, 98
transformation, 119

[Selenium (Se), *continued*]
transport, 119, 133–134, 390
volatilization (*see* Volatilization, selenium; Selenium, microbial transformation)
water-extractable (*see* Selenium, extraction methods)
Selenium dioxide, 3
Selenocysteine, 369, 374–375
Selenocystine, 36, 369, 374–375
Selenomethionine, 32, 36, 38, 369, 374–375
Selenosis, 3–5, 31, 35, 42–44
Shales, black, 152
Sludge, 17–18
Sodium selenite, 37–38, 42
Soils:
in China, 47–65
distribution of selenium in (*see also* Selenium, distribution), 32–35, 122–129, 328, 369–370
leaching of selenium from, 18
mobilization of selenium through, 18–20, 370
seleniferous, 7, 10, 32–34, 280–281, 328
selenium concentration, 370
selenium-contaminated
relationship to salinity and boron, 330–331
remediation of (*see* Kesterson Reservoir, bioremediation)
sludge-treated, 17–18
solubility of selenium in, 32
speciation of selenium in, 32, 60–65, 119–120, 130, 370
effects on animal health, 61, 63
effects on human health, 61, 63
at Kesterson Reservoir (*see also* Kesterson Reservoir), 107, 110, 331
Sulfate (*see also* Sulfur):
at California Coast Ranges, 141, 146–149, 151
competition with selenate, 20, 346, 349, 391–392, 397, 399–409, 422

[Sulfate, *continued*]
inhibition of selenium reduction, 261–262, 275
at Kesterson, 202, 204, 251, 253
reducing bacteria, 397–402
reduction of, 403
salts, 151, 153
selenium-tolerant plants, effects on, 290–298, 301, 304, 335–337, 339
selenium volatilization, effects on, 351–352, 355, 364
synergism with selenate, 347
Sulfide, 152, 214
Sulfur (*see also* Sulfate):
and blind staggers disorders, 31
at California Coast Ranges, 141
chemistry, 186
similarities to selenium, 186, 223, 369, 390, 422
correlations with selenium, 151, 153
deficiency, 41
oxidation states, 422
plant uptake of selenium, effect on, 20
relationship to selenium toxicity, 31, 347–348, 390
selenium bioavailability, effect on, 41
selenium dietary requirement, effect on, 43
selenium metabolism, effect on, 37–38

Thauera selenatis:
biological reactor studies with, 433–443
bioremediation of selenium oxyanions, 421–443
application to San Joaquin Valley drainage waters, 421–422, 433–443
importance of nitrates in process, 433, 443
effectiveness in detoxification of Se oxyanions, 433–443

[*Thauera selenatis, continued*]
  growth in drainage water, 433, 435, 442–443
  involvement of nitrate in selenate reduction, 426–429, 443
  involvement of nitrite in selenite reduction, 429–432
  isolation and characterization of, 422–426
  selenate reductase activity in, 426, 428–429, 431–432, 442
Trimethylselenonium, 3, 37–39, 374

U.S. Bureau of Reclamation, 69, 71, 76, 94, 109, 114, 136, 234
U.S. Department of Agriculture Soil Conservation Service, 247
U.S. Environmental Protection Agency, 31
U.S. Fish and Wildlife Service, 84
U.S. Food and Drug Administration, 41, 43

Vitamin E, 6–8, 29–30
Volatilization, selenium (*see also* Selenium, microbial transformation; Dimethylselenide; Dimethyl diselenide; Dimethylselenone), 12, 15, 56
  acceleration of, 374, 382
  by bacteria, 12, 371–373
  in China, 57–59, 61
  detoxification by, 311–312
  enhancement factors, 373–379
    microorganisms, 373, 382–384
    organic amendments, 374–378, 383, 394
    pH, 378–379
    selenium speciation, 374–375, 380–381
    soil moisture, 378, 383–384
    temperature, 375, 378, 383
    water-soluble Se, 374
  fate of methylated gas, 371

[Volatilization, selenium, *continued*]
  field experiments (*see also* Kesterson Reservoir, bioremediation), 382–384
  by fungi, 12, 371–373, 375
  gaseous products of, 3, 187, 369, 398, 414–415
  inhibitory factors, 379–381
    heavy metals, 379
    nitrate and nitrite, 379
  at Kesterson Reservoir (*see also* Kesterson Reservoir, bioremediation), 233, 311–312, 339, 382–385, 394
  by microbes, 12, 19, 20, 24, 56–58, 61, 339, 343–344, 351, 355, 356–360, 369–385, 393
    acceleration of, 382
    irrigation on, effects of, 383–384
    organic amendments on, effects of, 383
    production of dimethyl diselenide, 37, 311, 357–358, 372, 374, 393
    production of dimethylselenide, 37, 56, 311, 357–358, 372–375, 393–396
    production of dimethylselenone, 357–358, 374
    seasonal variation in, 382–383
    soil temperature on, effects of, 383
  in natural waters, 357–358
  by plants, 343–365
    comparison of roots and shoots, 354–356, 358–359, 364–365
    mechanisms of, 360–365
    nitrate on, effect of, 352
    products of, 3, 37, 344, 360–363
    rate of, 351
    rhizosphere microbes on, effect of, 356–360, 364–365
    sulfate on, effect of, 351–352, 355, 364
    temperature, pH, light, effect of, 353–354

[Volatilization, selenium, *continued*]
  [by plants, *continued*]
    use in vegetative management, 363–365
    relationship to Se deficiency diseases, 59
    seasonal variation in, 375, 378, 382–383
    by selenium-accumulator plants, 344, 360–363
    toxicity of methylated products, 373, 393

White muscle disease, 8–9, 33, 40, 85
WHR database (*see* California Department of Fish and Game)

X-ray absorption spectroscopy, 120–122, 136
X-ray fluorescence spectrometry, 122–123